138
Advances in Polymer Science

Springer
Berlin
Heidelberg
New York
Barcelona
Hong Kong
London
Milan
Paris
Singapore
Tokyo

Polymers in Confined Environments

Volume Editor: S. Granick

With contributions by
K. Binder, P.-G. de Gennes, E.P. Giannelis,
G.S. Grest, H. Hervet, R. Krishnamoorti, L. Léger,
E. Manias, E. Raphaël, S.-Q. Wang

This series presents critical reviews of the present and future trends in polymer and biopolymer science including chemistry, physical chemistry, physics and materials science. It is addressed to all scientists at universities and in industry who wish to keep abreast of advances in the topics covered.

As a rule, contributions are specially commissioned. The editors and publishers will, however, always be pleased to receive suggestions and supplementary information. Papers are accepted for „Advances in Polymer Science" in English.

In references Advances in Polymer Science is abbreviated Adv. Polym. Sci. and is cited as a journal.

Springer WWW home page: http://www.springer.de

ISSN 0065-3195
ISBN 3-540-64266-8
Springer-Verlag Berlin Heidelberg New York

Library of Congress Catalog Card Number 61642

Printed in Germany

Typesetting: Data conversion by MEDIO, Berlin
Cover: E. Kirchner, Heidelberg
SPIN: 10573665 02/3020 - 5 4 3 2 1 0 - Printed on acid-free paper

Editorial Board

Contents

Phase Transitions of Polymer Blends and Block Copolymer Melts in Thin Films
K. Binder 1

Flexible Polymers in Nanopores
P.-G. de Gennes 91

Polymer-Silicate Nanocomposites: Model Systems for Confined Polymers and Polymer Brushes
E.P. Giannelis, R. Krishnamoorti, E. Manias 107

Normal and Shear Forces Between Polymer Brushes
G.S. Grest 149

Surface-Anchored Polymer Chains: Their Role in Adhesion and Friction
L. Léger, E. Raphaël, H. Hervet 185

Molecular Transitions and Dynamics at Polymer/Wall Interfaces: Origins of Flow Instabilities and Wall Slip
S.-Q. Wang 227

Author Index Volumes 101 – 138 277

Subject Index 287

Phase Transitions of Polymer Blends and Block Copolymer Melts in Thin Films

K. Binder

Institut für Physik, Johannes Gutenberg-Universität Mainz, Staudingerweg 7,
D-55099 Mainz, Germany
e-mail: binder@chaplin.physik.uni-mainz.de

1 **Introduction** 1

2 **Theoretical Methods** 8

2.1 Mean-Field Theory of Phase Separation in Thin Films 8
2.2 Block Copolymers in the Weak Segregation Limit 22
2.3 Block Copolymers in the Strong Segregation Limit 37
2.4 Survey of Results Obtained with the Self-Consistent Field Theory . . 41
2.5 Concepts on Interfaces in Confined Geometry 47
2.6 Computer Simulation of Polymer Blends in Thin Films 51
2.7 Computer Simulation of Confined Block Copolymers 60
2.8 Dynamics of Phase Separation in Films: an Introduction 65

3 **Outlook on Pertinent Experimental Results** 69

3.1 Polymer Blends in Thin Film Geometry in Thermal Equilibrium . . 69
3.2 Ordering in Thin Films of Block Copolymers 73
3.3 Dynamics of Phase Separation 76

4 **Discussion and Conclusions** 79

5 **References** 82

1 Introduction

Polymer mixtures and block copolymers in thin film geometry find much recent attention, both theoretically [1–63] and experimentally [5,7,8,29,64–122]. This interest arises because of various applications of thin polymer films in materials science (adhesive properties, lubrication, coatings, etc.) [123,124], but also from the point of view of basic science: e.g., from suitable measurements of thin films of polymer mixtures one can extract information on bulk phase behavior (e.g. [70,81]), interfacial widths (e.g. [84,125]), and surface properties controlling surface enrichment or the formation of wetting layers (e.g. [5,8,65,69,74–82,126]). In such thin films, one may also observe very interesting kinetic phenomena and associated structure formation, e.g. growth of surface enrichment layers and adjacent depletion layers [66,69,83], dynamics of phase inversion in unmixed films [77], and – last but not least – surface-directed spinodal decomposition [5,127–158]. Of course, also the dynamics of ordering in block copolymer films is a topic of great current interest (e.g. [101]), but will not be considered further in this article.

We emphasize at the outset that this article deals with flexible linear chains only, neither branched polymers [54,159] nor the packing of stiff chains near surfaces [52,53] will find much attention. However, we also shall not cover films formed by end-grafted chains ("polymer brushes" [160–172]), although in brushes formed from two different types of chains A,B interesting phase separation behavior can occur [165,166] that is related to the phase separation in non-grafted films as treated here. Also films formed from strictly two-dimensional chains in a plane [173–175] are outside of our attention.,

However, the most important restriction in the scope of the present review is the fact that we consider only the case where both the lower surface of the thin film (provided by the substrate) and the upper surface (against air or another solid coating) are perfectly flat and structureless. The unmixing phase transition of a polymer blend in this geometry may lead to inhomogeneous structures as shown in Fig. 1 while structures connected to dewetting {e.g. [176–181]} of one or both of the constituents (Fig. 2) are not considered (see, however, Ref. [182]): so we here assume that either all contact angles in Fig. 2 are zero, or that we have the fluid film coated by another solid layer, so that only the possibilities shown in Fig. 1 occur. Thus we also do not consider the possibility that the unmixing of binary polymer films (A,B) leads to a roughening of its surface [182–184]. Similarly, for block copolymers we restrict attention mostly to symmetric diblock copolymers (chain lengths of the blocks $N_A=N_B=N/2$), and again assume that the film is confined between two flat walls, such that one only needs to consider a few types of lamellar arrangements in the ordered phase of the thin film (Fig. 3) [34–36,63]. On the other hand, if one considers the practically relevant case of thin films on substrates with free surfaces, again various types of inhomogeneous structures may form (Fig. 4b-d).

The questions that will be considered in this review will be concerned with a better understanding of the phase behavior of systems as considered in Figs. 1, 3 and the associated structure of the polymer coils under these conditions. E.g., for a polymer mixture confined by two symmetric walls (Fig. 1) we may ask how the phase diagram describing the unmixing of the binary mixture is different from the corresponding phase diagram in the bulk (Fig. 5); what about the nature of critical phenomena near the critical point of the thin film? Does one understand the concentration profiles across the thin film? Are polymer coils in the surface enriched layers adjacent to the walls deformed in comparison with the bulk? Etc. Of course, quite analogous questions can be asked for block copolymers [34–36,63]. One knows that already for block copolymer melts in the bulk

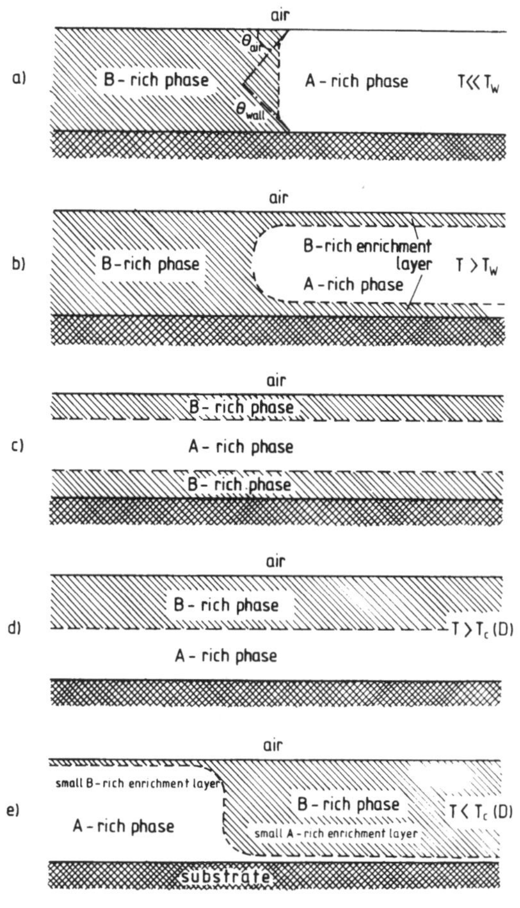

Fig. 1a–e. Schematic description of the equilibrium structure of phase separated binary polymer mixtures (*AB*) in thin film geometry, assuming that the lower surface is provided by a solid substrate, the upper surface being against air (or vacuum). Cases **a-c** refer to the case where the B-rich phase is energetically preferred both by the solid substrate and the air, while cases **d,e** refer to the situation where the B-rich phase is preferred by the air surface only, while the solid substrate prefers the A-rich phase. Note that case **c** occurs as a metastable state only, while in the two-phase coexistence region of thin films where both walls prefer the same phase in equilibrium, interfaces between the A-rich and the B-rich phase run perpendicular to the film. In the nonwet case the A-B interface meets the air surface (or the substrate surface, respectively) under a nonzero contact angle θ_{air} (θ_{wall}) {case **a**}, while in the wet case it bends over to a B-rich enrichment layer at the surface of the A-rich phase {case **b**}. In the "antisymmetric situation" where the two surfaces prefer different phases {cases **d,e**} the situation with a single interface in the center of the film {case **d**} still is in the one-phase region of the film, $T>T_c(D)$, D being the thickness of the film, with $T_c(D)\approx T_w$, the wetting transition temperature in semi-infinite geometry (for simplicity it is assumed that both air and solid substrate have the same wetting transition temperature). For $T\leq T_c(D)$ the A-B interface gets bound on the walls, only microscopically thin enrichment layers may remain at both walls, and the perpendicular part of the A-B interface ultimately (for $T<<T_w$) develops nonzero contact angles with both surfaces, as in part **a**. From Binder et al. [62]

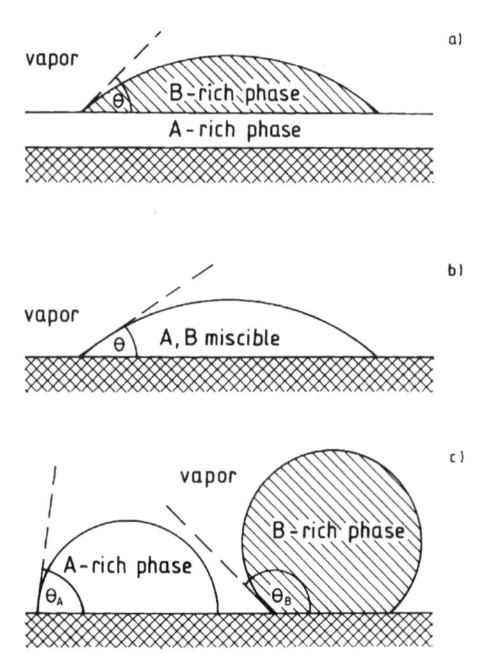

Fig. 2. Schematic description of a thin film containing two constituents *A,B* on a substrate (*cross-hatched*) indicating various possibilities of dewetting, depending on interfacial tensions between A-rich and B-rich phases and the solid substrate (γ_{AS}, γ_{BS}) and the vapor (γ_{AV}, γ_{BV}) and the interfacial tension between coexisting A-rich and B-rich fluid phases (γ_{AB}). In case **a**, the A-rich phases wets the substrate and forms a thin A-rich layer coating the substrate, while the B-rich phase on top does not wet this layer (contact angle θ being nonzero). In case **b**, the A,B mixture is miscible (γ_{AB}<0) but does not wet the substrate. In case **c**, neither phase wets the substrate

there is a subtle interplay between ordering and coil stretching [185–196]. Since the ordering of block copolymers in the bulk involves a non-zero wavelength λ_0 of mesoscopic dimensions (λ_0 is at least of the same order as the gyration radius of the coils [185–187,197]), and λ_0 in general will not be commensurate with the film thickness D, interesting "frustration effects" arise [8,34–36,63,119].

The present review intends to treat these topics from a tutorial point of view, emphasizing the insight that one has gained from both computer simulations [198] and from phenomenological, mean-field type theories [6,11,12,38,58,60–62]. Starting from the Flory-Huggins theory for a binary polymer blend [186,199–205] augmented by gradient-square terms accounting for long-wavelength volume fraction inhomogeneities [186,206–209] and a phenomenological term describing the wall effects [6,11,38,58,62], one obtains a useful qualitative description of the behavior alluded to in Fig. 5. Of course, for a quantitatively accurate description near the critical point mean-field theory is inadequate [55], and for strongly segregated states ($\chi^{-1} << \chi_{crit}^{-1}$) the long wavelength approximation is inadequate [13,186,207]. Thus, it is no surprise that there is clear

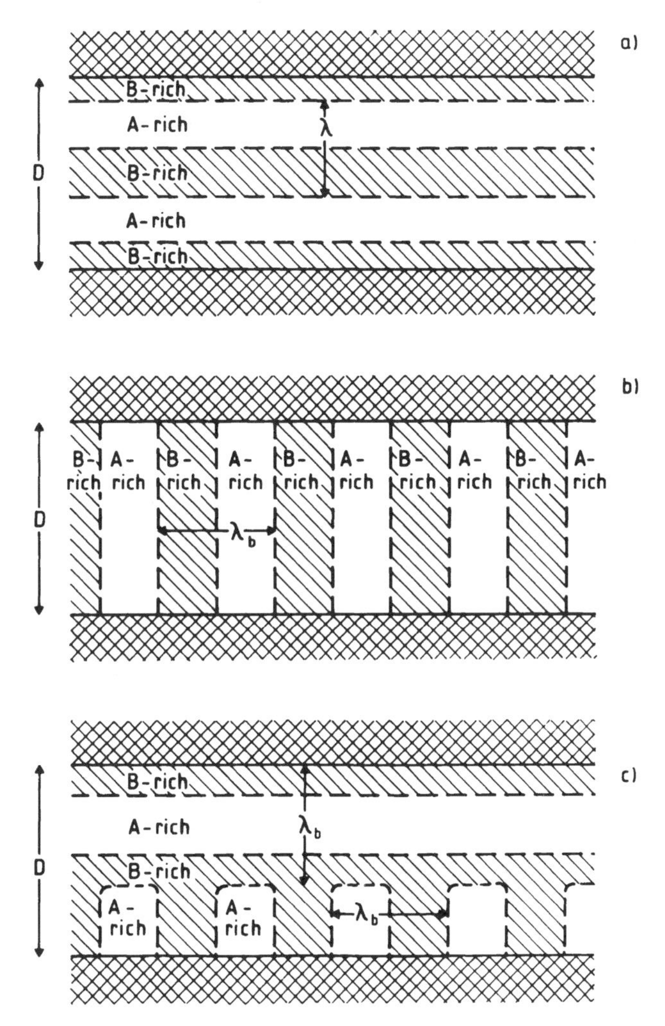

Fig. 3a–c. Possible lamellar structures of block copolymers in a thin film of thickness D in between two equivalent walls, assuming that the B-rich domains are preferred by the boundaries. For this symmetric situation, an arrangement with all AB-interfaces parallel to the walls (case **a**) involves necessarily some "frustration" of the order if the wavelength λ that is enforced by the thickness ($\lambda n=D$ with n integer) differs from its value λ_b in a bulk block copolymer melt under otherwise identical conditions. For a perpendicular arrangement (case **b**) the system can take its bulk wavelength, avoiding thus a compression (or expansion) of the ordered structure relative to the bulk state, but one loses favorable surface energy in the boundary regions of the A-rich domains. This loss can be reduced by a factor of two by choosing an inhomogeneous domain arangement of the type shown in **c** or its mirror image, on the cost of the interfacial energy between domains with perpendicular orientation.

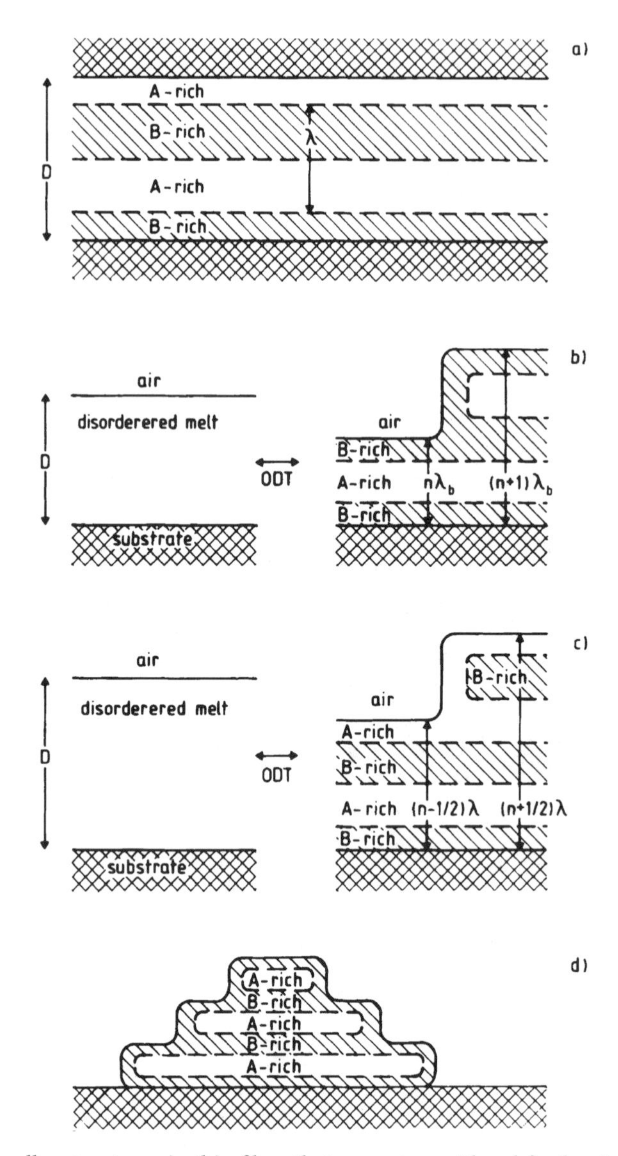

Fig. 4a–d. Lamellar structures in thin films that are not considered further in detail in the present article: **a** Thin film confined between inequivalent walls, where the lower one favors the B-rich domains and the upper one the A-rich domains. Then an arrangement where the interfaces run parallel to the walls requires that thickness D and wavelength λ are related as $D=(n+1/2)\lambda$, $n=0,1,2\ldots$ **b** Thin film on a substrate that favors B-rich domains undergo at the order-disorder transition (ODT) of the block copolymer melt a phase separation into a fraction x of thickness $n\lambda_b$ and a fraction 1–x of thickness $(n+1)\lambda_b$, such that $D=[xn+(1-x)(n+1)]\lambda_b$, if the air also favors B-rich domains. **c** If the air favors A-rich domains instead, the phase separation happens in a fraction x of thickness $(n-1/2)\lambda$ and a fraction 1-x of thickness $(n+1/2)\lambda$ with $n=1,2,3\ldots$ **d** If the block copolymer film undergoes dewetting at the substrate, droplets form with a step-pyramide like structure ("Tower of Babel" [30]).

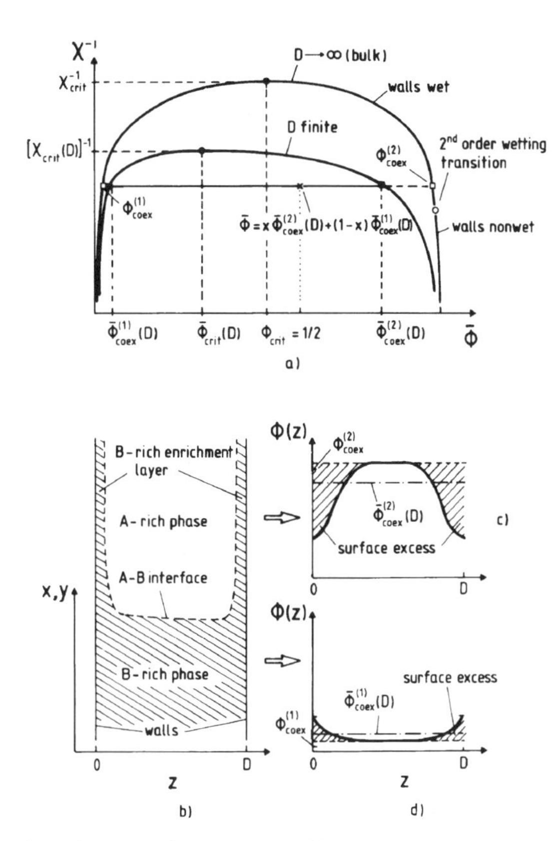

Fig. 5.a Qualitative phase diagram of a symmetrical binary (AB) mixture in both a thin film of thickness D and in semi-infinite geometry (D→∞), in the plane of variables inverse Flory-Huggins parameter χ^{-1} and average volume fraction $\bar{\phi}$ of A in the system. Assuming a symmetrical mixture in the bulk (D→∞), the critical volume fraction is ϕ_{crit}=1/2, and for $\chi^{-1} < \chi^{-1}_{crit}$ the volume fractions $\phi^{(1)}_{coex}, \phi^{(2)}_{coex}$ of the two coexisting phases are related by the symmetry relation $\phi^{(2)}_{coex} = 1 - \phi^{(1)}_{coex}$. Assuming a preferential attraction of one species (B) by the walls, this symmetry is broken, and both the critical volume fraction $\bar{\phi}_{crit}$ (D) and the volume fraction at the A-rich branch of the coexistence curve $\bar{\phi}^{(2)}_{coex}$(D) are shifted towards smaller volume fractions as compared to the bulk. With short range forces at the wall, and unchanged pairwise interactions between monomers of the mixture near the wall, one expects a second-order wetting transition at a temperature T_w<T_{cb} in the semi-infinite system, but this transition is rounded in the thin film geometry. **b** Schematic description of a state of the thin film for a volume fraction $\bar{\phi}$ inside the coexistence curve, $\bar{\phi}^{(1)}_{coex}(D) < \bar{\phi} < \bar{\phi}^{(2)}_{coex}(D)$ {e.g. the point marked by a cross in the phase diagram, part a}. Then the film is inhomogeneous in the x,y-directions parallel to the walls, the A-rich part of the film in equilibrium being separated from the B-rich part by a single A-B interface running perpendicular to the film (for free boundary conditions in lateral directions) or by two such interfaces (for periodic boundary conditions). The relative amounts of A-rich (x) and B-rich (1-x) phases are simply given by the lever rule, $\bar{\phi} = \bar{\phi}^{(2)}_{coex}(D) + (1-x)\bar{\phi}^{(1)}_{coex}(D)$, with 0<x<1. Here we assume that the film thickness is much larger than the interfacial thickness w and so the A-rich phase has enrichment layers of the B-rich phase "coating" the walls of the thin film. **c** Volume fraction profile ϕ(z) in the z-direction across the film in the A-rich phase. Here the shaded area denotes the surface excess ϕ_s. **d** Same as c but for the B-rich phase. From Flebbe et al. [58]

evidence {e.g. [74]} that for a quantitatively reliable description of surface enrichment near walls one needs better treatments, e.g. on the basis of the self-consistent field theory [13,20,32,57]. Since this theory to a large extent is of a numerical character, we shall treat it only rather briefly here, however. Also not much attention will be paid to the effect of long-range surface forces [16,31], although they surely will be present for most systems, at least in addition to (possibly stronger?) short range surface forces (e.g. due to packing effects near the hard walls, preferred steric interactions, etc.). Similarly, our treatment of block copolymers in thin films will either again stay on a mean-field level, based on (simplified) versions [12,61] of Leibler's [197] theory, or make use of order-of magnitude estimates valid in the strong segregation limit [35,36,63]. While rather sophisticated theories exist for block copolymers in the bulk, taking into account fluctuation effects [192,210,211] or unifying the strong and weak segregation limits in terms of self-consistent field theories [212,213], one still does not have an approach that includes fluctuation effects and is valid also for intermediate segregation strength, as is necessary for not too long chains. Neither of these theories have been extended to thin films yet, and since surfaces entail additional entropic effects [19,51–54,59] which may compete with energetic effects the development of quantitatively reliable theories clearly will be very difficult [61,167].

In Sect. 2 we hence give only a rather selective overview of theoretical methods and results – in such a rapidly developing field with still many controversial aspects the selection clearly is biased by the knowledge and interests of the author, but it is hoped that this survey nevertheless is a useful introduction to the field, and should stimulate further developments. The selection of experimental results that will be discussed in Sect. 3 again cannot aim at an authorative review of all work that has been done in the field, but should be rather viewed as an introduction again, based on some key examples. In Sect. 4, a brief summary will be given.

2 Theoretical Methods

2.1 Mean-Field Theory of Phase Separation in Thin Films

As emphasized in Figs. 1 and 5, the surfaces of a thin film introduce an inhomogeneity of the local volume fractions $\phi_A(\vec{r})$, $\phi_B(\vec{r})$ of the two species A,B at position $\vec{r}=(x,y,z)$, at least in the z-direction across the film. In the two-phase coexistence region, domains of A-rich and B-rich phases are formed, and this may lead to a lateral inhomogeneity (i.e., along the x,y-directions) as well. Since a treatment of this inhomogeneity in two spatial directions is difficult even on the mean-field level, we ignore the regions near those interfaces in Fig. 1 that are running perpendicular to the substrate. Allowing only an inhomogeneity of the local volume fractions $\phi_A(z)$, $\phi_B(z)$ in the z-direction, we treat only either the one-phase region of the thin film (the region above the coexistence curve in Fig. 5a) or – in the case of two-phase coexistence – the interior of one of the co-existing phases. The assumption that contributions from regions containing A-B- interfaces perpendicular to the walls can be ignored implies that the average volume fraction is related to the volume fraction of the two coexisting phases with the lever rule. Assuming strict incompressibility everywhere in the system

$$\phi_A(z) + \phi_B(z) = 1, \text{ for all } z \tag{1}$$

we reduce the problem to a single concentration field $\phi(z) \equiv \phi_A(z)$, $\phi_B(z) = 1 - \phi(z)$. Of course, Eq. (1) is rather restrictive – it neglects effects such as the well-known "layering" (density oscillations) near hard walls [214], the density reduction in strongly segregated A-B-interfaces [215], etc. In this case the average volume fraction $\bar{\phi}$ of monomers A is related to the volume fractions $\bar{\phi}^{(1)}_{coex}(D)$, $\bar{\phi}^{(2)}_{coex}(D)$ of A-monomers of the two phases coexisting in the thin film as

$$\bar{\phi} = x\bar{\phi}^{(2)}_{coex}(D) + (1-x)\bar{\phi}^{(1)}_{coex}(D) \tag{2}$$

x denoting the volume fraction taken by the A-rich phase. Denoting the profiles $\phi(z)$ in the two coexisting phases by $\phi^{(1)}(z)$ and $\phi^{(2)}(z)$, we simply have

$$\bar{\phi}^{(1)}_{coex}(D) = (1/D)\int_0^D \phi^{(1)}(z)dz \qquad \text{B-rich phase} \tag{3}$$

$$\bar{\phi}^{(2)}_{coex}(D) = (1/D)\int_0^D \phi^{(2)}(z)dz \qquad \text{A-rich phase} \tag{4}$$

As shown schematically in Fig. 5c,d, the profile $\phi^{(2)}(z)$ is expected to be significantly depressed near the walls, due to the preferential attraction of B-monomers to the walls B-rich surface enrichment layers form. In contrast, there is at best a weak enhancement of the volume fraction of A in the B-rich phase $\phi^{(1)}(z)$ near the walls: this enhancement opposing the energetic preference of the walls for B, is entropically driven (due to the "missing neighbor effect" at the walls entropy of mixing is more efficient at the walls than in the bulk [56,216]). This effect is important only at temperatures rather close to the critical temperature $T_c(D)$ {which is proportional to $\chi^{-1}_{crit}(D)$, if the Flory-Huggins parameter χ is predominantly of enthalpic origin}.

A quantity of particular interest is the surface excess ϕ_s induced by the walls, which we define separately for the left wall and right wall in Fig. 5b)-d), to allow for a general asymmetric situation,

$$\phi_s^{\ell} = \int_0^{D/2} [\phi(z) - \phi(D/2)]dz, \quad \phi_s^{r} = \int_{D/2}^{D} [\phi(z) - \phi(D/2)]dz \tag{5}$$

Per definition we have

$$\bar{\phi} = \phi(D/2) + (1/D)(\phi_s^{\ell} + \phi_s^{r}) \tag{6}$$

In the limit of large D we have $\phi(D/2) \to \phi_b$, the volume fraction of A in a bulk mixture. In the following, we shall for the most part consider only the situation with symmetric walls, such that $\phi_s^{\ell} = \phi_s^{r} = \phi_s$.

The surface excess ϕ_s is the basic quantity to characterize the wetting behavior of the (semi-infinite) mixture (Fig. 6). Many excellent reviews exist [217–223] that describe various aspects of the theory of wetting. Therefore, we shall not duplicate these efforts here, but simply recall in Fig. 6 the basic notions. Note that no prewetting transition line exists if the wetting transition is second order, as has been assumed in Fig. 5a in contra-distinction to Fig. 6a. Although experimental studies of wetting phenomena for polymer mixtures always deal with films of finite (albeit large) thickness D [69,76,81–83], we maintain the well-known [62,224,225] fact that in a strict sense wetting transitions occur in the limit $D \to \infty$ only, while for finite D they are the more rounded off the smaller D (only the first-order prewetting transition can survive in a thin film). Unfortunately, this point is sometimes missed in the recent literature {e.g. [26,40]} where the phase transition in a symmetrical film leading to lateral inhomogeneity (Fig. 5) was ignored, by requiring that only an inhomogeneity $\phi(z)$ in z-direction can occur and then "transitions" are located in the unstable part of the phase diagram.

The phenomenological discussion given so far can be made more explicit on the mean-field level, combining a Flory-Huggins free energy density [199–206] with the "gradient square"-approximation [206–209], supplemented by local surface terms [11,58]. That is, the excess free energy of the mixture (relative to pure B) per unit wall area is written [58]

$$\frac{\Delta F}{k_B T} = \frac{1}{b}\int_0^D dz \left\{ f[\phi(z)] + \kappa(\phi)\left[\frac{d\phi(z)}{dz}\right]^2 \right\} + f_s^{(bare)}(\phi_0) + f_s^{(bare)}(\phi_D) \tag{7}$$

Here b is the effective lattice spacing of the underlying Flory-Huggins lattice [199–206], and the coefficient $\kappa(\phi)$ of the gradient energy term for a symmetric mixture (effective monomeric units have linear dimensions $\sigma_A = \sigma_B = b$, chain lengths are $N_A = N_B = N$) is [186,206,207]

$$\kappa(\phi) = b^2 / \{36\phi(z)[(1-\phi(z)]\} \tag{8}$$

For the free energy density $f(\phi)$ in the bulk we assume the standard expression

$$f(\phi) = \frac{1}{N}\,[\phi \ell n \phi + (1-\phi)\ell n(1-\phi)] + \chi\phi(1-\phi) - \Delta\mu\phi \tag{9}$$

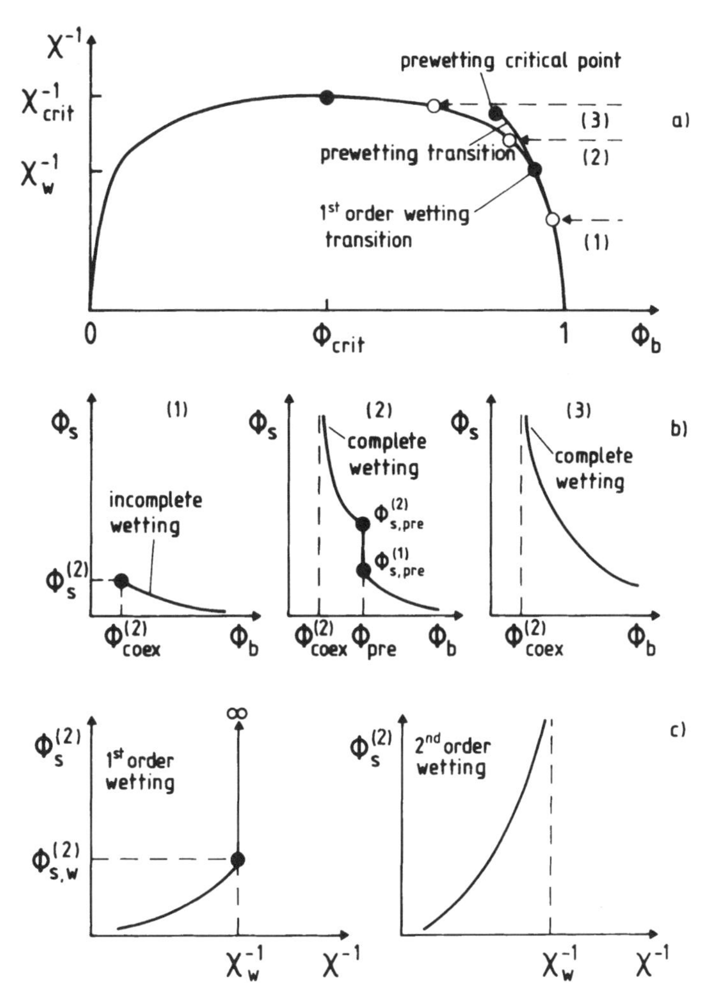

Fig. 6. a Schematic phase diagram of a semi-infinite polymer mixture exhibiting a first-order wetting transition on the A-rich side of the coexistence curve for $\chi=\chi_w$. In this case there exists a prewetting transition line in the A-rich one-phase region, beginning at the first-order wetting transition at the coexistence curve and ending in a prewetting critical point. Three paths (*1*), (*2*) and (*3*) are indicated in the one-phase region, where one approaches the coexistence curve at constant χ by reducing the bulk volume fraction ϕ_b of the mixture. **b** Variation of the surface excess ϕ_s plotted vs ϕ_b for the three paths (*1*), (*2*) and (*3*) {schematic}. For $\chi^{-1} < \chi_w^{-1}$ and $\phi_b \to \phi_{coex}^{(2)}$ the surface excess reaches a finite limit $\phi_s^{(2)}$ ("incomplete wetting"), while for $\chi^{-1} > \chi_w^{-1}$ the surface excess diverges at phase coexistence ("complete wetting"). For $\phi=\phi_{pre}$ a finite jump of ϕ_s from $\phi_{s,pre}^{(1)}$ to $\phi_{s,pre}^{(2)}$ occurs ("first-order prewetting transition"). This jump $\phi_{s,pre}^{(2)} - \phi_{s,pre}^{(1)}$ smoothly vanishes as the prewetting critical point is reached. **c** Variation of $\phi_s^{(2)}$ with χ^{-1} at phase coexistence (i.e., ϕ_b is kept at $\phi_{coex}^{(2)}(\chi)$). At the first order wetting transition, $\phi_s^{(2)}$ jumps from a finite value $\phi_{s,w}^{(2)}$ discontinuously to infinity, while a critical divergence of $\phi_s^{(2)}$ is encountered for the second order wetting transition

where $\Delta\mu$ is the normalized chemical potential difference between effective subunits, $\Delta\mu=(\mu_A-\mu_B)/k_BT$. Choosing χ (which for simplicity is chosen independent of ϕ) and $\Delta\mu$ as independent thermodynamic parameters, we have implied that it is most convenient to work in the semi-grandcanonical ensemble rather than in the canonical ensemble of the thin film. Assuming a macroscopic extent in lateral directions, there is still a well-defined thermodynamic limit, and then the various ensembles of statistical mechanics yield equivalent results. In the one-phase region of the thin film (above the coexistence curve in Fig. 5a), there is a smooth relation between $\Delta\mu$ and $\bar{\phi}$, the average volume fraction in the thin film, while in the two phase region $\Delta\mu=\Delta\mu_{coex}(D)$=const and the relation between $\bar{\phi}$ and the volume fraction x of the A-rich phase is given by the lever rule, Eq. (2). Of course, due to the symmetry of Eq. (9) against an interchange of A and B ($\phi \leftrightarrow 1-\phi$) for $\Delta\mu=0$ we have phase coexistence at $\Delta\mu=0$ in the bulk, $\Delta\mu_{coex}(D\to\infty)=0$, but this symmetry typically is broken in a thin film due to surface effects, as anticipated already in the qualitative phase diagram, Fig. 5a).

The last terms $f_s^{(bare)}(\phi_0)$, $f_s^{(bare)}(\phi_D)$ in Eq. (7) are the bare contributions to the surface excess free energy. Since we consider here only symmetric walls, we assume the same functional form for both walls at z=0 and at z=D, assuming the quadratic approximation [11] for simplicity,

$$f_s^{(bare)}(\phi_0)=-\mu_1\phi_0-\frac{1}{2}g\phi_0^2,\quad f_s^{(bare)}(\phi_D)=-\mu_1\phi_D-\frac{1}{2}g\phi_D^2 \tag{10}$$

Of course, some general aspects of our treatment could be easily extended to a general form of $f_s^{(bare)}(\phi)$, as in the semi-infinite case [226], but for explicit numerical work a specific form of $f_s^{(bare)}(\phi)$ is needed. Equation (10) can be justified for Ising-type lattice models near the critical point [216,220], i.e. when ϕ is near $\phi_{crit}=1/2$, as well as in the limits $\phi\to 0$ or $\phi\to 1$ [11]. The linear term $-\mu_1\phi$ is expected due to the preferential attraction of component B to the walls, and to "missing neighbors" for the pairwise interactions near the walls while the quadratic term can be attributed to changes in the pairwise interactions near the walls [144,216,227]. We consider Eq. (10) only as a convenient model assumption to illustrate the general theoretical procedures – there is clear evidence that Eq. (10) is not accurate for real polymer mixtures [74,81,82,85].

Also we have implicitly assumed in Eqs. (7) and (10) that all effects of the walls on the mixture are strictly of short range, and thus $f_s^{(bare)}$ depends on the local concentrations ϕ_1, ϕ_D only. Because of long range van der Waals forces this is not realistic [220], and in fact for the semi-infinite case extensions to long range surface forces are available [16,31,228]. However, since the mathematics is more cumbersome, and in the semi-infinite geometry the differences to the short range case are rather minor [31], this extension is disregarded here.

We also emphasize that $f_s^{(bare)}$ is not the total surface excess free energy f_s that in principle is accessible to measurements, but only a part thereof. For large D Eq. (7) can be cast to the form

$$\Delta F / k_B T = (D / b)\, f(\phi) + 2f_s \tag{11}$$

and in the limit $D \rightarrow \infty$ the distinction between ϕ and $\bar{\phi}$ vanishes, and Eq. (11) can be considered as a definition of f_s. Note that the gradient energy contributions under the integral in Eq. (7) contribute to f_s also, and thus f_s differs from $f_s^{(bare)}$.

One solves the problem posed by Eqs. (7)-(10) by minimizing the free energy functional, Eq. (7), $\delta\, \Delta F\{\phi(z)\}/\delta\phi(z)=0$. This yields an Euler-Lagrange equation together with two boundary conditions at the walls. Since in thermal equilibrium the resulting profiles $\phi(z)$ are symmetrical around the center of the film, it is convenient to take the origin of the coordinate system not at the left wall as in Fig. 5, but rather at this midpoint, so henceforth the coordinates of the walls are $z=\pm D/2$. Putting $b\equiv 1$ the Euler-Lagrange equation reads

$$\partial f(\phi) / \partial\phi - \partial\kappa(\phi) / \partial\phi\, (d\phi / dz)^2 - 2\kappa(\phi) d^2\phi / dz^2 = 0 \tag{12}$$

and the boundary conditions

$$2\kappa(\phi) d\phi / dz \Big|_{z=-D/2} = -\mu_1 - g\phi\ (z = -D/2) \tag{13}$$

$$2\kappa(\phi) d\phi / dz \Big|_{z=-D/2} = \mu_1 + g\phi\ (z = D/2) \tag{14}$$

Imposing the condition that in equilibrium the profiles are symmetrical around $z=0$ yields further

$$d\phi / dz \Big|_{z=0} = 0 \tag{15}$$

and it suffices to consider one half of the film only ($z\geq 0$).

Eq. (15) is analogous to the situation in the bulk of a semi-infinite system [11],

$$\lim_{z\rightarrow\infty} d\phi / dz = 0 \qquad \phi(z \rightarrow \infty) = \phi_b \tag{16}$$

ϕ_b being the solution of the Euler-Lagrange equation in the bulk,

$$\partial f(\phi) / \partial\phi \big|_{\phi=\phi_b} = 0 \qquad \Delta\mu = (1/N)\ell n\left[\phi_b / (1-\phi_b)\right] + \chi(1-2\phi_b) \tag{17}$$

While in the semi-infinite case we hence know the value ϕ_b at which the profile $\phi(z)$ saturates, for the thin film the volume fraction $\phi^*\equiv\phi(z=0)$ in the center of the film is not known in beforehand, but must also be found from the minimization procedure (i.e., the condition that both boundary conditions Eqs. (14) and (15) are satisfied).

Such inhomogeneous mean field theories are solved by noting an analogy to classical mechanics [216]: one considers the motion of a point particle in a potential – $f(\phi)$, i.e. ϕ plays the role of a spatial coordinate and z the rôle of "time". Then the integral in Eq. (7) corresponds to the action, and the integrand $\alpha(\phi, d\phi/dz) \equiv f(\phi) + \kappa(\phi)\,(d\phi/dz)^2$ is the Lagrangian, i.e. $2\kappa(\phi)$ a "mass" that depends on the "position" of the point particle. This analogy readily yields that there exists a conserved quantity, the total energy E of the point particle:

$$H(\phi, d\phi/dz) \equiv \kappa(\phi)(d\phi/dz)^2 - f(\phi) = E = \mathrm{const} = -f(\phi^*) \quad . \tag{18}$$

Equation (18) yields an implicit for the profile $\{G(\phi) \equiv f(\phi) — f(\phi_b)\}$

$$z(\phi) = \pm \int_{\phi^*}^{\phi} \sqrt{\kappa(\phi')/\left[G(\phi') - G(\phi^*)\right]}\, d\phi' \tag{19}$$

In the semi-infinite problem [11] $E = -f(\phi_b)$ and thus Eq. (19) simplifies to

$$z(\phi) = \pm \int_{\phi_0}^{\phi} \sqrt{\kappa(\phi')/\left[f(\phi') - f(\phi_b)\right]}\, d\phi' \tag{20}$$

choosing the surface at z=0 then. The shape of the profile in Eq. (20) is clearly independent of the boundary condition – the latter just acts as a cutoff for the profile $z(\phi)$, one simply looks for the point $\phi = \phi_0$ along the profile where $2\kappa(\phi_0)/(dz/d\phi)_{\phi=\phi_0} = -\mu_1 - g\phi_0$ holds.

This property is no longer true for thin films, where ϕ^* and hence the shape of the profile also depend on the boundary condition. Obviously, Eqs. (14), (19) then define a non-trivial self-consistency problem, which requires an iterative numerical treatment [58]. We note, however, that Eqs. (14), (18) yield a simple relation between ϕ^* and $\phi_0 \equiv \phi(z = \pm D/2)$,

$$\mu_1 = -g\phi_0 \pm 2\sqrt{\kappa(\phi_0)\left[G(\phi_0) - G(\phi^*)\right]} \tag{21}$$

In Eqs. (19)-(21), always the sign has to be chosen which yields the absolute minimum of the free energy. As usual, mean field theory yields metastable branches, and the phase separation from one state $\bar{\phi}^{(1)}_{coex}(D)$ to the other state $\bar{\phi}^{(2)}_{coex}(D)$ in this treatment shows up as an intersection of two branches for $\Delta F/k_BT$ when plotted vs $\Delta\mu$ or the conjugate variable $\Delta\phi \equiv \phi_b - \bar{\phi}^{(2)}_{coex}$ (cf. Fig. 5a), see Fig. 7a. This example is for a case where $R_{gyr} = b\sqrt{N/6} \approx 4.082$, D=100, i.e. $D/R_{gyr} \approx 25$, i.e. a relatively thick film. This calculation also yields "isotherms" for ϕ^* as a function of $\Delta\phi$ (Fig. 7b), and the corresponding profiles $\phi(z)$, and thus ϕ_0 as function of ϕ^* is also known explicitly. It is interesting to examine the volume fraction profiles at phase coexistence (Fig. 7c). For the parameters shown, the volume fraction $\phi_0$at the surface is almost the same in both coexisting phases.

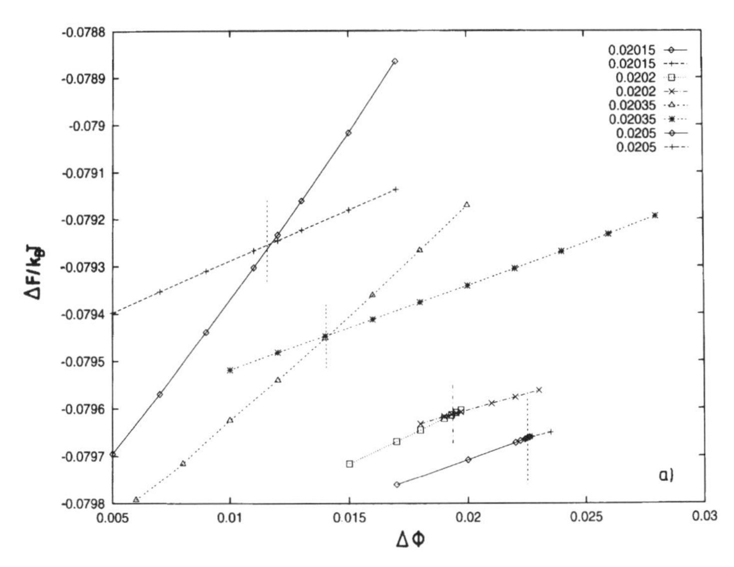

Fig. 7a Free energy of mixing $\Delta F/k_BT$ of the symmetric binary polymer mixture, described by the model Eqs. (7)-(10) plotted as a function of $\Delta\phi = \phi_b - \phi^{(2)}_{coex}$, for the choice of parameters b=1, D=100, N=100, μ_1=0.2, g=–0.5. Several "isotherms" labelled by their Flory-Huggins parameter χ are shown (note χ_{crit}=0.0200 for N=100 in the bulk since χ_{crit}=2/N [186,200]. *Vertical broken lines* indicate the location of the first-order transition. Note that the A-rich branch has a steeper variation with $\Delta\phi$ than the B-rich branch. A third branch is an unstable solution to the mean field equations and not shown at all.

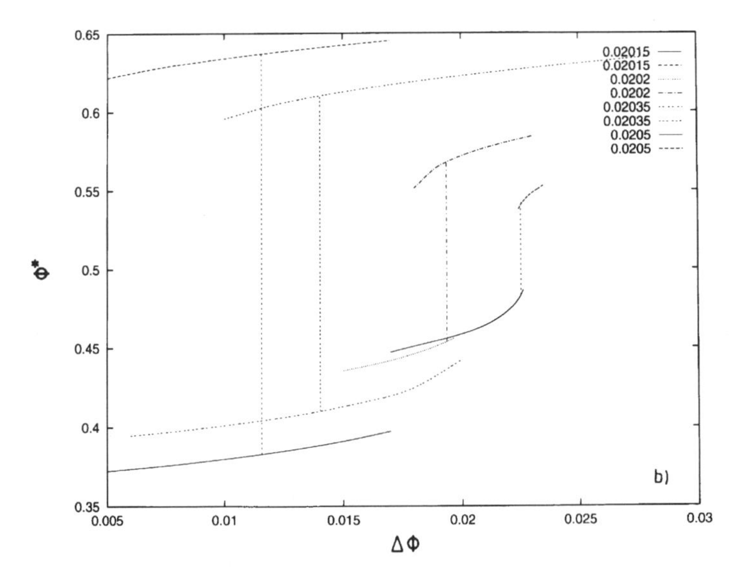

Fig. 7b Isotherms of ϕ^* vs $\Delta\phi$, for the same case as in part a).

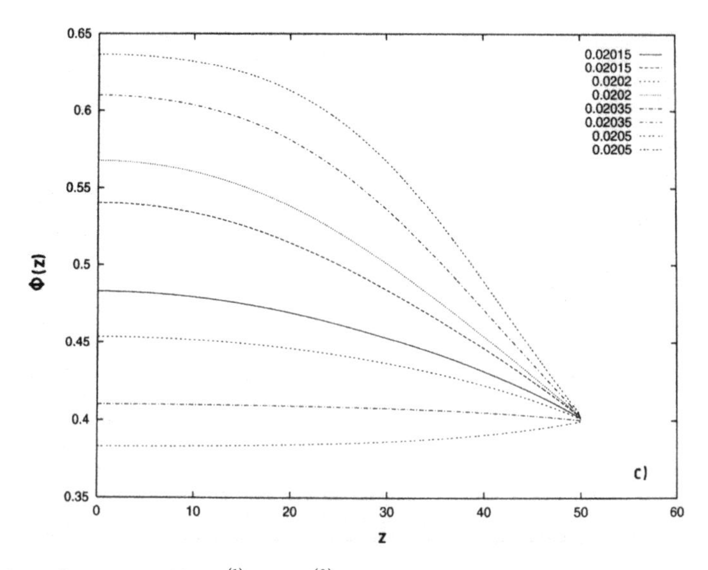

Fig. 7c Volume fraction profiles $\phi^{(1)}_{coex}(z)$, $\phi^{(2)}_{coex}(z)$ at phase coexistence. From Flebbe et al. [58]

For the temperature farthest away from T_c (χ=0.0205), the volume fraction $\phi^*=\phi(z=0)$ of A in the B-rich phase in the center of the film is clearly smaller than at the surface: i.e., there is an entropically driven enrichment of the minority component at the surface. For χ near $\chi_{crit}(D)$, however, both coexisting profiles have a similar shape, $\phi(z)$ is maximal at z=0 and decreases towards the walls. Of course, this must happen because as $\chi \rightarrow \chi_{crit}(D)$ both profiles must merge into a common profile, for $\chi < \chi_{crit}(D)$ there exists a unique profile (at any value of ϕ^* or $\bar{\phi}$) only.

From the profiles, Eq. (20), it is straightforward to obtain other quantities of physical interest such as the average value of the volume fraction in the film,

$$\bar{\phi} = (1/D) \int_{-D/2}^{+D/2} \phi(z) dz = \phi_0 + (2/D) \int_{\phi_0}^{\phi^*} z(\phi') d\phi' \tag{22}$$

as well as the surface excess ϕ_s {Eqs. (5) and (6)}. At phase coexistence, one finds $\bar{\phi}^{(1)}_{coex}(D)$, $\bar{\phi}^{(2)}_{coex}(D)$ by using Eq. (22) for the two coexisting profiles, of course, and if desired one can find the fractions x, 1–x of the two phases for a given $\bar{\phi}$ from Eq. (2) and then also construct an average concentration profile $\phi^{av}(z)=x\phi^{(1)}(z)+(1-x)\phi^{(2)}(z)$ in the thin film [62].

Since experiments on polymer mixtures in thin film geometry are often carried out in the intention to examine wetting behavior [69,71,81–83], we discuss here also the behavior of ϕ_s as function of ϕ^*. In the semi-infinite case, we have a logarithmic divergence when phase coexistence is approached, (cf. Fig. 6b)

$$|\phi_s| \propto \left| \ell n \left(\phi_b - \phi^{(2)}_{coex} \right) \right| \tag{23}$$

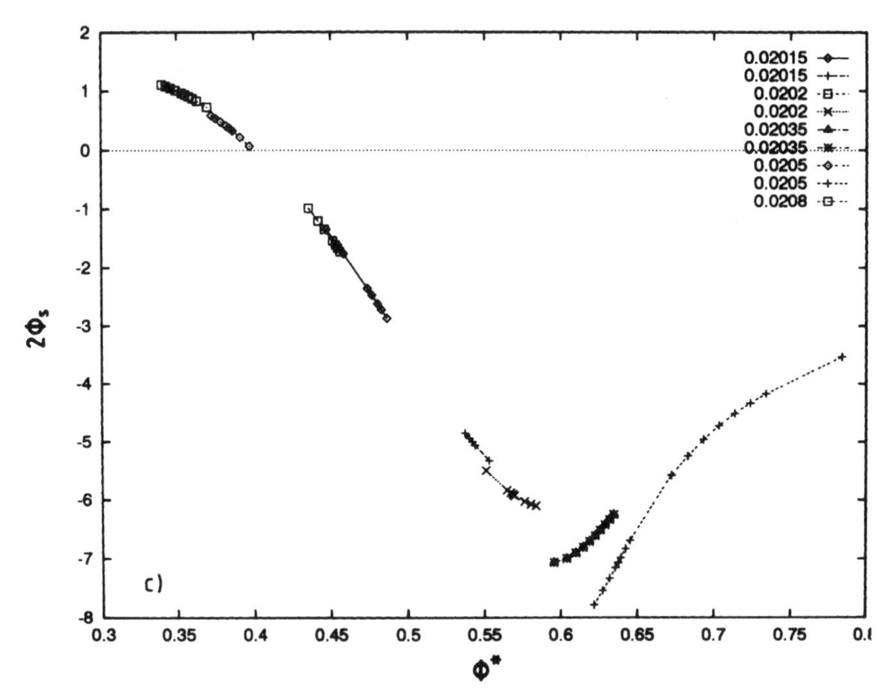

Fig. 8. Plot of the surface excess $2\phi_s$ vs the volume fraction ϕ^* in the center of the thin film, for the same model as shown in Fig. 7. Note that there are two branches at each value of χ, corresponding to the A-rich state (larger ϕ^*) and to the B-rich state (smaller ϕ^*) of the thin film. From Flebbe et al. [58].

In our notation, ϕ_s is typically negative (Fig. 8), corresponding to surface enrichment of B in the A-rich phase (only on the B-rich side positive values can be observed, corresponding to the situation sketched in Fig. 5d). It is seen that the wetting transition (a second-order wetting transition occurs for χ_w=0.02027 for the corresponding semi-infinite system [58]) is completely blurred: a rather strong decrease of $2\phi_s$ with decreasing ϕ^* is observed for χ=0.0205, where the system at phase coexistence ($\phi^*\approx0.636$, cf. Fig. 7c) is still non-wet. At χ=0.0235 the decrease of $2\phi_s$ is weaker as one passes phase coexistence ($\phi^*\approx0.611$), and for still smaller χ the curves even have bent over, ϕ_s gets less negative with decreasing ϕ^* when one passes phase coexistence. Thus we conclude that analyses of surface enrichment in polymer mixtures should take the finite film thickness into account, if reliable results on $f_s^{(bare)}(\phi)$ are to be obtained.

As a final example in this subsection, Fig. 9 examines the distortion of the phase diagram due to these wall effects, for three choices of the film thickness [58]. One notes that the shape of the coexistence curves is always parabolic near the critical point, as expected, since this shape reflects the mean field order parameter exponent, β=1/2 [186]. In real systems, one expects – except for extremely long

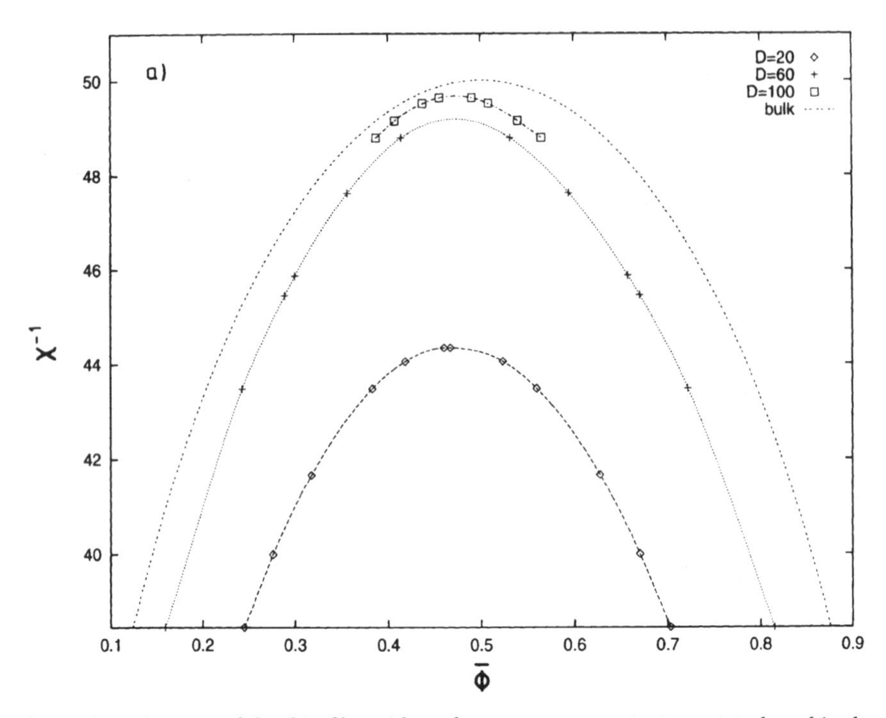

Fig. 9. Phase diagram of the thin film with surface parameters μ_1=0.2, g=–0.5 plotted in the plane of variables χ^{-1}, $\bar{\phi}$ for polymers of chain length N=100 and for three choices of film thicknesses: D=20 (*diamonds*), D=60 (*crosses*) and D=100 (*squares*). *Broken curve* shows the bulk phase diagram of the underlying Flory-Huggins model for comparison. Remember that lengths are measured in units of the size b of an effective monomer. From Flebbe et al. [58]

chains, $N\to\infty$ [186,200] that the shape of the coexistence is flatter, corresponding to a three-dimensional Ising exponent $\beta(d=3)\approx 0.325$ [229,230] in the bulk, and a two-dimensional Ising exponent β (d=2)=1/8 [229] for thin films. We shall return to this problem in Sect. 2.6.

As anticipated in Fig. 5a), the critical volume fraction $\bar{\phi}_{crit}(D)$ gets shifted to smaller values, for a case where the walls preferentially attract monomers of the other kind (i.e. B-monomers when $\bar{\phi}$ denotes the average volume fraction of A). One also sees that χ_{crit}^{-1} gets more depressed the smaller D is, i.e. confinement makes the polymer mixture more compatible.

Assuming that $\bar{\phi}_{crit}(D)$ is still close to ϕ_{crit}=1/2, and that the surface perturbations are weak enough so that for large D the profile $\phi(z)$ stays close to ϕ_{crit} throughout, the shift of the critical coupling $\chi_{crit}(D)$ can be calculated analytically, by expanding in Eq. (12) $\phi(x)$ around ϕ_{crit}=1/2, and keeping only the leading terms of the reduced "order parameter" $m(z)\equiv 2\phi(z)-1$ [186],

$$m(z)\left(1-\frac{\chi}{\chi_{crit}}\right)+\frac{1}{3}m^3(z)-\frac{N}{18}\frac{d^2m}{dz^2}=H \qquad H=N\Delta\mu/2 \tag{24}$$

and the boundary conditions become $\{H_1 \equiv (\mu_1 + g/2)N/2, \lambda \equiv -2/(9\,g)\}$

$$dm/dz \Big|_{z=-D/2} - m(z=-D/2)/\lambda = -H_1/C \tag{25}$$

$$dm/dz \Big|_{z=D/2} + m(z=D/2)/\lambda = H_1/C \tag{26}$$

where the constant $C \equiv N/18$. Eqs. (24)-(26) are identical to the analogous problem of the shift of T_c in ferromagnetic thin films, where m(z) denotes the magnetization profile {cf. Eqs. (2.5) and (2.6) of Ref. [216]}, where H_1 is a "surface magnetic field", and the "extrapolation length λ" again accounts for "missing bonds" near the surface. For $T>T_c$ ($\chi<\chi_{crit}$) the non-linear term $1/3\, m^3(z)$ can be neglected, and then the problem is easily solved, using the results for the order parameter and correlation length in the bulk,

$$m_b = H/(1-\chi/\chi_{crit}) \qquad \xi_b = \sqrt{N/\left[18\left(1-\chi/\chi_{crit}\right)\right]} \tag{27}$$

$$m(z) = m_b - (m_b - m_1)\cosh(z/\xi_b)/\cosh\,[D/(2\xi_b)] \qquad -\frac{D}{2} \le z \le \frac{D}{2} \tag{28}$$

with the order parameter m_1 at the surface (Eqs. 2.18 a,b of ref. [216])

$$m_1 = \{(\lambda/\xi_b)\, m_b \tanh[D/(2\xi_b)] + H_1\lambda C\} \;/\; \{1+(\lambda/\xi_b)\tanh[D/(2\xi_b)]\} \tag{29}$$

The average order parameter $\overline{m}$ then becomes

$$\overline{m} \equiv \int_{-D/2}^{+D/2} m(z)dz/D = m_b - \frac{2}{D}\frac{\xi_b\left(m_b - H_1\lambda/C\right)}{\coth\left[D/\left(2\xi_b\right)\right] + \lambda/\xi_b} \tag{30}$$

As is well-known, the critical point can be found from locating the divergence of the response function ("susceptibility") $\overline{S} \equiv \partial\overline{m}/\partial H$, which signals the onset of symmetry breaking and spontaneous order. The simplest case occurs for $H_1=0$, and then

$$\overline{S} = \frac{\lambda\xi_b/C + \left(1-\chi/\chi_{crit}\right)^{-1}\coth\left[D/\left(2\xi_b\right)\right] - 2\xi_b/\left[D\left(1-\chi/\chi_{crit}\right)\right]}{\coth\left[D/\left(2\xi_b\right)\right] + \lambda/\xi_b} \tag{31}$$

In the further discussion, we must distinguish whether λ is positive or negative. In the latter case, the ordering tendency of the thin film is enhanced as compared to the bulk, i.e. the critical temperature $T_c(D)$ in the thin film exceeds the critical temperature T_{cb} of the bulk, ξ_b being still finite at $T_c(D)$. The transition simply results from the vanishing of the denominator, $\{x \equiv D/(2\xi_b)\}$

$$\coth[D/(2\xi_b)] = -\lambda/\xi_b, \text{ or } \coth(x) = 2\lambda x/D \tag{32}$$

For large $D/|\lambda|$ this equation has a solution for large x, where $\coth(x)\to 1$, i.e. $x=D/(2\xi_b)\approx -D/(2\lambda)$, i.e. D cancels and the transition occurs when $\xi_b=-\lambda$. This means that the film orders at the "surface transition" of the corresponding semi-infinite system (phase separation occuring in two dimensions only, namely in a surface region of thickness ξ_b. For small $D/|\lambda|$, however, the solution of Eq. (32) occurs for small x, where $\coth x\approx 1/x$ and hence

$$2\xi_b / D \approx -\lambda / \xi_b \quad \xi_b^{-2} \approx -2/(D\lambda) \quad \chi_{crit}(D)/\chi_{crit} = 1 + N/(9D\lambda) \tag{33}$$

i.e. the shift is proportional to the chain length and inversely proportional to the film thickness.

When we consider the case $\lambda>0$, we need to look for a singularity in the region $\chi>\chi_{crit}$, where ξ_b when calculated still from Eq. (27) would be imaginary. It turns out, however, that formally Eqs. (27)-(31) can still be used, we must,however, note relations such as $\cot x = i \coth ix$ where $i=\sqrt{-1}$. Then the condition that the denominator in Eq. (31) vanishes yields

$$\cot\left(D\sqrt{\left|1-\chi/\chi_{crit}\right| 18/N}\,/2 \right) = \lambda\sqrt{\left|1-\chi/\chi_{crit}\right| 18/N} \tag{34}$$

and defining now the auxiliary variable x as $x \equiv D\sqrt{\left|1-\chi/\chi_{crit}\right| 18/N}\,/2$ yields

$$\cot x = 2\lambda x/D \qquad x\to\pi/2 \text{ for } D\to\infty \tag{35}$$

and hence for thick films we get

$$\frac{\chi}{\chi_{crit}} - 1 = -\frac{N\pi^2}{18D^2} \tag{36}$$

while for very thin films ($\lambda/D>>1$) the solution occurs for small x where $\cot x\approx 1/x$ and hence $x^2=D/2\lambda$, or

$$\frac{\chi}{\chi_{crit}} - 1 = -\frac{2N}{9D\lambda} \tag{37}$$

Note (from the relations between H_1, λ, μ and g) that a useful order of magnitude of μ_1 and g is order 1/N, because only then is H_1 (a field normalized by temperature) of order unity {this is already clear from Eqs. (7) and (9), because $f(\phi)$ is of order 1/N, and in order to have a non-trivial competition between bulk terms and the "bare" surface free energy the latter also should be small of order 1/N, for the treatment in Eqs. (24)-(37) to be valid}. Since g is of order 1/N, we write it as $g=g'/N$, g' being a constant of order unity, and thus $\lambda=-2\,N/9g'$. Now the condition $|\lambda|/D=1$, where crossover from Eq. (36) to Eq. (37) occurs, implies a crossover at a large thickness D_{cross}

$$D_{cross}\approx 2\,N/(9g') \tag{38}$$

While the spirit of this treatment hence closely parallels correponding treatments of phase transition shifts in thin films in general [216], the existence of the large factor N in Eqs. (36), (38) is special for polymers, as emphasized by Tang et al. [38].

While the situation $H_1=0$ considered above implies that we consider a "neutral" film, i.e. there is no enthalpic preference for any component at the walls, the results obtained in Eqs. (36)-(38) are in fact more general. As an example, we consider the case $H=0$, $H_1\neq 0$. Then Eq. (30) yields for the "susceptibility" $\bar{S}_1 = \partial \bar{m} / \partial H_1$

$$\bar{S}_1 = \frac{\lambda}{C} \frac{2\xi_b / D}{\coth\left[D/(2\xi_b)\right] + \lambda/\xi_b} \tag{39}$$

Since the denominators of $\bar{S}_1$ and of $\bar{S}$ are the same, and the divergence of $\bar{S}$ was caused by the vanishing of this denominator, it readily follows that the $\chi_{crit}(D)$ as located above also yields a divergence of $\bar{S}_1$.

Of course, it would be of interest to consider the full Eq. (24), where the nonlinear term $m^3(z)/3$ is kept: then the full behavior both for $\chi>\chi_{crit}(D)$ and for $\chi<\chi_{crit}(D)$ can be analyzed, but such a treatment is already fairly involved [224,225], and not given here. We only note that on the level of Eq. (24)-(26), the only reference that is made to polymers (rather than mixtures of small molecules) is via factors of N in various coefficients. In particular, the factor N in front of the term $d^2 m/dz^2$, which gives rise to the $\sqrt{N}$ prefactor in the correlation length ξ_b, Eq. (27), explicitly illustrates the fact that the gyration radius is a natural length scale of the problem. It is already clear from Eqs. (7)-(10) that the dependence on chain length can be completely eliminated, if D is measured in units of $b\sqrt{N}$ (here we restore again the unit of length b) and if the surface term is written in terms of normalized coefficients $\mu_1' = \mu_1\sqrt{N}$, $g' = \sqrt{N}g$, since then all N-dependence apart from a scale factor $\sqrt{N}$ for ΔF disappears,

$$\sqrt{N}\Delta F / k_B T = \int_0^{D/b\sqrt{N}} dz' \left\{ \phi \ell n\phi + (1-\phi)\ell n(1-\phi) + 2\frac{\chi}{\chi_{crit}}\phi(1-\phi) - \Delta\mu'\phi + \frac{1}{36\phi(1-\phi)}\left(\frac{d\phi}{dz'}\right)^2 \right\}$$

$$-\mu_1'\phi_0 - \frac{1}{2}g_1'\phi_0^2 - \mu_1'\phi_D - \frac{1}{2}g'\phi_D^2 \quad \text{with } \Delta\mu' \equiv \Delta\mu N \tag{40}$$

Equation (40) may help to convert the examples presented in Figs. 7-9 to other cases of physical interest. Equation (40) makes it clear once more that the present mean field theory is too crude to describe the interplay between the configurational statistics of flexible polymers and surface enrichment in thin films:

the only remnant of the gaussian statistics of coils in the present theory is the prefactor $[36\phi(1-\phi)]^{-1}$ of the $(d\phi/dz')^2$ term.

2.2 Block Copolymers in the Weak Segregation Limit

In a diblock copolymer (AB), a chain of N_A subunits of type A is at one end covalently bonded to a chain of N_B subunits of type B. Therefore, a positive χ parameter (describing the net repulsive interaction between A and B monomers) cannot lead to macroscopic phase separation but only to microphase separation [185–187]: at an order-disorder transition concentration waves are formed spontaneously, with a wavelength of the same order as the gyration radius of the coils. The type of long range order that forms depends on the composition $f=N_A/N$, $N \equiv N_A+N_B$, of the copolymers, and on the Flory-Huggins parameter χ (ignoring also effects due to disparity in the size and shape of the monomeric subunits [231], due to the finite chain length, etc.). Here we emphasize the simplest case, symmetrical composition $f=1/2$, for which the lamellar mesophase forms [185–187,197,210–213]. We take as an order parameter density $\Psi(\vec{x})$ the deviation of the local volume fraction $\phi_A(\vec{x})$ at point $\vec{x}$ from the average composition, $\Psi(\vec{x}) \equiv \phi_A(\vec{x}) - f$. As in Sect. 2.1, the system is assumed to be incompressible, and hence $\phi_A(\vec{x}) + \phi_B(\vec{x}) \equiv 1$.

For this problem already the simple mean field approximation becomes rather involved [197,213]. Therefore, we describe here only an approach, which is even more simplified, appropriate for wavenumbers q near the characteristic wavenumber q^*, but strictly correct neither for $q \to 0$ nor for large q: {the spirit of our approach is similar to the long wavelength approximation encountered in the mean field theory of blends, Eq. (7)}. That is, we write the effective free energy functional as an expansion in powers of ψ and include terms $(\nabla\psi)2$ as well as $(\nabla^2\psi)^2$, as in the related problem of lamellar phases of microemulsions [232,233], namely [234]

$$\frac{N\Delta F}{k_B T} = \int d\vec{x} \left\{ \frac{1}{2} \Psi(\vec{x}) \left[\tau_0 + e_0 \left(\nabla^2 + q^{*2} \right)^2 \right] \Psi(\vec{x}) + \frac{u_0}{4!} \left[\Psi(\vec{x}) \right]^4 \right\} \tag{41}$$

From the random phase approximation [200] the characteristic wavenumber q^* of lamellar ordering can be calculated [197],

$$q^* = 2\pi / \lambda^* \approx 1.9456 / \sqrt{\left\langle R_g^2 \right\rangle} \tag{42}$$

where $\left\langle R_g^2 \right\rangle$ is the mean square gyration radius of the chains. The coefficient τ_0 of the quadratic term is the temperature-like variable, which changes sign when the Flory-Huggins parameter χ reaches a critical value χ_c [197],

$$\tau_0 = 2\rho_c N(\chi_c - \chi) \qquad \chi_c = 10.495 / N \tag{43}$$

Note that on the mean field level at χ_c a second-order transition is predicted for f=1/2, while taking fluctuations into account renders the transition first order [192,210,211], as also found experimentally [231]. Although the nature of surface effects on this transition is quite different for the first-order case [6] than for the second-order case [12], we discuss mostly the second-order case here. The constant ρ_c in Eq. (43) is the density of the chains (ρ_c=1/N if a Flory-Huggins lattice with lattice spacing unity is invoked), and the constants e_0 and u_0 can be derived [197] from the random phase approximation as

$$e_0 = 1.5\rho_c c^2 \left\langle R_g^2 \right\rangle / q^{*2}, u_0 = \rho_0 \Gamma \tag{44}$$

with c and Γ being constants of order unity [197].

Assuming now a solution of the form of a simple standing wave,

$$\Psi(\vec{x}) = 2A\cos(q^* \vec{n} \cdot \vec{x}) \tag{45}$$

where $\vec{n}$ is a unit vector specifiying the orientation of the lamellae, Eq. (41) reduces to the standard free energy density $f_L(A)$ of the Landau theory of second order transitions, V being the total volume of the system

$$f_L(A = N\Delta\mathcal{F}/(\mathbf{k}_b TV) = \tau_0 A^2 + \frac{1}{4} u_0 A^4 \tag{46}$$

which implies that for $\tau_0<0$ there is a non-zero amplitude $A = \sqrt{-2\tau_0 / u_0}$ while for $\tau_0 \geq 0$ the spontaneous order is zero, $A \equiv 0$.

When we now consider a thin film of thickness D, Eq. (41) must be supplemented by boundary conditions of the same type as in the polymer blend case, Eqs. (7) and (10), i.e. we add a (bare) surface free energy contribution to the free energy that accounts for preferential attraction of one kind of monomers to the walls, "missing neighbors" in the pairwise interactions, and possible changes in the pairwise interactions near the surface. As in the blend case, this surface contribution is taken locally at the walls only and expanded to second order in the local order parameter $\psi(z)$. Per unit area of the wall, this free energy is written as

$$NF_s^{(\text{bare})} / k_B T = \int_{-D/2}^{+D/2} dz \left[\delta(z - D/2) + \delta(z + D/2)\right] f_s\{\psi(z)\} \tag{47}$$

with

$$f_s(\psi) = -H_1\psi + 1/2a\,\psi^2 \tag{48}$$

We now make the restrictive assumption that only lamellar ordering with $\vec{n}$ perpendicular to the walls needs to be considered, i.e. $\Psi(\vec{x})$ in Eq. (41) is a function of the z-coordinate in the direction perpendicular to the walls only.

Then the free energy that needs to be minimized becomes, per unit area of the walls,

$$\frac{N\Delta F_{\text{film}}}{k_B T} = \int_{-D/2}^{+D/2} dz \left\{ \frac{1}{2} \Psi(z) \left[\tau_0 + e_0 \left(\frac{d^2}{dz^2} + q^{*2} \right)^2 \right] \Psi(z) + \frac{u_0}{4!} \Psi^4(z) \right\} + \frac{N F_s^{(\text{bare})}}{k_B T} \tag{49}$$

and the condition that this free energy is minimal leads to the following Euler-Lagrange equation

$$\frac{N\delta\Delta F_{\text{film}}}{k_B T \delta\Psi(z)} = \left[\tau_0 + e_0 \left(\frac{d^2}{dz^2} + q^{*2} \right) \right] \Psi(z) + \frac{u_0}{3!} \Psi^3(z) = 0 \tag{50}$$

supplemented by the boundary conditions ($\Gamma \equiv 2\left\langle R_g^2 \right\rangle / N$)

$$a\Psi(z = -D/2) - H_1 - \Gamma \frac{d\Psi}{dz} \Big|_{z=-D/2} = 0 \tag{51}$$

$$a\Psi(z = D/2) - H_1 + \Gamma \frac{d\Psi}{dz} \Big|_{z=D/2} = 0 \tag{52}$$

Due to the symmetry between the boundary conditions at $z=-D/2$ and at $z=+D/2$ the solution in the center of the film must be flat, i.e. $d\Psi/dz=0$ for $z=0$. Thus $\Psi(z)$ must be an even function of z.

For $\tau_0>0$ and sufficiently weak forces from the walls it is legitimate to neglect the non-linear term $u_0\Psi^3(z)/3!$ in Eq. (50). Then it is easy to find the solution of Eqs. (50)-(52) as [61]

$$\psi(z) = A[e^{z/\xi} \cos(q^* z - \phi) + e^{-z/\xi} \cos(q^* z + \phi)] \tag{53}$$

with the amplitude A determined by the parameters a, H_1 of the boundary condition

$$\begin{aligned} A = \xi H_1 / \{ & a\xi \exp(D/2\xi) \cos(-q^* D/2 + \phi) + a\xi \exp(-D/2\xi) \\ & \cos(q^* D/2 + \phi) + q^* \xi \Gamma \exp(D/2\xi) \sin(-q^* D/2 + \phi) \\ & - q^* \xi \Gamma \exp(-D/2\xi) \sin(q^* D/2 + \phi) + \Gamma \exp(D/2\xi) \\ & \cos(-q^* D/2 + \phi) - \Gamma \exp(-D/2\xi) \cos(q^* D/2 + \phi) \} \end{aligned} \tag{54}$$

In Eqs. (53) and (54) ξ denotes the correlation length of order parameter fluctuations in the disordered phase,

$$\xi = \hat{\xi}\left(1-\chi/\chi_c\right)^{-1/2} \tag{55}$$

the amplitude $\hat{\xi}$ being of the order of the gyration radius of the coils [12,61,234]. Finally, the phase ϕ is determined by the condition that for a symmetric block copolymer melt the total number of A-monomers equals the total number of B-monomers, and hence the integral over the order parameter must vanish [12],

$$\int_{-D/2}^{+D/2} \psi(z)dz = 0 \tag{56}$$

For a bulk block copolymer melt, there is translational invariance and hence the phase ϕ could be omitted in Eq. (45) by a suitable choice of the coordinate origin. In a thin film, the translational invariance in z-direction is broken, and hence it is indispensable to allow for a non-zero phase ϕ in Eq. (53). In fact, Eq. (56) then yields [61]

$$\tan\phi = \frac{e^{\frac{D}{2\xi}}\left[\cos\frac{q^*D}{2} + q^*\xi\sin\frac{q^*D}{2}\right] - e^{-\frac{D}{2\xi}}\left[\cos\frac{q^*D}{2} - q^*\xi\sin\frac{q^*D}{2}\right]}{e^{\frac{D}{2\xi}}\left[q^*\xi\cos\frac{q^*D}{2} - \sin\frac{q^*D}{2}\right] - e^{-\frac{D}{2\xi}}\left[q^*\xi\cos\frac{q^*D}{2} + \sin\frac{q^*D}{2}\right]} \tag{57}$$

Of course, the semi-infinite geometry ($D\to\infty$) is a special case of this treatment, and hence we first demonstrate that the solution obtained by Fredrickson [12] is a special case of Eqs. (53)-(57). Being interested in the solution near $z=-D/2$, it is clear that the second term in Eq. (53) dominates, and the first term may be neglected. In terms of a shifted coordinate $z'=z+D/2$ this yields

$$\psi(z) \approx A\exp((D/2-z')/\xi)\,\cos(q^*z'-Dq^*/2+\phi) \qquad D\to\infty \tag{58}$$

with

$$A \approx \xi H_1 \exp(-D/2\xi)/\{(a\xi+\Gamma)\cos(-q^*D/2+\phi) + q^*\xi\Gamma\sin(-q^*D/2+\phi)\} \tag{59}$$

In terms of a redefined phase $\varphi=\phi-q^*D/2$ Eqs. (58) and (59) indeed yield the result of Fredrickson [12] ($z'=z+D/2$)

$$\psi(z) = \xi H_1 \exp(-z'/\xi)\,\cos(q^*z'+\varphi)/\{(q\xi+\Gamma)\cos\varphi + q^*\xi\Gamma\sin\varphi\} \tag{60}$$

noting that Eq. (57) in this limit $D \to \infty$ reduces to [61]

$$\tan\varphi = (q^*\xi)^{-1} \tag{61}$$

From Eq. (60) one sees that the preferred attraction of one species to the wall causes a local lamellar ordering near the wall. The amplitude function $B(z') \equiv \Psi(z') \frac{\cos\varphi}{\cos(q^* z' + \varphi)} = \xi H_1 \exp(-z'/\xi)/\{a\xi + \Gamma + q^*\xi\Gamma\tan\varphi\} = \xi H_1 \exp(-z'/\xi) / (a\xi + 2\Gamma)$ decays exponentially on the scale of the correlation length ξ towards the bulk. Since this (Leibler-type [197]) mean field theory of block copolymer ordering implies at $\tau_0=0$ ($\chi=\chi_c$) a second-order transition, cf. Eq. (55), the surface induced ordering becomes long-range as $\chi \to \chi_c$ (for $a>0$). If $a<0$, however, we encounter a divergence of the denominator of $\psi(z)$ in Eq. (60) already in the disordered phase before the bulk orders, namely for

$$a\xi + 2\Gamma = 0, \text{i.e. for } 1 - \chi_s/\chi_c = a^2\hat{\xi}^2 / \left(4\Gamma^2\right) \qquad a<0 \tag{62}$$

The singularity at $\chi=\chi_s$ means that spontaneous two-dimensional long range order sets in at the surface even in the absence of a preferential attraction ($H_1=0$). This "surface transition" has been discussed extensively in the context of surface effects on magnetic systems [216] and related phase transition problems. Although one can draw a similar "surface phase diagram" here (Fig. 10), this phase transition is of less interest here: (i) it would occur as a sharp transition only for perfectly neutral walls, $H_1=0$, while for non-neutral walls it is rounded. (ii) Fluctuation effects [210,211] change the phase transition in the bulk from second order to first order even in a symmetric mixture, and this fact also significantly alters the surface phase diagram (Fig. 10).

Returning now to the case of thin films, we note that in general our treatment implies a rounded crossover from an inhomogeneous lamellar ordering with largest amplitudes at the walls, as described by Eqs. (53)-(57) and illustrated in Fig. 11, to a nearly homogeneous lamellar ordering. Note, however, that there is an interesting problem in the thin film that is absent for the semi-infinite geometry: in general, the thickness D is incommensurate with the wavelength λ^* of the lamellar ordering. This causes interesting "finite size"-effects: in addition to factors $\exp(\pm D/2\xi)$ in Eqs. (54) and (57), which are common in the theory of finite size scaling [236–239] ("the linear dimension scales with the correlation length"), there is an oscillatory size-dependence via the terms $\cos(q^*D/2)$, $\sin(q^*D/2)$. Proceeding as in the case of unmixing (Sect. 2.1), we may again consider neutral walls and ask when spontaneous ordering would set in (the signal of spontaneous lamellar ordering parallel to the walls would again be the vanishing of the denominator of the amplitude A in Eq. (54), i.e. a divergent response to a weak biasing field H_1 at the walls). This condition $A^{-1}=0$ yields [61]

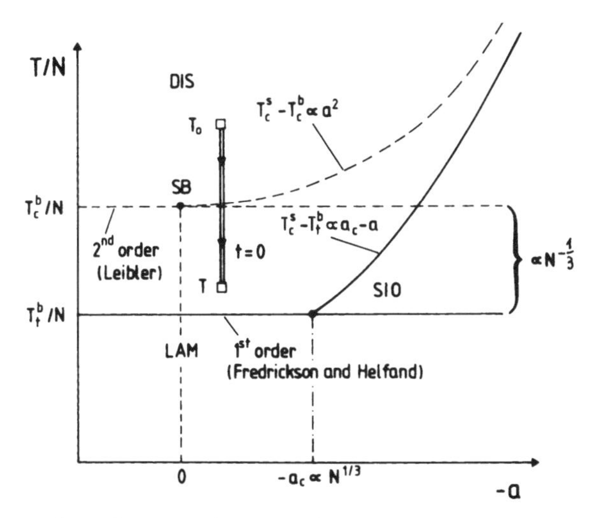

Fig. 10. Schematic phase diagram of a semi-infinite block copolymer melt for the special case of a perfectly neutral surface (H_1=0). Variables chosen are the surface interaction enhancement parameter (–a) and the temperature T rescaled by chain length (assuming $\chi \propto 1/T$ the ordinate hence is proportional to χ_c/χ). While according to the Leibler [197] mean-field theory a symmetric diblock copolymer transforms from the disordered phase (DIS) at $T_c^b \propto N$ in a second-order transition to the lamellar phase (LAM), according to the theory of Fredrickson and Helfand [210] the transition is of first-order and depressed by a relative amount of order $N^{-1/3}$. In the second-order case, the surface orders before the bulk at a transition temperature $T_c^s (\propto 1/\chi_s)$ as soon as a is negative [216], and the enhancement of T_c^s is quadratic in a, $T_c^s - T_c^b \propto a^2$. In the first-order case surface ordering before the bulk occurs only if a exceeds a threshold value, $-a > -a_c$, with $-a_c \propto N^{1/3}$, and the enhancement of T_c^s then scales linear with a ($T_c^s - T_c^b \propto a_c - a$) [61]. For $T_t^b < T < T_c^s$ one then should observe formation of an ordered "wetting layer" which grows in thickness as $T \to T_t^b$ ("surface induced ordering", SIO [6,235]). Note, however, that the physically natural scale for both parameters H_1 and a is N, cf. Eqs. (47) and (48). From Binder et al. [61]

$$\exp(-D/\xi) = (2 - \xi a/\Gamma)^{-1} \left\{ \begin{array}{l} 2\cos(q^* D) + \dfrac{a}{\Gamma q^*}\sin(q^* D) \\ + \left[\left(\dfrac{\xi a}{\Gamma}\right)^2 - 4\sin^2(q^* D) + \dfrac{2a}{\Gamma q^*} \right. \\ \left. \sin(2q^* D) + \left(\dfrac{a}{\Gamma q^*}\right)^2 \sin^2(q^* D) \right]^{1/2} \end{array} \right\} \tag{63}$$

which shows that superimposed on the power law {Eq. (36)} there is an oscillatory dependence of the critical point on the thickness of the film,

$$\frac{\chi_c(D)}{\chi_{crit}} - 1 = D^{-2} f\{\cos(q^* D)\} \tag{64}$$

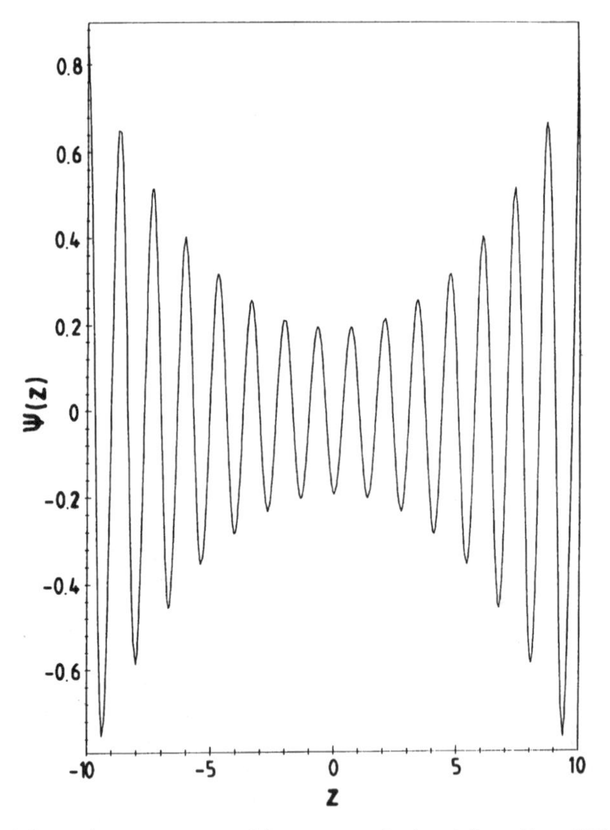

Fig. 11. Plot of the order parameter $\psi(z)$ vs z, as calculated from Eqs. (53)-(57), for the choice of parameters $\xi=5$, $D=20$, $H_1=0.2$, $a=0.2$, and $\Gamma=0.08$ (note that all lengths are measured in units of the gyration radius R_g here, and hence $q^*=1.9456\sqrt{6}=4.7656$). From Binder et al. [61]

where $f\{\cos(q^*D)\}$ is a function that can be calculated from Eq. (63) numerically. Analysis is easiest for the case $q^*D=n\pi$, n=1,2,3,..., i.e. film thicknesses commensurate with the lamellar ordering: then there is no finite size effect on the phase φ in the thin film, Eq. (61) holds also for these finite values of D. For $a<0$ Eq. (63) then reduces to

$$D/\xi=\ell n\{(2\Gamma+|a|\xi)/(2\Gamma-|a|\xi)\} \tag{65}$$

For large D we note that $\chi_c(D)$ converges to the surface transition χ_s, where $2\Gamma=|a|\xi$, while for small D the ordering occurs for even smaller values of χ. For $a>0$, however, the thin film does not order at smaller values of χ than the bulk would order ($\chi=\chi_{crit}$). In fact, one finds that for odd multiples n $\chi_c(D)=\chi_{crit}$, there is no shift of the critical point in this case. However, for even multiples n Eq. (63) leads to

$$\xi \coth(D/2\xi) = -2\Gamma / a \text{ or } (\chi > \chi_{crit})\ x^{-1} \cot x = 4\Gamma / (aD) \qquad x \equiv D(2/\xi) \quad (66)$$

where in the last step coth ix=–i cot x $\left(i \equiv \sqrt{-1}\right)$ was used. For D→∞ this equation,

Eq. (66), is solved by x→π/2, and hence one obtains a shift

$$\frac{\chi_c(D)}{\chi_{crit}} - 1 = \hat{\xi}^2 \pi^2 / D^2 \quad (67)$$

As D increases one hence has an oscillatory dependence of the amplitude of the D^{-2} term between zero as a minimum value and $\hat{\xi}^2\pi^2$ as a maximum value, explicitly illustrating Eq. (64).

We emphasize, however, that these predictions only apply in the framework of a treatment where one insists on a lamellar arrangement parallel to the walls from the outset, and ignores deliberately any lateral inhomogeneities of the possible ordering. Monte Carlo studies [34,35], in fact, have given the first indication that this assumption is too restrictive: for thin films and weak surface fields, the system can deal better with the misfit arising from $D \neq (2n+1)\lambda$, with n integer, by forming arrangements of the lamellae perpendicular to the walls rather than parallel. The same conclusion emerges from treatments based on the strong segregation theory [35,36,63], as will be discussed below. The simplification of the free energy functional, Eq. (41), in terms of ∇^2 and ∇^4 terms can be avoided [12,42] by using a vertex function $\Gamma_2(q)$ that has in reciprocal space the correct limits both for small and for large wavenumbers, namely [12]

$$\Gamma_2(q) = 24 / \left(q^2 \langle R_g^2 \rangle \right) + 2q^2 \langle R_g^2 \rangle - \bar{\chi} \qquad \bar{\chi} = 2N(\chi - \chi_c) + 2\sqrt{3} \quad (68)$$

However, the Euler-Lagrange equation then becomes an integro-differential equation [12]

$$\frac{24}{\langle R_g^2 \rangle} \int_{-D/2}^{+D/2} dz G(z - z')\psi(z) - 2\langle R_g^2 \rangle \frac{d^2\psi}{dz^2} - \bar{\chi}\psi(z) - \theta = 0 \quad (69)$$

where θ is a Lagrange multiplier (necessary to ensure $\int_{-D/2}^{+D/2} \psi(z) dz = 0$) and G(z-z') is the Green's function for the one-dimensional Poisson equation, $G(z-z') = D^{-1} \sum' q^{-2} \exp[iq(z-z')]$, the prime on the sum indicating that the q=0 term is excluded. For the semi-infinite problem (D→∞), this more elaborate approach yields again a solution of the form $\psi(z) = \psi_1 \exp(-z/\xi) \cos(\tilde{q}z + \phi) / \cos\phi$ as above, but it predicts a weak temperature dependence of the wavenumber of the oscillations, $\tilde{q}=q^*\left[\left(1+\overline{\chi}/2\sqrt{3}\right)/2\right]^{1/2}$. For films of finite thickness, however, a numerical treatment is required [42]. From a stability analysis of the disordered phase an oscillatory variation of the transition temperature $\chi_c(D)$ with film thickness D is found for $H_1=0$, as in our simplified treatment above. Interestingly, in the presence of non-zero surface fields $H_1\neq 0$, where one expects the sharp transition from the disordered phase to the lamellar phase to be rounded because there is a gradual transition from a surface-induced ordered state for $\chi>\chi_c$, the treatment of Wong et al. [42] does predict sharp second-order transitions to occur for certain regimes of $D/\sqrt{\langle R_g^2\rangle}$. These transitions have nothing to do with the transitions between different orientations of the lamellae (Fig. 3a,b), since the treatment suppresses any lateral inhomogeneity from the start, treating the order parameter as a function $\psi(z)$ of the coordinate z perpendicular to the walls only.

Another criticism of the present treatment in the case of $H_1\neq 0$ is that for χ near χ_c the surface term induces such a strong local order near the surface that it is no longer legitimate to neglect the quartic term $u_0[\psi(\vec{x})]^4/4!$ in Eq. (41) [23]. Tang and Freed [23] argue that even for $\chi\rightarrow\chi_c$ the decay length ξ of the oscillatory profile of the order parameter is distinct from the correlation length ξ_b in the bulk that is defined in the framework of the Leibler theory [197] from the expansion of the structure factor S(q) at q^*, $S^{-1}(q)=S^{-1}(q^*)\,[1+(q-q^*)^2\,\xi_b^2\,]\}$ and ξ does not diverge when $\chi\rightarrow\chi_c$ while ξ_b does {Eq. (55)}. Tang and Freed [23] also discuss the case of thin films but do not constrain the system to have the same thickness D everywhere, but rather assume that a lateral phase separation in domains with n lamellae and with (n+1) lamellae is possible (Fig. 4).

The result that the decay length ξ of the order parameter oscillations remains finite at χ_c is also suggested by Shull [20] using a numerical self-consistent field treatment. On the other hand, for not too strong H_1 and $\chi<\chi_c$ his treatment is in good numerical agreement with the result of Fredrickson [12], and hence one can conclude that the extension to thin films [61] described above is sensible.

It must be remembered, however, that the Leibler-type mean field theory [197] is believed to be accurate for the limit of infinite chain length, $N\rightarrow\infty$; for finite N effects of order parameter fluctuations are important and change the character of the transition from second order to first order even for symmetric composition [185,186,192,210,211]. With a self-consistent Hartree approximation that one believes to be valid for large N, Eq. (41) gets replaced by [234]

$$\frac{N\Delta F}{k_BT}=\int d\vec{X}\left\{\frac{1}{2}\overline{\psi}(\vec{X})\left[\tau_R+e\left(\nabla^2+q_0^2\right)^2\right]\overline{\psi}(\vec{X})+\frac{U_R}{4!}\overline{\psi}^4+\frac{W_R}{6!}\overline{\psi}^6\right\} \tag{70}$$

where we have introduced rescaled variables, $\vec{X}=\vec{x}/\sqrt{6\langle R_g^2\rangle}$, $\overline{\psi}=c\overline{N}^{1/4}\psi$, $\overline{N}$ being an effective chain length $\left\{ \overline{N} \equiv \rho_c^2 \left(6 \left\langle R_g^2 \right\rangle^3 \right) \right\}$. The parameters e_0, q^* of Eq. (41) then also get rescaled into $e \equiv \left[24 q^{*2} \left\langle R_g^2 \right\rangle \right]^{-1}$, $q_0^2 \equiv 6 q^{*2} \left\langle R_g^2 \right\rangle$. However, the key point of this calculation is that the parameters τ_0, u_0 get a much more complicated dependence on χ, and near χ_c the coefficient U_R is negative, and this is the mechanism by which a first-order transition arises [234]. These renormalized parameters τ_R, U_R, W_R in Eq. (70) are given by the implicit equations

$$\begin{aligned} &\tau_R = \tau + du\tau_R^{-1/2}, \; U_R = u\left(1 - du\tau_R^{-3/2}/2\right)/\left(1 + du\tau_R^{-3/2}/2\right) \\ &W_R = 4.5du^3 \tau_R^{-5/2}\left(1 + du\tau_R^{-3/2}/2\right)^3 \end{aligned} \tag{71}$$

where $\tau \equiv 2N(\chi_c - \chi)/c^2$ and the constants d, u being $d \equiv 3q^{*2} \left\langle R_g^2 \right\rangle / (2\pi)$, $u \equiv u_0 / (\rho_c c^4 \overline{N}^{1/2})$. If one again tries to describe the ordering in terms of a single wave, $\overline{\psi}(\vec{X}) = 2A\cos(q_0 \vec{n} \cdot \vec{X})$, Eq. (70) can be reduced to a simple free energy density which exhibits three minima in the transition region,

$$f_H(A) = \left(6 \left\langle R_g^2 \right\rangle \right)^{3/2} \Delta F / (k_B T V) = \tau_R A^2 + \frac{U_R}{4} A^4 + \frac{W_R}{36} A^6 \tag{72}$$

because for small τ_R the coefficient U_R is negative. The first-order transition occurs when all three minima of the polynomial are equally deep, which leads to a shift of the transition of order $\overline{N}^{-1/3}$ [210],

$$\chi_t N = 10.495 + 37.823 \left(\overline{N} \right)^{-1/3} \qquad \overline{N} \to \infty \tag{73}$$

and at this value $\chi = \chi_t$ the amplitude A of the order parameter jumps discontinuously from zero for $\chi < \chi_t$ to $A_t \propto (\overline{N})^{-1/6}$ at $\chi \geq \chi_t$. While in the disordered phase it is now the parameter τ_R which controls the χ-dependence of the correlation length [234],

$$\xi = \sqrt{6 \left\langle R_g^2 \right\rangle / \tau_R} \tag{74}$$

at the transition point ($\chi = \chi_t$) we have τ_R non-zero (of order $\overline{N}^{-1/3}$ as well) and hence $\xi \propto N^{2/3}$ for $\chi = \chi_t$. If one neglects non-linear effects in the disordered phase, the above mean field treatment of surface-induced ordering remains valid for $\chi < \chi_t$, the only significant distinction being that the temperature (or χ)-dependence of ξ now is different, and ξ is no longer divergent at the transition [61].

However, this is not the whole story, since near χ_t the non-linear terms in the Euler-Lagrange equation

$$\frac{N}{k_B T}\frac{\delta \Delta F}{\delta \overline{\psi}(\vec{X})} = \left[\tau_R + e\left(\nabla^2 + q_0^2\right)^2\right]\overline{\psi}(\vec{X}) + \frac{U_R}{3!}\left[\overline{\psi}(\vec{X})\right]^3 + \frac{W_R}{5!}\left[\overline{\psi}(\vec{X})\right]^5 = 0 \quad (75)$$

must not be neglected [6,60], and actually the surface terms may stabilize a layer of the ordered phase at the surface, whose thickness ℓ diverges as χ approaches its value χ_t at the first-order transition [6,220,235]

$$\ell \propto \xi_t \ln(1 - \chi/\chi_t) \quad (76)$$

For distances z from the surface smaller than ℓ the order parameter amplitude A(z) basically takes already its value A_t at the transition, while for $z \approx \ell$ there is a rather well developed interfacial profile, that resembles the interfacial profile between bulk ordered and disordered phases coexisting at χ_t [5], see Figs. 12,13. Writing $\mathfrak{z}=z/\xi$ and using a solution in terms of waves cos (q^*z), sin (q^*z) as

$$\overline{\psi}(Z) = \left(8\tau_R / |U_R|\right)^{1/2}\left[B(\mathfrak{z})\exp(iq^*z) + B^*(\mathfrak{z})\exp(-iq^*z)\right] \quad (77)$$

one can derive from Eq. (75) a simplified amplitude equation for the complex amplitude $B(\mathfrak{z})$,

$$\left(1 - \frac{d^2}{d\mathfrak{z}^2}\right)B - 4B|B|^2 + \frac{16W_R\tau_R}{3U_R^2}B|B|^4 = 0 \quad (78)$$

Note that in deriving Eq. (78) terms of order ξ^{-1} have been omitted (which is legitimate for χ near χ_t only), and the constraint $\int dZ\overline{\psi}(Z) = 0$ has not yet been considered. Since at the order-disorder transition we have $16W_R\tau_R/(3U_R^2)=3$, it is convenient to rewrite this coefficient in general as $16W_R\tau_R/(3U_R^2) \equiv 3(1+\varepsilon)$, where near the transition point $\tau_R = \tau_R^t$ we have $\varepsilon \approx (\tau_R - \tau_R^t)/\tau_R^t$. Equation (78) allows 5 uniform solutions: the disordered phase being B=0, two solutions are unstable, corresponding to maxima of $f_H(A)$ rather than minima, and two solutions $\pm B_0$ correspond to the ordered phase, with

$$B_0^2 = \left[2 + (1 - 3\varepsilon)^{1/2}\right] / \left[3(1+\varepsilon)\right] \approx 1 - \frac{3}{2}\varepsilon \quad \text{for } \varepsilon \to 0 \quad (79)$$

Considering now the problem of an interface between bulk coexisting phases, we have $\varepsilon=0$, and must choose boundary conditions that ensure that the appro-

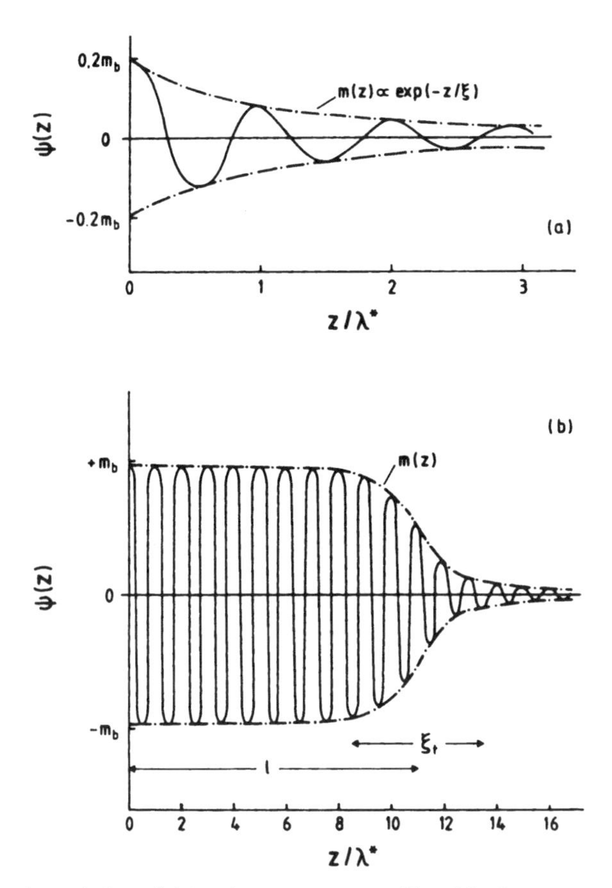

Fig. 12. Schematic variation of the order parameter profile $\psi(z)$ of a symmetric (f=1/2) diblock copolymer melt as a function of the distance z from a wall situated at z=0. It is assumed that the wall attracts preferentially species A. Case (a) refers to the case $\chi<<\chi_t$, where non-linear effects are still negligible, correlation length ξ and wavelength λ^* are then of the same order of magnitude, and it is also assumed that the surface "field" H_1 is so weak that at the surface it only induces an order parameter 0.2 m_b if m_b is the order parameter amplitude that appears for $\chi=\chi_t$ at the first-order transition in the bulk. Case (b) refers to a case where χ is only slightly smaller than χ_t, such that an ordered "wetting layer" of thickness l [Eq. (76)] much larger than the interfacial thickness {which is of the same order as ξ [Eq. (74)]} is stabilized by the wall, while the bulk is still disordered. The envelope (denoted as m(z) in the figure) of the order parameter profile is then essentially identical to an interfacial profile between the coexisting ordered phase at $T=T_t$ for (z<l) and the disordered phase (for z>l). The quantitative form of this profile [234] is shown in Fig. 13. From Binder [6]

priate bulk order parameter is reached at either side of the interface,

$$\lim_{\mathfrak{z}\to-\infty} B(\mathfrak{z}) = 1, \lim_{\mathfrak{z}\to+\infty} B(\mathfrak{z}) = 0, \lim_{\mathfrak{z}\to\pm\infty} \frac{d}{dZ} B(\mathfrak{z}) = 0 \tag{80}$$

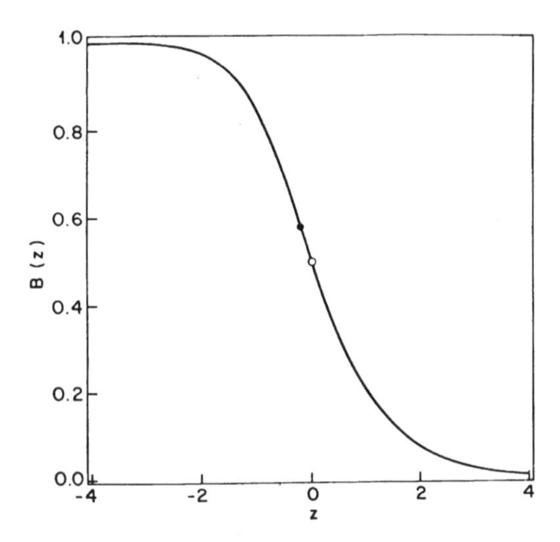

Fig. 13. Plot of the amplitude (envelope) function B(Z) vs Z, for a planar interface between coexisting disordered ($Z\to\infty$) and lamellar ($Z\to-\infty$) phases of block copolymer melts in the bulk. The midpoint of the profile (Z,B)=(0,1/2) is denoted by an *open circle*, while the inflection point (*solid circle*) is at [—(1/2) ln(3/2), $1/\sqrt{3}$]. From Fredrickson and Binder [234]

The solution of Eq. (78) with these boundary conditions is [234]

$$B(\mathfrak{z})=[1+3\exp(2\mathfrak{z})]^{-1/2} \tag{81}$$

Since there is no symmetry between the ordered and disordered phases of block copolymers, this profile is not antisymmetric around its midpoint. Now for $\varepsilon>0$ ($\chi<\chi_t$) Eq. (79) shows that the amplitude of the profile B_0 at which saturation would occur is somewhat reduced, and also the finite distance ℓ of the midpoint of the profile from the wall (Fig. 12) implies that the left part of the envelope $B(\mathfrak{z})$ is 'cut off'.

So the description of surface-induced ordering at the surface of a semi-infinite lamellar block copolymer melt is the gradual unbinding of an interface between a thin ordered layer at the surface and the disordered bulk: as $\chi\to\chi_t$ the distance ℓ of this interface from the surface diverges, Eq. (76). At this point, we note two generalizations:

(i) for non-symmetric block copolymers {composition $f=N_A/(N_A+N_B)\neq 1/2$} there is a term proportional to ψ^3 in the free energy functionals, Eqs. (41) and (70), respectively, and this term leads to a first-order transition in the absence of fluctuation corrections as well. It has been suggested to combine these effects by generalizing Eq. (74) for $\chi=\chi_t$ as follows [6]

$$\xi_t \propto \sqrt{\langle R_g^2\rangle}\left[\bar{N}^{-1/3}+\text{const}(f-1/2)^2\right]^{-1/2} \tag{82}$$

(ii) For long-range surface perturbations (such as van der Waals forces), one expects that the growth of the ordered surface layer is faster than logarithmic, namely a power law [220]

$$\ell \propto \xi_t \left(1 - \chi / \chi_t\right)^{-1/3} \tag{83}$$

The situation is fully analogous to complete wetting at the surfaces of fluids or fluid mixtures [220], of course. Perhaps the closest analogy occurs between surface-induced lamellar ordering and the surface melting [220] of crystals – the distinction being, of course, that in the latter case it is the disordered rather than the ordered phase that is stabilized by the surface.

This wetting picture (Fig. 12 [6]) of surface-induced ordering in block copolymer melts has been considered recently by Milner and Morse [60]. They considered the transition from the state of weak surface-induced order (Fig. 12a) to the case of strong-surface induced order (Fig. 12b) and pointed out that typically a first-order transition may occur between these states, in analogy to the "prewetting transition" first proposed by Cahn [226] (Fig. 14a). This prewetting-type first-order transition may persist in a thin film (Fig. 14b), but it ends in a triple point where the surface excess (m_s^+) is still finite, of course, since no divergence of excess quantities is possible for finite thickness D. In the disordered state of the thin film for $H_1 < H_1^{trip}(D)$ both surfaces have weak surface-induced oscillations of the type of Fig. 12a, while for $H_1 > H_1^{trip}(D)$ both walls are coated with ordered layers of the type of Fig. 12b, only the bulk of the film still being disordered. The first-order transition where the two interfaces between the ordered surfaces and the disordered bulk of the film disappear, such that for $\chi > \chi_t$ $(D;H_1)$ the surface-induced order propagates throughout the film, can be [60] viewed as the analogue of "capillary condensation", encountered for the binary mixtures already in Sect. 2.1. But the analogy is not complete, since for the mixtures we have the average concentration (or the conjugate variable, the chemical potential difference betwen the two species A and B) as a "control variable", while no such variable exists for the block copolymers (a field conjugate to their order parameter is not physically meaningful). Thus for a suitable chosen average concentration we encounter the critical point in the thin film {at $\bar{\phi}_{crit}(D, H_1), \chi_{crit}(D, H_1)$}, where there is a second-order unmixing transition, and this transition exists beyond the mean field approximation (Fig. 5a). The line of first-order prewetting transitions (Fig. 6) does not end at the critical line $\chi_{crit}(D,H_1)$ in Fig. 14c, of course, it rather hits the coexistence curve at some value of $\chi_{trip}(D) > \chi_{crit}(D,H_1)$. At this point we have coexistence between a B-rich phase, and A-rich phase with a small surface excess, and an A-rich phase with a large surface excess in the thin film.

Thus, although the analogy between wetting phenomena in mixtures and surface-induced ordering in thin block copolymer films is not complete, the analogy does allow to extend the mathematical methods to study wetting phenomena to the present case, at least approximately. In particular, Milner and Morse [60]

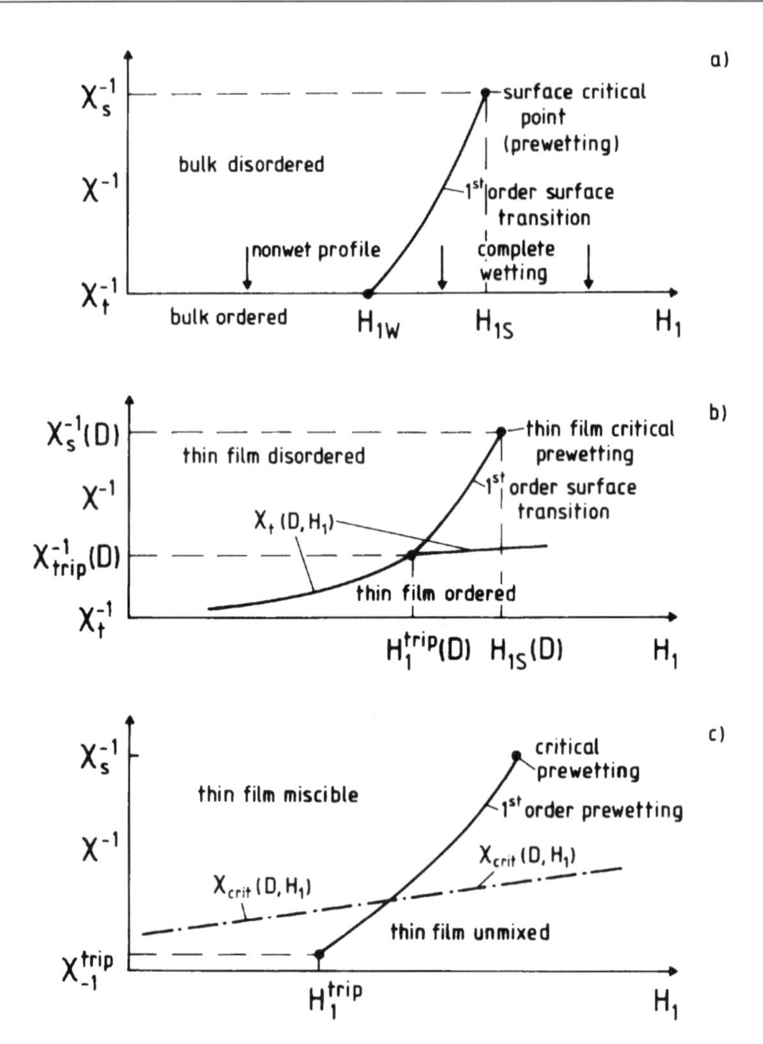

Fig. 14. Schematic phase diagram of the surface of a semi-infinite block copolymer (**a**) and a related system in a thin film geometry, assuming two walls a distance D apart, at which both the same surface field H_1 acts **b.** In the semi-infinite system for $H_1<H_{1w}$ the profile of the surface-induced order stays of the non-wet type for $\chi \rightarrow \chi_t$ (Fig. 12a), while complete wetting occurs for $H_1>H_{1w}$ (Fig. 12b). While for $H_1 \geq H_{1s}$ the transition between the profiles in Figs. 12a,b is completely gradual, for $H_{1w}<H_1<H_{1s}$ a 1st order surface transition occurs, where the surface excess (e.g. defined as $m_s = \int_0^\infty m(z)dz$ in Fig. 12) jumps from a small value (m_s^-) to a larger value (m_s^+). This is analogous to the prewetting transition sketched in Fig. 6b. At the surface critical point this jump vanishes, while at H_{1w} it diverges. This transition may persist in a thin film (**b**), but the distinction is that this surface order transition line will meet the line $\chi_t(D)$, where the thin film exhibits a first-order transition to the ordered phase, at a triple point $(H_1^{trip}(D),\ \chi_{trip}(D))$ where m_s is still finite. Case (**c**) draws the analogous phase diagram of a polymer mixture in thin film geometry that has a prewetting transition in semi-infinite geometry (Fig. 6a).

discuss in detail the complications due to the constraint $\int_{-D/2}^{+D/2} \psi(z)dz = 0$, which in the above linear treatment we have taken into account by introducing a suitable phase ϕ. Milner and Morse [60] show that in the non-linear case one needs to consider a slow variation of the phase $\phi(z)$ across the film.

Finally, we return to the case H_1=0, i.e. consider a surface that does not prefer one species as compared to the other. In the framework of the linear theory, we have seen that still surface-induced order occurs if the constant a in the surface term {Eq. (48)} is negative. As shown already in Fig. 10, this transition can only occur if -a exceeds a critical value -a_c, if fluctuations are taken into account. This fact is simply seen considering the response function χ_{11} [216]

$$\chi_{11} = \partial\psi_1 / \partial H_1 \propto \left(1/\xi + a\left\langle R_g^2 \right\rangle\right)^{-1} \tag{84}$$

since the surface transition can be located via a divergence of χ_{11} {remember that ψ_1 is the local order parameter $\psi(z=0)$ right at the wall}. From Eq. (84) we find $a_c = -\sqrt{\left\langle R_g^2 \right\rangle \tau_R^t / 6} \propto N^{1/3}$, as anticipated in Fig. 10 and near a_c the line of the surface transitions then is described by $a - a_c \approx -\frac{1}{2}\sqrt{\tau_R^t}\sqrt{\left\langle R_g^2 \right\rangle / 6}\left(\tau_R / \tau_R^t - 1\right)$, see Fig. 10, i.e. $a - a_c \propto N^{1/3}\left(\tau_R / \tau_R^t - 1\right)$ [60].

Thus, while in the literature [20] sometimes the view is expressed that surface fields in thin block copolymer films strongly suppress fluctuations and hence the simple mean field theory is adequate, we see that the inclusion of fluctuation effects leads to rich new physics, like analogues of wetting and prewetting phenomena, capillary condensation etc. [60].

2.3 Block Copolymers in the Strong Segregation Limit

In this section, we consider confined symmetric diblock copolymers in the limit $\chi N \to \infty$: then the lamellar phase consists of a sequence of essentially pure A domains and pure B domains, separated by sharp interfaces (of width $w \approx b / \sqrt{6\chi}$ [185,240,241]). Effect of confinement (Fig. 3) now simply can be accounted for by a phenomenological description, where one considers the balance of the bulk energy in the domains (there occurs an "elastic" energy due to chain stretching [185,186]), the interfacial free energy (which is of oder $\gamma_{AB} = \sqrt{\chi / 6}\, b^{-2}$ per unit area) and the free energy at the walls, which involves interfacial tensions γ_{AW}, γ_{BW}. Such an approach was first attempted by Turner

[24], who considered only structures with lamellae parallel to the walls, however, treating (for symmetric walls) the competition between symmetric (Fig. 3a) and anti-symmetric (Fig. 4a) structures. His conclusions are somewhat misleading, however, because the anti-symmetric structure for symmetric walls is never favored when one allows also a perpendicular arrangement of the lamellae [36]. The competition between such parallel (Fig. 3a) and perpendicular (Fig. 3b) lamellar arrangements was treated in the framework of such a free energy model by Walton et al. [36] and, independently, by Kikuchi and Binder [35]. Since the latter work does ignore prefactors of order unity, we follow Walton et al. [36] here. First we recall the free energy in the bulk, which for a system containing p copolymers of length N can be expressed in terms of the lamellar period λ as follows [241]

$$F/k_BT = \frac{3p}{8Nb^2}\lambda^2 + \frac{2pNb}{\lambda}\sqrt{\frac{\chi}{6}} \tag{85}$$

The first term is the entropic contribution in the bulk ("gaussian springs"!), the second term accounts for the A-B interfaces. Minimization of Eq. (85) with respect to λ yields the corresponding equilibrium period λ_b and free energy F_b,

$$\lambda_b = 2\left(\sqrt{\chi/6}/3\right)^{1/3} N^{2/3} b,\; F_b/k_BT = \frac{3p}{2}\left(\chi/2\right)^{1/3} N^{1/3} \tag{86}$$

Now the free energy of thin films is modified due to the appearance of surface terms γ_{AW}, γ_{BW}, and also the period may be expanded (or compressed) relative to λ_b. For parallel (or "horizontal", respectively) arrangements, we can write the free energy in terms of the normalized period $\tilde{\lambda} = \lambda/\lambda_b$ as [36]

$$F_h/F_b = \frac{1}{3}\left\{\tilde{\lambda}^2 + \frac{2}{\tilde{\lambda}} + \frac{1}{m\tilde{\lambda}}\left[2\Gamma + \delta\left(\frac{1-(-1)^{2m}}{2}\right)\right]\right\} \tag{87}$$

with (assuming $\gamma_{AS}<\gamma_{BS}$ the A-rich phase is preferred by the walls, unlike Fig. 3)

$$\Gamma \equiv \gamma_{AW}/\gamma_{AB},\; \delta \equiv (\gamma_{BW} - \gamma_{AW})/\gamma_{AB} \tag{88}$$

and m=n for a horizontal symmetric morphology (D=nλ) while m=n+1/2 for a horizontal anti-symmetric morphology (D=(n+1/2)λ), cf. Fig. 3.

In the perpendicular (or "vertical") morphology, Fig. 3b, the imposed thickness constraint acts perpendicular to the lamellar ordering, and hence the bulk period λ_b is still realized. Now the amount of interface between A-rich phase and the walls and of B-rich phase and the walls is exactly the same for the anti-symmetric arrangement (Fig. 4a) and the vertical arrangement (Fig. 3b), if one neglects the (relatively very small) regions of width w where AB-interfaces meet the walls (in case of Fig. 3b). Since the sample volume is fixed by the amount of copolymer present, assuming a constant monomer density we can express the wall separation D as D=mλ, where both m and λ are the values that would be realized by the horizontal system [36]. This yields for the reduced free energy of the vertical lamellar arrangement

$$F_v / F_b = \frac{1}{3}\left[3 + \frac{1}{m\tilde{\lambda}}(2\Gamma + \delta)\right] \tag{89}$$

The vertical arrangement is favored over the horizontal one whenever $F_v<F_h$. By subtracting Eqs. (89), (87), one finds that a vertical morphology is favored over a horizontal anti-symmetric one whenever $\tilde{\lambda}^3 - 3\tilde{\lambda} + 2 = (\tilde{\lambda}-1)^2(\tilde{\lambda}+2) > 0$, which is realized for all $\tilde{\lambda}$ except $\tilde{\lambda}=1$ where the two structures are degenerate. The vertical morphology is favored over a horizontal symmetric one when

$$\tilde{\lambda}^3 - 3\tilde{\lambda} + 2 - \delta / n > 0 \tag{90}$$

which is realized for $\tilde{\lambda}$ in the regions $\tilde{\lambda} < \tilde{\lambda}_-$ and $\tilde{\lambda} > \tilde{\lambda}_+$, $\tilde{\lambda}_-$, $\tilde{\lambda}_+$ being the two positive roots of Eq. (90), with $\tilde{\lambda}_+ > \tilde{\lambda}_-$. Since the change in the lamellar period (with corresponding chain deformation) is less than half of the bulk period (further deformation can be avoided by a change in the number n of the lamellae), we have

$$-\frac{1}{2} < n(\tilde{\lambda} - 1) < \frac{1}{2} \tag{91}$$

The combination of Eqs. (90) and (91) yields critical numbers n_+,n_- of layers

$$n_\pm = \left[\left[\left(3 + \sqrt{9 \pm 8\delta}\right) / 8\delta\right]\right] \tag{92}$$

[[...]] denoting the largest integer less than or equal to its argument, and n_+,n_- refer to chains expanded and contracted from their bulk equilibrium configuration, respectively. For n greater than this critical number of layers, a horizontal symmetric arrangement of lamellae (Fig. 3a) is always favored, and below it, a horizontal symmetric or a vertical morphology can occur, depending on the value of $\tilde{\lambda}$ as in Eq. (90). Equation (92) can be rewritten to examine the critical values of δ as a function of the number of layers as

$$\delta_\pm = (6n \pm 1) / 8n^2 \tag{93}$$

subtracting the lower bound from the upper one yields $\delta_+ - \delta_- = 1/(4n^2)$, i.e. the range over which the vertical morphology is stable becomes very small for large n [35,36]. As an example, Fig. 15 shows the reduced lamellar period $\tilde{\lambda}$ as a function of the reduced distance between the walls for a typical case. Every vertical line corresponds to a first-order transition (when one of the states has $\tilde{\lambda}=1$, it

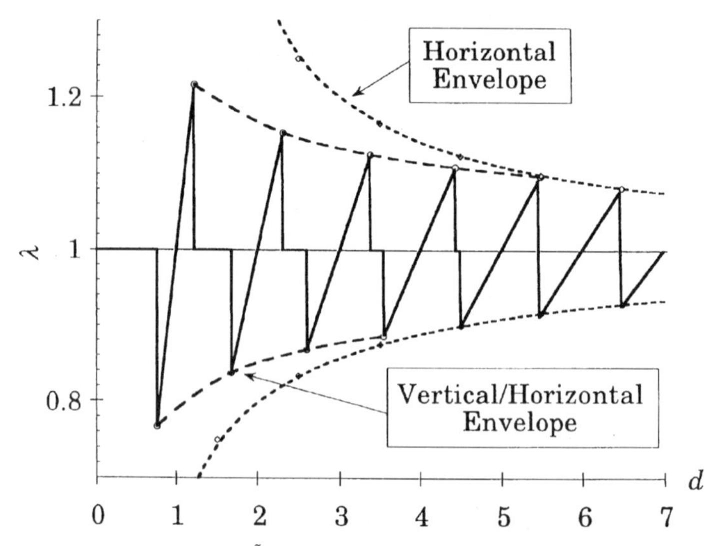

Fig. 15. Reduced lamellar period $\tilde{\lambda}=\lambda/\lambda_b$ as a function of reduced distance between the symmetric walls, $d=D/\lambda_b$, for $\delta=(\gamma_{BW}-\gamma_{AW})/\gamma_{AB}=0.15$. Wherever the vertical arrangement (Fig. 3b) is favored, the bulk equilibrium lamellar period occurs, and hence $\tilde{\lambda}=1$. From Walton et al. [36]

is a horizontal to vertical transition, while transitions in between the two horizontal envelopes mean changes in the number of lamellae, $n\rightarrow n+1$).

This treatment can easily be extended to two non-equivalent walls 1,2, which means we have to distinguish the parameters $\Gamma_1=\gamma_{AW}^{(1)}/\gamma_{AB}$, $\delta_1=\left(\gamma_{BW}^{(1)}-\gamma_{AW}^{(1)}\right)/\gamma_{AB}$ for wall 1 from those of wall 2 (Γ_2, δ_2). Then Eq. (87) is replaced by [36]

$$F_h/F_b=\frac{1}{3}\left\{\tilde{\lambda}^2+\frac{2}{\tilde{\lambda}}+\frac{1}{m\tilde{\lambda}}\left[\Gamma_1+\Gamma_2+\delta_K\left(\frac{1-(-1)^{2m}}{2}\right)\right]\right\} \tag{94}$$

k being the index of the substrate which is adjacent to the B block for the antisymmetric horizontal case. Equation (89) is replaced by

$$F_v/F_b=\frac{1}{3}\left\{3+\frac{1}{m\tilde{\lambda}}\left[\left(\Gamma_1+\frac{\delta_1}{2}\right)+\left(\Gamma_2+\frac{\delta_2}{2}\right)\right]\right\} \tag{95}$$

One finds [36] that the vertical morphology is favored over the horizontal anti-symmetric morphology when

$$\tilde{\lambda}^3 - 3\tilde{\lambda} + 2 + (-1)^k (\delta_2 - \delta_1) / (2n+1) > 0 \tag{96}$$

which is always true if k=1, while for k=2 there exists now a region where the anti-symmetric structure (Fig. 4a) is the stable one. The vertical morphology is favored over the horizontal symmetric one when

$$\tilde{\lambda}^3 - 3\tilde{\lambda} + 2 - (\delta_1 + \delta_2) / 2n > 0 \tag{97}$$

i.e. the same condition as Eq. (90) when we interpret δ as the average of δ_1 and δ_2. For a more detailed discussion of asymmetric walls see [48].

A very interesting situation also occurs when one considers asymmetric block copolymers $\{f=N_A/(N_A+N_B)$ sufficiently distinct from $f=1/2\}$, for which the bulk ordering may be no longer lamellar, but, for instance, hexagonal [44]. Then, nevertheless, the walls may favor a lamellar surface-induced ordering, separated from the bulk by an undulated interface between the lamellar domains and the hexagonal arrangement of cylinders. This phase competes with a phase where no lamellae exist near the surface, rather the surface is coated by "half-cylinders" [44]. Such non-symmetric situations have also been studied in the weak segregation limit by numerical techniques [44,45].

2.4 Survey of Results Obtained with the Self-Consistent Field Theory

As emphasized already in Sects. 2.1,2.2, the square gradient-type theories {Eqs. (7), (41)} are only a good description for long wavelength phenomena, as occurring in the weak segregation limit, but are inappropriate in all cases where on length scales much smaller than a coil size steep concentration variations occur, as is typical for the strong segregation limit. While the extreme case $\chi N \to \infty$ can be treated directly, as exemplified in Sect. 2.3, many systems of practical interest have χN large but finite, and hence need a different treatment even on the mean-field level. Such a treatment is provided by the self-consistent field theory, which has been developed both for interfaces in blends [242–255], free surfaces of blends [13,32,57,59,74], and block copolymers [20,63,240,256–259]. The theory exists in many different variations – formulations on a lattice [20,41,260] exist as well as in the continuum [13,57,63,246–249,253], versions both assuming compressible [57,59,255] and strictly incompressible [63] melts are used, versions assuming strictly gaussian chains [242–245] or allowing for local chain stiffness [253] have been proposed, etc. We cannot attempt to describe all these developments in detail, but rather intend to only sketch the spirit of the approach, and then summarize some of the results. A general drawback of the self-consistent field approach, of course, is that usually it requires rather extensive numerical calculations. Here we follow Schmid [57] in briefly describing the theory for surfaces of compressible symmetric blends (another example, interfaces in incompressible fully segregated blends, can be found in [6]).

The basic quantities of the theory are monomer density operators

$$\hat{\rho}_{A,B}(\vec{r}) = N \sum_{i=1}^{n_{A,B}} \int_0^1 ds \delta(\vec{r} - \vec{r}_i(s)) \tag{98}$$

where n_A, n_B are the numbers of polymers of type A,B in the system, and each chain molecule is described by a space curve $\vec{r}(s)$, with s varying from 0 to 1. If instead we want to deal with block copolymers of composition f, the corresponding expression simply would be [63] (n is the total number of polymers in the system)

$$\hat{\rho}_A(\vec{r}) = N \sum_{i=1}^{n} \int_0^f ds \delta(\vec{r} - \vec{r}_i(s)) \qquad \hat{\rho}_B(\vec{r}) = N \sum_{i=1}^{n} \int_{1-f}^{1} ds \delta(\vec{r} - \vec{r}_i(s)) \tag{99}$$

The partition function then is written as a path integral [245], for the blend {Eq. (98)}

$$Z = \frac{1}{n_A! n_B!} \int \prod_{i=1}^{n_A} D(\vec{r}_i) \exp\left[-\frac{3}{2Nb_i^2} \int_0^1 ds \left| \frac{d\vec{r}_i}{ds} \right|^2 \right] \int \prod_{j=1}^{n_B} D(\vec{r}_j) \exp\left[-\frac{3}{2Nb_j^2} \int_0^1 ds \left| \frac{d\vec{r}_j}{ds} \right|^2 \right] \exp\left[-F\{\hat{\rho}_A, \hat{\rho}_B\} / k_B T \right] \tag{100}$$

Here the exponential weighting factors represent gaussian statistics (gyration radii $R_{g,i}^2 = b_i^2 N/6$, where we allow different effective segment lengths b_A, b_B for both species). A normalization factor of the path integrals is ignored for simplicity. The free energy functional is split in a part that depends on the total density $\rho = \rho_A + \rho_B$ only and an interaction part,

$$\mathcal{F}\{\rho_A, \rho_B\} = \mathcal{F}_0\{\rho\} + \mathcal{F}_{inter}\{\rho_A, \rho_B\} \tag{101}$$

where we also invoke a mean field approximation, i.e. the monomer density operators $\hat{\rho}$ in Eq. (100)are replaced by the average monomer density, $\rho_i(\vec{r}) \equiv \langle \hat{\rho}_i(\vec{r}) \rangle$, ignoring monomer-monomer correlations. The chains are thus treated as independent random walks in the external field $W_i(\vec{r}) = \delta(\mathcal{F}/k_B T)/\delta\rho_i(\vec{r})$. The densities $\hat{\rho}_i(\vec{r})$ can be expressed in terms of probability densities $Q_i(\vec{r}, s)$ that describe the statistical weight for a part of a polymer of type i with chain length sN and one end fixed at $\vec{r}$.

$$\rho_i(\vec{r}) = \rho_b \int_0^1 ds Q_i(\vec{r},s) Q_i(\vec{r},1-s) \tag{102}$$

$$Q_i(\vec{r},s) = \int \mathfrak{D}(\vec{r}') \exp\left[-\frac{3}{2Nb_i^2}\int_0^s ds' \left|\frac{d\vec{r}'}{ds'}\right|^2\right] \exp\left[-\int_0^s ds' W_i(\vec{r}'(s'))\right] \delta(\vec{r}-\vec{r}'(s)) \tag{103}$$

The Q_i simply satisfy a diffusion equation [245]

$$\left[\frac{1}{N}\frac{\partial}{\partial s} - \frac{1}{6} b_i^2 \nabla_{\vec{r}}^2 + W_i\right] Q_i(\vec{r},s) = 0 \tag{104}$$

with initial condition $Q_i(\vec{r},0)=1$. The configurational entropy of a polymer of type i then is

$$S_i^{conf} = k_B \ell n z_i + \frac{1}{n_i} \int d\vec{r} W_i(\vec{r}) \rho_i(\vec{r}) \tag{105}$$

where z_i is the single chain partition function, $z_i = \int d\vec{r} Q_i(\vec{r},1) = N n_i / \rho_b$, ρ_b being the bulk density of the system. The free energy of this model is then

$$F = \mathcal{F}\{\rho_A, \rho_B\} - k_B T\left(n_A S_A^{conf} + n_B S_B^{conf}\right) \tag{106}$$

For explicit calculations, one needs to specify the form of the free energy functional $\mathcal{F}\{\rho_A, \rho_B\}$ in this self-consistent formulation in detail. Schmid [57] uses for $\mathcal{F}_0$ an extension of the generalized Flory theory [261]

$$\mathcal{F}_0\{\rho\}/k_B T = \int d\vec{r} \rho(\vec{r}) \left\{ f[\rho(\vec{r})] - \mu \right\} \tag{107}$$

where μ is the chemical potential and $f(\rho)$ the bulk free energy per monomer, which is written in term of the packing fraction $\eta \equiv b^3 \rho \pi / 6$ (we assume $b_A = b_B$ here),

$$f(\rho) = C_0 \eta \, (4 - 3\eta + c_1 \eta^3)/(1-\eta)^2 \tag{108}$$

with constants $C_0=0.6583$, $c_1=3$, $b=1.96\, a_0$ (a_0 being the lattice spacing) in order to numerically fit the equation of state of the bond fluctuation model [262]. Equation (108) basically describes the hard-core repulsion between the effective monomers (in the bond fluctuation model each effective monomer blocks the eight sites of an elementary cube from further occupation [263–265]).

The pairwise interactions $U_{AA}(\vec{r})$, $U_{BB}(\vec{r})$ and $U_{AB}(\vec{r})$ between monomers then enter the second term on the right hand side of Eq. (101),

$$\mathcal{F}_{inter} = \int d\vec{r} d\vec{r}' \frac{1}{2} \left[\sum_{i,j} \rho_{ij}^{(2)}(\vec{r},\vec{r}') U_{ij}(\vec{r}-\vec{r}') \right] \tag{109}$$

where the pair density $\rho_{ij}^{(2)}(\vec{r},\vec{r}') = \gamma(\vec{r}-\vec{r}')\rho_i(\vec{r})\rho_j(\vec{r}')$ in mean field approximation is put proportional to the product of the local densities. The function $\gamma(\vec{r}-\vec{r}')$ accounts for local effects due to chain connectivity (e.g. $\gamma(\vec{r})=0$ inside the excluded volume of the central monomer, and $\gamma(\vec{r})=1-p(\vec{r})$ outside, where $p(\vec{r})$ is the probability that direct neighbor monomers along the chain block point $(\vec{r})$ for occupation. For short range interactions one can relate Eq. (109) to the familiar Flory-Huggins interaction via a term $\chi = \rho_b \int d\vec{r} \gamma(\vec{r}) \left[U_{AB}(\vec{r}) - \left(U_{AA}(\vec{r}) + U_{BB}(\vec{r}) \right)/2 \right] / k_B T$.

Finally some fields describing enthalpic preference of the wall to one component could be added, but this case is not considered now: rather we follow Schmid [57] in addressing the case of a "free" surface of the polymer blend (against air or vacuum, respectively). Figs. 16a,b show typical results, for a choice of parameters that correspond to the Monte Carlo studies of the bond fluctuation model by Rouault et al. [56]. While the scale for the profile of the total density ρ is given by the bond length b, and hence much smaller than the gyration radius R_g, the profile describing the surface enrichment of the minority component decays on the scale of R_g. But Fig. 16a shows that right at the surface, in the region where the total density ρ decays to zero, the volume fraction profile $\phi_A(z)$ is flattened out. This flat part is missing in the case where the range of the pairwise interaction tends to zero (dashed curve in Fig. 16a), and is also not predicted by calculations for incompressible blends. It is also interesting that the (entropically driven!) surface enrichment of the minority species is about twice as large if the finite range of the interactions is considered than for the limit of zero range. In view of these differences, it is clear that it is very difficult to extract weak surface fields from an analysis of the surface enrichment of blends.

For such free surfaces with no energetic preference of the surface for one of the components (because of the arbitrary choice of a purely symmetric choice of interactions, $U_{AA}=U_{BB}$!) the surface enrichment effects are typically rather weak (Fig. 16b), even if one considers asymmetry of chain lengths. Even for $\chi=0$ the shorter chains are then attracted to the surface, but this purely entropic effect is particularly weak. For $\chi\neq 0$ the enthalpy is less effective near the free surface in comparison with the entropy of mixing, and this effect is for small volume fraction $\phi_{A,b}$ of the minority component nearly independent of the asymmetry in chain length (Fig. 16b): i.e., even if the minority chains are much longer ($N_A=2N_B$) they segregate to the surface, contrary to the non-interacting case.

Another interesting driving force for surface enrichment in a blend is conformational asymmetry, described e.g. by different statistical segment lengths b_A,

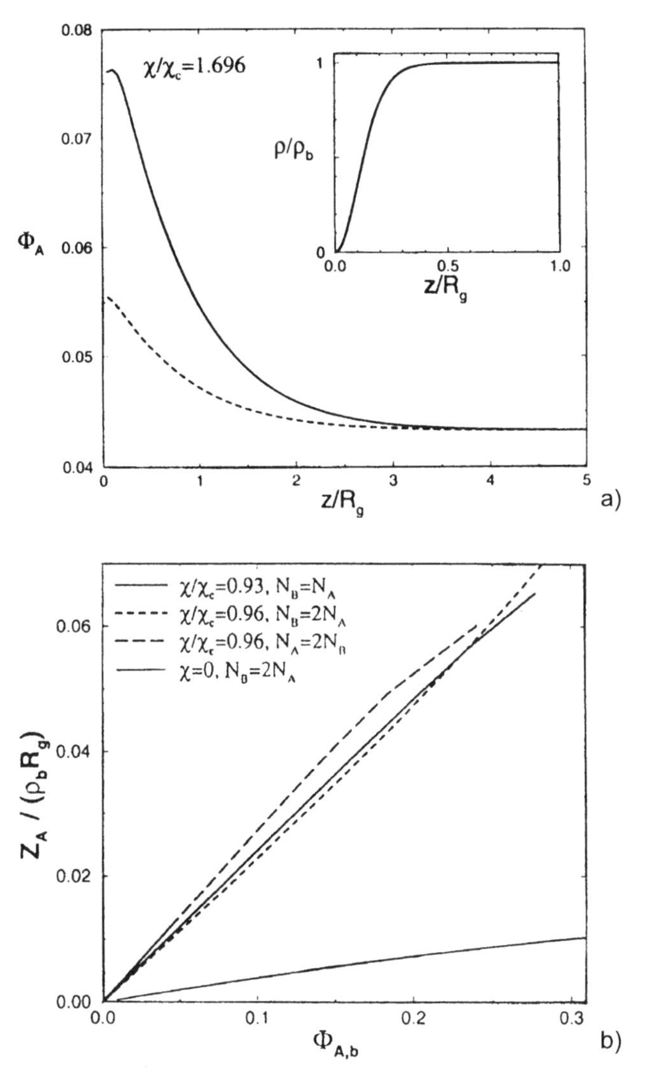

Fig. 16. a Volume fraction profile of the minority component monomers at χ/χ_c=1.696 in a symmetric polymer mixture. *Dashed line* shows the volume fraction profile for the case where monomers are assumed to have pure contact interactions with each other, while the *full curve* corresponds to a finite interaction range (chosen as in the bond fluctuation model, where each monomer interacts with neighbors at distances 2, $\sqrt{5}$ or $\sqrt{6}$ lattice spacings apart). The *inset* shows the total density profile, which is independent of χ, for the choice N=32, and symmetric interactions $U_{AA}=U_{BB}=-U_{AB}$. From Schmid [57].
b Surface excess $Z_A^* = \int dz\rho(z)(\phi_A(z)-\phi_{Ab})$ in units of $\rho_b R_g$ plotted vs the bulk volume fraction of A, ϕ_{Ab}, in the one phase region, for symmetric mixtures ($N_A=N_B$, *thick solid line*) and asymmetric mixtures, where A polymer chains are either longer (*long-dashed line*) or shorter (*short-dashed line*) than B polymer chains. The *thin line* indicates results for asymmetric mixtures, with interactions turned off (χ=0). From Schmid [57]

b_B. Fredrickson et al. [19,59] suggested that the relevant parameter is $\beta_k^2 = b_k^2/(6v_k)$, where k=A,B and v_k is the segment volume for type-K segments. Fredrickson and Donley [19] suggested that the chains with the smaller values of β are entropically favored at the surface. However, for such chains with different segment lengths there is also no physical reason to assume that the cohesive energies U_{AA}, U_{BB} are equal. Yethiray [54] suggested that the chains with the weakest cohesive forces are preferred at the surface in order to minimize the overall interaction energy, i.e. surface enrichment would have enthalpic origin. A systematic treatment of both effects together seems rather difficult.

We now turn to the case of block copolymers. Early work by Shull [20] considered a lamellar arrangement parallel to the walls only. From such calculations, one obtains a unified treatment of both the weak and the strong segregation limit within mean field theory, and a detailed description of the order parameter profiles across the thin film emerges.

Matsen [63] has considered more general orderings, including the perpendicular (Fig. 3b) and mixed parallel-perpendicular phase (Fig. 3c). While the latter structure is unstable for symmetric diblocks, it can become stable for asymmetric composition. Figure 17 shows a phase diagram obtained from this treatment, considering thin films of symmetric diblocks with symmetric walls. The phase boundaries obtained in this treatment differ only slightly from a corresponding treatment by the simple strong segregation theory (Sect. 2.3).

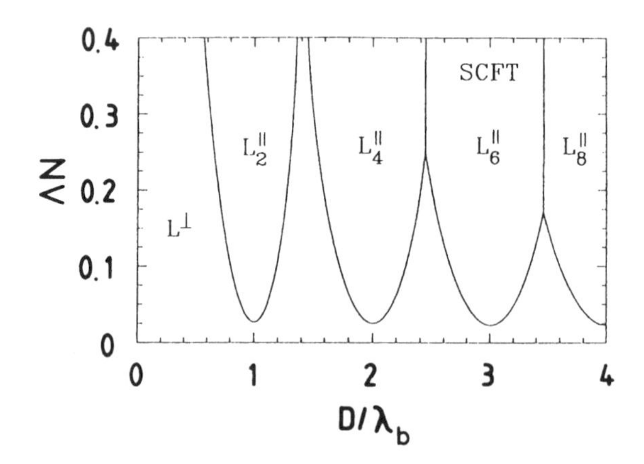

Fig. 17. Phase diagram for intermediately segregated (χN=20) symmetric diblock (f=0.5) films confined between identical walls calculated from self-consistent field theory. The film thickness D is normalized by the bulk lamellar period λ_b, and the ordinate ΛN is a measure of the surface field (which has the functional form $H(z)=\Lambda_1\,(+\cos(\pi z/\varepsilon))\;b\sqrt{N}/\varepsilon$ for $0 \le z \le \varepsilon$, and an analogous form at the other wall). The index ν of the phases $L_\nu^{\parallel}$ denotes the number of A/B interfaces parallel to the walls. From Matsen [63]

2.5 Concepts on Interfaces in Confined Geometry

In this section we return to the situation sketched in Fig. 1d, namely a polymer mixture (A,B) confined between two different planar surfaces, such that the lower surface prefers species A, and the upper surface prefers species B. We are interested in the temperature region in between the critical temperature of phase separation in the bulk and the wetting transition temperatures of the two surfaces (i.e., $\chi_w^{-1} < \chi < \chi_{crit}^{-1}$ in Fig. 6; in principle, one must for film thicknesses $D \to \infty$ deal also with the general asymmetric situation that wetting transitions of the A-rich phase at the lower surface and of the B-rich phase at the upper surface occur at different Flory Huggins parameters $\chi_w^{(\ell)}, \chi_w^{(u)}$, but for simplicity we deal with $\chi_w^{(\ell)}, \chi_w^{(u)} = \chi_w$ here only). Then we expect that the surfaces stabilize an interface which (apart from fluctuations) runs in parallel to the flat walls.

While the average position $\bar{h}$ of the interface (height above the lower surface) again follows from the average volume fraction $\bar{\phi}$ of species A in the thin film, using the lever rule, Eq. (2), it is a problem of current interest to discuss the interfacial profile in this confined geometry [84,266]. This geometry is used to study both the compositions of coexisting unmixed phases [70] and interfacial widths [71,81–84,125]. Studying this problem theoretically for the simpler case of the Ising model for binary mixtures [267–271], it was found that in this case interfacial profiles may significantly deviate from "intrinsic profiles" [216,272,273], because they are anomalously broadened by fluctuations. Recently it has been recognized [84,266] that this problem is very relevant for polymer mixtures as well. We sketch here the treatment of Werner et al. [266] and summarize their main conclusions.

Starting point is the description of the interface in terms of the "drumhead model" [273], i.e. we disregard for the moment the fact that the interface has a non-trivial intrinsic profile with width w_0, and treat it like a sharp kink in terms of the local interfacial height h(x,y) only, x,y being coordinates along the lower surface (at z=0). Then the effective free energy functional $H_{eff}\{h\}$ associated with interfacial fluctuations can be written as [220]

$$\mathcal{H}_{eff}\{h\}/k_BT = \int dx \int dy \left\{ \frac{\sigma}{2} \left[\left(\frac{\partial h}{\partial x} \right)^2 + \left(\frac{\partial h}{\partial y} \right)^2 \right] + V(h) \right\} \tag{110}$$

where σ is the interfacial tension for a flat interface between coexisting A-rich and B-rich phases, and V(h) is an effective potential due to the surfaces at z=0 and z=D, stabilizing the average interface position in the center of the film, z= D/2, if we consider the most symmetric situation, where $\bar{\phi} = 1/2$. The gradient square term in Eq. (110) just accounts for the enhancement of the interfacial area due to long wavelength interfacial fluctuations [216,272,273]. In the following, it is convenient to measure the origin of h at the average height D/2, so $\langle h \rangle \equiv 0$.

For a short range potential acting on the interface due to the wall, one considers an exponentially decaying form, i.e. for the lower wall

$$V_{(\ell)}(h)=a_0\left(\frac{T-T_w}{T_w}\right)\exp\left\{-\kappa\left(\frac{D}{2}+h\right)\right\}+b_0\exp\left\{-2\kappa\left(\frac{D}{2}+h\right)\right\} \tag{111}$$

where a_0, b_0 are phenomenological (positive) constants, and κ^{-1} is the range of the potential. Note that we have allowed here the first term to change sign at the wetting transition temperature T_w, and thus a term proportional to exp $(-2\kappa(D/2+h))$ is required also. Taking $V_{(u)}(h)=V_{(\ell)}(-h)$ for the potential due to the upper wall, the total potential $V(h)=V_{(\ell)}(h)+V_{(u)}(h)$ becomes

$$\begin{aligned} V(h)&=2a_0\left(\frac{T-T_w}{T_w}\right)\exp(-\kappa D/2)\cosh(\kappa h)+2b_0\exp(-\kappa D)\cosh(2\kappa h)\approx \\ &\approx\left[a_0\left(\frac{T-T_w}{T_w}\right)+4b_0\exp(-\kappa D/2)\right]\exp(-\kappa D/2)\kappa^2h^2 \end{aligned} \tag{112}$$

where in the last step we have expanded the potential quadratically, which is possible if h stays small, i.e. for $T>T_c(D)$ where

$$T_c(D)=T_w\left[1-4\frac{b_0}{a_0}\exp(-\kappa D/2)\right] \tag{113}$$

Note that at $T_c(D)$, where the square bracket in Eq. (112) changes sign, a transition from the state of the film as shown in Fig. 1d) to the state shown in Fig. 1e) is predicted to occur [268–271], with interfaces bound to the walls, rather than a freely fluctuating interface unbound in the center of the film.

Using the potential $V(h)\propto h^2$ in Eq. (110), one recognizes that Eq. (110) is formally identical to a Ginzburg-Landau theory of a second-order transition for $T>T_c(D)$, with h(x,y) the order "parameter field" [186,216]. Therefore, it is straightforward to read off the correlation length ξ associated with this transition at $T_c(D)$, namely

$$\xi_{\parallel}^{-2}=\frac{1}{\sigma}\left(\frac{\partial^2 V(h)}{\partial h^2}\right)_{h=0}=\frac{2a_0}{\sigma}\kappa^2\exp\left(-\frac{\kappa D}{2}\right)\left(\frac{T-T_c(D)}{T_w}\right) \tag{114}$$

Even if one is at a temperature $T>>T_c(D)$, where the factor $(T-T_c(D))/T_w$ is not small, $\xi_{\parallel}\propto\exp(\kappa D/4)$ results from Eq. (114). In other words, for large D the potential V(h) is so weak that $H_{eff}\{h\}/k_BT$ is almost identical to the free energy functional of a free interface,

$$\mathcal{H}_{\rm eff}^{\rm free}\{h\}/k_BT=\int dx\int dy\left\{\frac{\sigma}{2}\left[\left(\frac{\partial h}{\partial x}\right)^2+\left(\frac{\partial h}{\partial y}\right)^2\right]\right\}=\frac{\sigma}{2}\sum_{q}q^2\left|h(\vec{q})\right|^2 \tag{115}$$

where $h(\overline{q})$ is the Fourier component of h(x,y) for wavevector $\overline{q}$. Equation (115) is the well-known "capillary wave" Hamiltonian, and because of the harmonic character of this Hamiltonian one can immediately conclude that the mean square value of $h(\overline{q})$ is [216,272,273]

$$\left\langle\left|h(\vec{q})\right|^2\right\rangle=1/\left(\sigma q^2\right)\quad \text{free interface.} \tag{116}$$

The mean square displacement now is

$$\left\langle h^2(x,y)\right\rangle=\sum_{\vec{q}}\left\langle\left|h(\vec{q})\right|^2\right\rangle=\frac{1}{4\pi^2}\int d\vec{q}\left\langle\left|h(\vec{q})\right|^2\right\rangle=\frac{1}{2\pi\sigma}\int dq/q=\\ \left(2\pi\sigma\right)^{-1}\ell n\left(q_{max}/q_{min}\right)\quad \text{free interface,} \tag{117}$$

where in the last step the divergences of the integral $\int dq/q$ were eliminated by introducing a cutoff q_{min} near q=0 and q_{max} for large q {q_{max} is believed to be related to the intrinsic width w_0, namely [274] $q_{max}=2/w_0$}. For a free interface in a macroscopic system, the cutoff q_{min} is either the lateral sample size L (if the measurement accesses interfacial fluctuations over the full size of the sample) or a lateral resolution length L^{res}, and then $q_{min}=2\pi/L$ or $q_{min}=2\pi/L^{res}$, respectively.

Comparing now Eqs. (112) and (115), we see that for a confined interface Eq. (116) gets replaced by $\left\langle\left|h(\vec{q})\right|^2\right\rangle=1/\left[\sigma\left(q^2+\xi_{||}^{-2}\right)\right]$, i.e. the divergence of the integral in Eq. (117) for $q\to 0$ is removed, and we have instad of Eq. (117)

$$<h^2(x,y)>\approx\left(2\pi\sigma\right)^{-1}\ell n\left(q_{max}/2\pi\right)\quad \text{confined interface} \tag{118}$$

since now $q_{min}\approx 2\pi\xi_{||}$ instead of $q_{min}=2\pi/L$, if $\xi_{||}<L$ or L^{res}.

In practice, it is important, of course, to take into account that the width of the interface is not due to capillary waves alone, but also the intrinsic width needs to be accounted for. This can be done by a convolution approximation [272,273] for the total profile $\phi(z)$ of the order parameter,

$$\phi(z)=\int_{-\infty}^{+\infty}dh\rho^{(int)}(z-h)P(h)\quad P(h)=\left(2\pi\left\langle h^2\right\rangle\right)^{-1/2}\exp\left(-h^2/2\left\langle h^2\right\rangle\right) \tag{119}$$

where the intrinsic profile usually is approximated to be of the tanh form [6,186,207–209,272]

$$\phi^{(int)}(z)=\frac{1}{2}\left\{\phi_{coex}^{(2)}+\phi_{coex}^{(1)}+\left(\phi_{coex}^{(2)}-\phi_{coex}^{(1)}\right)\tanh\left[(z-D/2)/w_0\right]\right\} \tag{120}$$

Approximating the total profile in Eq. (119) by the same tanh profile but with a broadened width w rather than the intrinsic width w_0, one finds [84,266]

$$w^2 \approx w_0^2 + \frac{\pi}{2}\langle h^2 \rangle = w_0^2 + (4\sigma)^{-1} \ell n\left(\xi_{\|}\, q_{max}/2\pi\right) = w_0^2 + \frac{\kappa D}{16\sigma} + \text{const} \tag{121}$$

where in the last step $\xi_{\|} \propto \exp(\kappa D/4)$ was used. Note that $\kappa^{-1}=\xi_b\,(1+\omega/2)$ [275,276] where ω is the constant appearing in the theory of critical wetting [220,277,278]

$$\omega = \left(4\pi\xi_b^2\sigma\right)^{-1} = \left(\pi w_0^2 \sigma\right)^{-1} \tag{122}$$

where in the last step it was used that in the critical region the intrinsic interfacial width w_0 is twice the bulk correlation length ξ_b, $w_0=2\xi_b$ [6,186]. The last form of Eq. (122) is believed to hold also in the strong segregation limit, where $\xi_b \approx b\sqrt{N}/6$ but $w_0 = \sqrt{6\chi}$ [242–245] is much smaller than ξ_b. Since in the strong segregation limit [242–245]

$$\sigma \approx b^{-2}\sqrt{\chi/6} \qquad \chi/\chi_{crit} >> 1 \tag{123}$$

while in the mean field critical region of the polymer mixture {i.e., for $N|1-\chi/\chi_{crit}|>>1$ but $|1-\chi/\chi_{crit}|<<1$}

$$\xi_b = \frac{b}{6}\sqrt{N}\left(1-\chi_{crit}/\chi\right)^{-1/2}, \sigma = \left\{8/\left(3b^2\sqrt{N}\right)\right\}\left(1-\chi_{crit}/\chi\right)^{3/2} \tag{124}$$

we conclude that the parameter ω {Eq. (122)} has an intricate temperature dependence,

$$\omega \approx 6\sqrt{6\chi}\pi, \;\; \chi/\chi_{crit} >> 1 \qquad (\text{remember}\; \chi_{crit} = 2/N) \tag{125a}$$

$$\omega \approx (27/8\pi)\left[N\left(1-\chi_{crit}/\chi\right)\right]^{-1/2} \qquad N|1-\chi_{crit}/\chi|>>1 \tag{125b}$$

while in the asymptotic (Ising-like [186]) critical region ω tends to a universal constant [220,279]

$$\omega \approx 0.86 \qquad N\left|1-\chi/\chi_{crit}\right|<<1 \tag{126}$$

This behavior is sketched in Fig. 18. Note that the parameter ω directly controls the anomalous broadening of interfacial widths in confined geometry, since Eq. (121) can be rewritten as {using $\kappa=2/[w_0(1+\omega/2)]$ in the general case} [266]

$$w^2 = w_0^2 + \frac{\pi\omega}{2+\omega}\frac{w_0 D}{4} + \text{const} \qquad \xi_{\|} << L \tag{127}$$

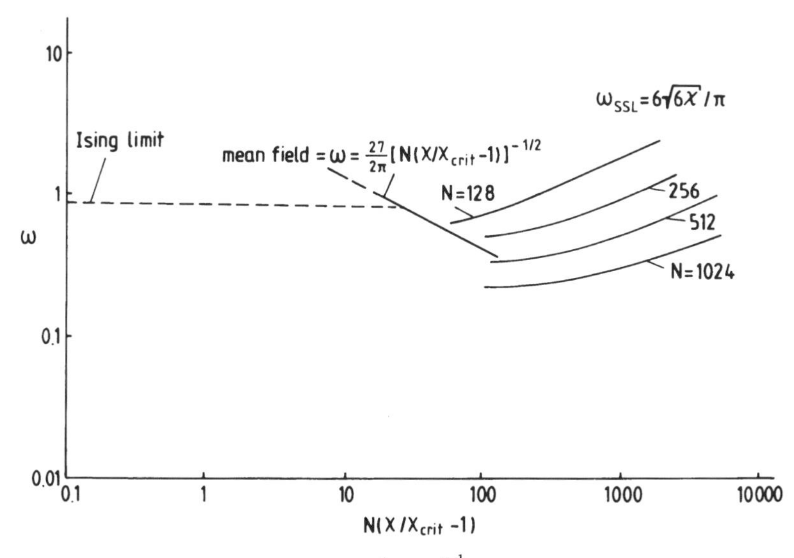

Fig. 18. Log-log plot of the parameter $\omega=\left(\pi w_0^2\sigma\right)^{-1}$, that controls the behavior of critical wetting [11,220,277,278], vs χ/χ_{crit} -1, for symmetrical polymer mixtues (N_A=N_B=N) with chain lengths ranging from N=128 to N=1024, showing the predictions Eqs. (124)–(126). From Werner et al. [266]

It must be emphasized that the result $w^2 \propto D$ holds for short range forces on the interface due to the walls. However, the treatment is easily extended to the case of long range forces, replacing Eq. (111) by a power law with an exponent p, e.g. $V_{(1)}(h) \approx a(D/2+h)^{-p}$, which then yields $V_{(\ell)}(h) \approx ap(p+1)(D/2)^{-p-2}h^2$, and Eq. (114) then gets replaced by $\xi_{||}^{-2} \propto (D/2)^{-p-2}$ rather than $\xi_{||}^{-2} \propto \exp(-\kappa D/2)$. Using again q_{min}=$\xi_{||}$; in Eq. (117) the result is that w^2 depends on D only in logarithmic form,

$$\begin{aligned} w^2 &= w_0^2+\frac{\pi}{2}\left\langle h^2\right\rangle = w_0^2+\left(p/2+1\right)\left(4\sigma\right)^{-1}\ell nD = \\ &w_0^2\left[1+\frac{\pi}{4}\left(\frac{p}{2}+1\right)\omega\ell nD\right]+\text{constant.} \end{aligned} \tag{128}$$

2.6
Computer Simulation of Polymer Blends in Thin Films

The first Monte Carlo study of phase transitions in thin films of symmetric binary polymer mixtures confined between two “neutral” hard walls (to which neither component is preferred energetically) was presented by Kumar et al. [39].

They used both a lattice model, choosing simple cubic lattices of size 30×30xD with D=5 and D=10 lattice spacings, respectively, and an off-lattice model, where chains are modelled as freely jointed rigid links, the beads of which interact with a standard Lennard-Jones potential which was truncated at 2.5σ, choosing the range parameter the same for both types of monomers, $\sigma_{AA}=\sigma_{BB}=\sigma_{AB}=\sigma$. Phase separation is caused by choosing the well depth distinct for like and unlike pairs, i.e. $\varepsilon_{AB}/\varepsilon_{AA}=0.9$, while $\varepsilon_{AA}=\varepsilon_{BB}$ to keep the mixture symmetric. The considered film thicknesses then were D=5σ and 10.5σ, for a chain length N=20, choosing a reduced density $\rho^*=\rho\sigma^3=0.7$ which is believed to correspond to a bulk polymer melt, while for the lattice model the polymers were modelled as self-avoiding walks, where nearest-neighbor interactions were chosen as $\varepsilon_{AA}=\varepsilon_{BB}=-\varepsilon$, $\varepsilon_{AB}=0$, and a volume fraction $\phi=0.8$ of occupied sites was chosen. While in the case of these thin film calculations one chooses periodic boundary conditions in the directions parallel to the walls, and confinement is simulated by requiring that the z-coordinates of all beads lie in the interval $0<z<D$, also simulations of cubic boxes without confining walls and periodic boundary conditions also in z-directions were carried out, to obtain the corresponding bulk behavior of the studied models [39].

Here we do not give any details on simulation methodologies, since these have been reviewed recently elsewhere [186,198], and just reproduce typical results on the change of the phase diagram due to confinement (Fig. 19 [39]). A note-

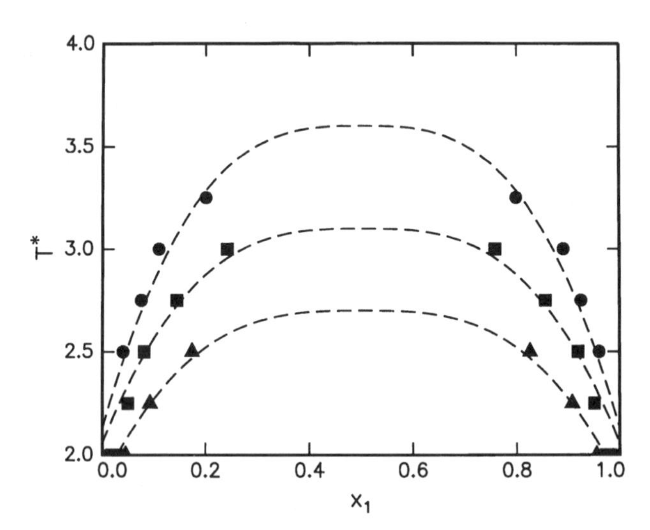

Fig. 19. Monte Carlo result for the phase diagrams of an off-lattice bead rod model of a symmetric binary polymer mixture with N=20, in the plane of variables reduced temperature $T^*=k_BT/\varepsilon_{AA}$ and volume fraction of component A, denoted here as x_1. Data are for bulk systems (*full dots*), and for confined films of thicknesses D=10.5σ (*squares*) and 5σ (*triangles*), respectively. *Dashed curves* represent fits to $|x_1-x_{1c}|\propto|T-T_c|^{\beta_I}$, where the Ising model exponent [229,230] was chosen as $\beta_I=1/3$. From Kumar et al. [39]

worthy conclusion is that the critical temperature is suppressed significantly with decreasing film thickness, i.e. the confinement enhances the compatibility of both components in a mixture. This result is understandable [39] already in terms of the simple mean field theory outlined in Sect. 2.1., since the "missing neighbors" at the hard wall are responsible for a non-zero term μ_1 in Eq. (10) [28,144,216]. Therefore, it is no surprise that the minority component gets segregated to the walls [39,56]: due to this "missing neighbor" effect, entropy of mixing is more effective near the walls than in the bulk, and thus the "order parameter" near the walls is reduced in comparison with the bulk. Rouault et al. [56] have studied this phenomenon in more detail, using the bond fluctuation model [198,263–265] of polymer chains, a chain length N=32, and interactions $\varepsilon_{AB}=-\varepsilon_{AA}=-\varepsilon_{AB}\equiv\varepsilon$ non-zero if the distance between the monomers does not exceed $\sqrt{6}$. For this model, one knows rather precisely both the bulk critical temperature ($k_BT_{cb}/\varepsilon\cong69.35$) [280] and the critical temperature of thin films [55], e.g. for D=20 lattice spacings we have $k_BT_c(D)/\varepsilon\cong58.5$. Note that this linear dimension is about 3 times larger than the bulk gyration radius of this model, $\sqrt{\langle R_g^2\rangle}\approx7$ lattice spacings. Figure 20 shows resulting order parameter profiles at four inverse temperatures [56]. The data clearly indicate that the profiles flatten out at both walls, and no longer reach the bulk value even in the center of the thin film. We emphasize that the mean-field treatment of Sect. 2.1 (cf. Fig. 7c) does not predict such a flattening; profiles of the type of Eq. (28) have their maximal slopes at the surfaces of the thin film. Rouault et al. [56] attributed these flat regions of the profiles near the walls to the orientation of the polymer coils at the walls: the instantaneous shape of a Gaussian polymer coil is not a sphere, but rather a "flattened egg-shaped" object, with a gyration tensor that has three distinct eigenvalues. The axes giving the eigendirections of this tensor are randomly oriented in the bulk, and then on average the coil is spherically symmetric. At the surfaces, however, best packing of the monomers is achieved if the axis corresponding to the largest eigenvalue is parallel to the wall (then the end-to-end distance of a polymer close to the wall is parallel to the wall), and the axis corresponding to the smallest eigenvalue is oriented along the z-axis perpendicular to the wall. Thus the polymer coil lies essentially flat on the wall, but without major distortion of its internal shape (which would be very costly in terms of the entropic elastic forces of the coil). Only when one uses strong binding potentials to the wall, as done in [17,18], it becomes possible to get a significant distortion of the shape of such a polymer coil attached to the wall, which ultimately takes a flat pancake-like shape for very strong wall-monomer interaction [281].

Although in the present case the smallest eigenvalue R_g^{min} is rather small, $R_g^{min}=1.75$ lattice spacings [215] {since the first layer which can be taken by monomers is z=1, we have marked the position $R_g^{min}+1$ in Fig. 20}, this explanation implies that for large N the region of the profile where a flattening occurs should scale like $\sqrt{N}$. Alternatively, Schmid [57] from self-consistent field cal culations suggested that a flattening of the profile near the wall extends over a region comparable to the range of interactions (which is $\sqrt{6}$ in our case), cf.

Fig. 16. In order to discriminate between these interpretations, calculations for larger N would be needed for which R_g^{min} is larger while the interaction range stays the same: however, this would require to choose both linear dimensions D, L significantly larger (note the finite size effects for $\varepsilon/k_BT=0.018$ in Fig. 20), and thus this problem has not yet been solved.

Since in Fig. 19 critical temperatures have been estimated by a rough extrapolation only, this study [39] could not make too definitive statements about the shift of the critical temperature due to confinement. A more extensive study has been possible for the bond fluctuation model [55], extracting both critical temperatures and the coexistence curves by use of the finite size scaling technique [236–239,280]. Figs. 21, 22 present the results for chain length N=32. While mean field theory {Eqs. (36) and (37) in Sect. 2.1} has predicted a crossover from

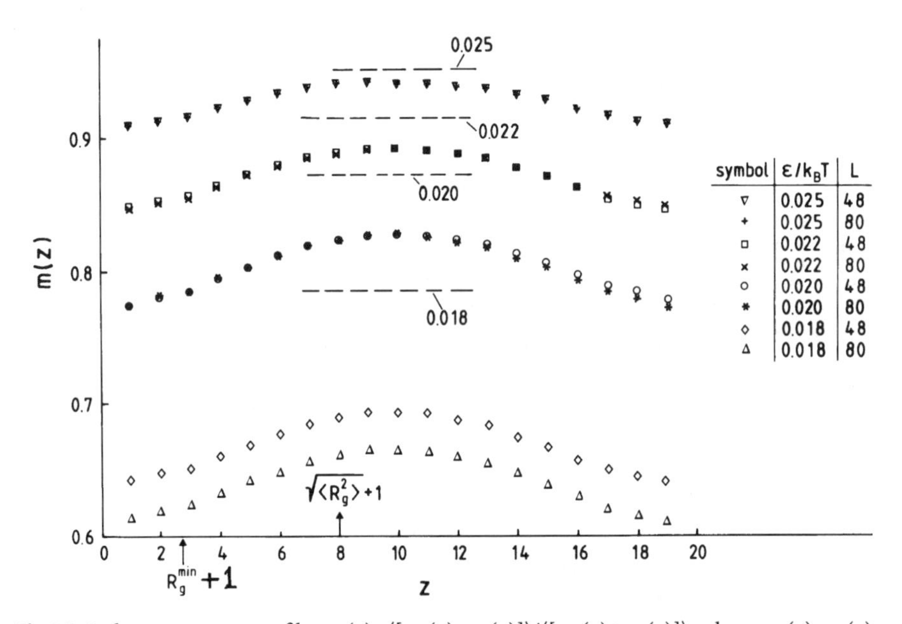

Fig. 20. Order parameter profiles $m(z)=\langle[\rho_A(z)-\rho_B(z)]\rangle/\langle[\rho_A(z)+\rho_B(z)]\rangle$, where $\rho_A(z)$, $\rho_B(z)$ are densities of A-monomers or B-monomers at distance z from the left wall, for LxLx20 films confining a symmetric polymer mixture, polymers being described by the bond fluctuation model with N=32, $\varepsilon_{AB}=-\varepsilon_{AA}=-\varepsilon_{BB}=\varepsilon$ and interaction range $\sqrt{6}$. Four inverse temperatures are shown as indicated. In each case two choices of the linear dimension L parallel to the film are included. While for $\varepsilon/k_BT\geq 0.02$ differences between L=48 and L=80 are small and only due to statistical errors (which typically are estimated to be of the size of the symbols), data for $\varepsilon/k_BT=0.018$ clearly suffer from finite size effects. Broken straight lines indicate the values of the bulk order parameters m_b in each case [280]. Arrows show the gyration radius and its smallest component in the eigencoordinate system of the gyration tensor [215]. Average volume fraction of occupied sites was chosen as 0.5. From Rouault et al. [56].

$T_c(\infty)$—$T_c(D) \propto 1/D$ to $1/D^2$, as D increases, one finds here (Fig. 21) a crossover from $T_c(\infty)-T_c(D) \propto 1/D$ at small thicknesses to the finite size scaling [236–239] relation

$$T_c(\infty) - T_c(D) \propto D^{-1/\nu} \tag{129}$$

at large thicknesses. A remarkable crossover also occurs in the shape of the coexistence curve, which near $T_c(D)$ is described by a power law [55,229,230]

$$M \equiv \phi_{coex}^{(2)}(D) - \phi_{coex}^{(1)}(D) = \hat{B}(D)\left[1 - T/T_c(D)\right]^{\beta} \quad \text{for } T \le T_c(D) \tag{130}$$

where $\hat{B}(D)$ is a critical amplitude, and the exponent β for $T \to T_c(D)$ for all finite thicknesses D should have the value of the two-dimensional Ising model [229,230], namely $\beta_2=1/8$, while for $D \to \infty$ the three-dimensional value {$\beta \approx 0.325$ [230]} applies. In practice {where the limit $1-T/T_c(D) \to 0$ is hard to study, one rather works in the regime $10^{-3}<1-T/T_c(D)<10^{-1}$} the two-dimensional value $\beta=1/8$ is only observed for ultrathin films, where D is of the same order as the polymer gyration radius, while for larger D one observes "effective exponents" in between the asymptotic values (Fig. 22). This crossover was also ignored in the study by Kumar et al. [39], and thus the critical temperatures that one could ex-

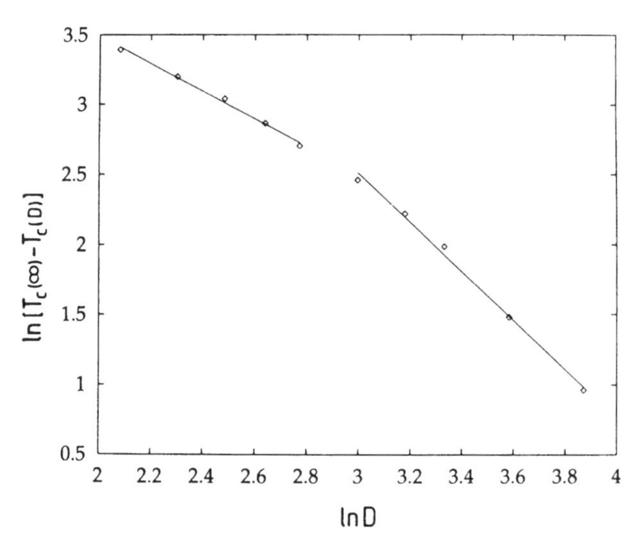

Fig. 21. Log-log plot of $T_c(\infty)-T_c(D)$ versus D, for the bond-fluctuation model of a symmetric polymer mixture with $N_A=N_B=N=32$. For small D, the straight line corresponds to a shift $T_c(\infty)-T_c(D) \propto 1/D$, while the second straight line for larger D shows the result $T_c(\infty)-T_c(D) \propto D^{-1/\nu}$, with $\nu=0.63$ being the critical exponent of the three-dimensional Ising model correlation length [229,230]. From Rouault et al. [55]

tract from Fig. 19 presumably suffer from some systematic errors. In any case, the data shown in Fig. 22 do exhibit a pronounced flattening of the coexistence curve, particularly for thin films, and look rather different from their Flory-Huggins counterparts, Fig. 9 (the parabolic shape there implies $\beta=1/2$).

A more complete description of the crossover from two- to three-dimensional critical behavior, that incorporates the shift of T_c, Eq. (129), can be written down in terms of crossover scaling functions $\tilde{M}$ {for the order parameter M, Eq. (130)} and $\tilde{S}$ {for the collective scattering intensity S_{coll} for scattering wavenumber $q \to 0$, which can be accessed by small angle scattering techniques [186]}. Defining $t_\infty \equiv 1-T/T_c$ $(D \to \infty)$ we can write

$$M = D^{-\beta/\nu}\tilde{M}\left(D^{1/\nu}t_\infty\right) \qquad S_{coll} = D^{\gamma/\nu}\tilde{S}^{\pm}_{coll}\left(D^{1/\nu}t_\infty\right) \tag{131}$$

where $\gamma \approx 1.24$ [229,230] and the two signs $\pm$ refer to the signs of t_∞. Note that all exponents β,γ,ν appearing in Eq. (131) are just those of the three-dimensional Ising model, describing criticality of polymer mixtures for $t_\infty \to 0$ in the bulk. Note that the standard power laws [229,230] $M \propto t_\infty^\beta$, $S_{coll} \propto |t_\infty|^{-\gamma}$ result from this description because the scaling functions $\tilde{M}(Y), \tilde{S}^{\pm}(Y)$ tend to power laws for large arguments, $\tilde{M}(Y \to \infty) \propto Y^\beta, \tilde{S}^{\pm}(Y \to \pm\infty) \propto |Y|^{-\gamma}$. In this description, two-dimensional criticality emerges as the singularity of the scaling functions when

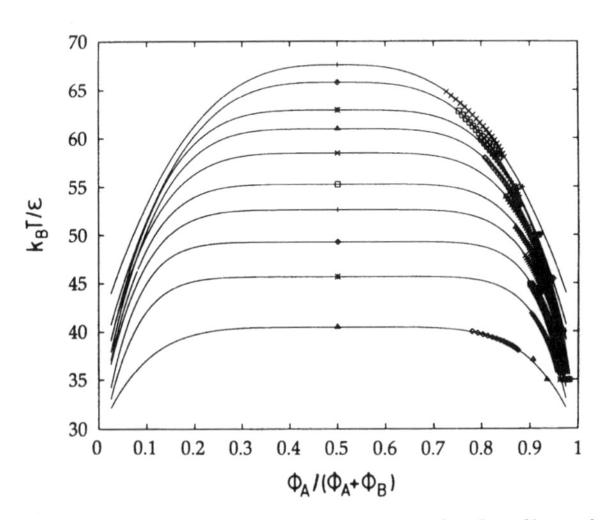

Fig. 22. Phase diagrams of the confined polymer mixtures for thin films of various thicknesses D, using the bond fluctuation model for symmetric polymer mixtures for $N_A=N_B=N=32$. The symbols refer to different film thicknesses: D=8,10,12,14,16,20,24,28,36 and 48 (from the bottom to the top). From Rouault et al. [55]

the argument $Y \equiv D^{1/\nu} t_\infty$ reaches a critical value Y_c that corresponds to $T=T_c(D)$ [55,236–239,282],

$$\tilde{M}(Y) \propto (Y-Y_c)^{\beta_2}, \tilde{S}^- \propto |Y-Y_c|^{-\gamma_2} \tag{132}$$

where β_2, γ_2 are the exponents of the two-dimensional Ising model ($\beta_2=1/8$, $\gamma_2=7/4$), and the equation for criticality, $Y=D^{1/\nu} t_\infty^{crit}=Y_c$, immediately yields $t_\infty^{crit}=1-T_c(D)/T_c(\infty)=Y_c D^{-1/\nu}$, i.e. Eq. (129). From Eq. (132) we immediately obtain the D-dependence of the critical amplitudes $\hat{B}(D)$ {defined in Eq. (130)} and $\hat{S}(D)$ {defined via $S_{coll}=|t-t_\infty^{crit}|^{-\gamma_2}\hat{S}(D)$}, namely [55]

$$\hat{B}(D) \propto D^{(\beta_2-\beta)/\nu} \qquad \hat{S}(D) \propto D^{(\gamma_2-\gamma)/\nu} \tag{133}$$

This description of critical behavior in thin films, by Eqs. (129)-(133) is not the whole story, however: it addresses only the regime of Ising-like critical behavior [186]. Now it is well-known that for symmetrical polymer mixtures the Ising critical region becomes very narrow when the chain length N becomes very large [186,200,283–287], i.e. Ising critical behavior is only observed for $N|t_\infty| \ll 1$, while mean-field critical behavior is predicted for $N|t_\infty| \gg 1$ (provided $|t_\infty|$ is still very small). However, for practically accessible molecular weights one often observes that the bulk critical behavior falls in the cross-over region between these two extreme limits [286,287]. In a thin film, there is then an interplay between two crossovers that one must deal with, namely both crossover from three-dimensional to two-dimensional critical behavior and crossover from meanfield to Ising behavior [55]. The situation can be best understood in terms of a comparison of lengths, namely the film thickness D, the correlation length ξ of concentration fluctuations, and the crossover correlation length ξ_{cross} that designates the value of ξ in the center of the mean-field to Ising crossover, say for $N|t_\infty|=1$. Since [186,200] $\xi = b/6\sqrt{N t_\infty^{-1/2}}$, cf. Eq. (123), we immediately conclude that $\xi_{cross} \propto bN$. Figure 23 shows that four regimes can be distinguished. If ξ_{cross} is rather small, as in the Monte Carlo simulation of Rouault et al. [55], basically no mean field behavior can be observed, and one has just a dimensional crossover from 3-dim Ising to 2-dim Ising, as discussed above {Eqs. (129)-(133), Figs. 21, 22}. If ξ_{cross} is very large, however, one stays essentially in the mean field regime and the treatment of Sect. 2.1 applies. The condition $D=\xi=b/6\sqrt{N t_\infty^{-1/2}}$ then yields $t_\infty^{cross} \propto N/D^2$, i.e. the temperature region where crossover occurs is of the same order of magnitude as the shift of T_c in this mean field region {Eq. (36)}. For $\xi_{cross}=D$, i.e. $D=D_{cross} \propto bN$, cf. Eq. (38), Fig. 23 predicts a crossover from mean field to 2-dim Ising behavior, whithout intermediate 3-dim Ising region. Note that for $D<D_{cross}$ the behavior of the shift of

T_c does not yet follow the asymptotic laws $D^{-1/\nu}$ {Eq. (129) or D^{-2} {Eq. (36)}, but a simpler 1/D behavior {Eq. (37)}, as confirmed also by the simulations [55] (Fig. 21).

All work reviewed so far in this subsection concerns thin films with neutral surfaces, but we feel that the general scaling description {Eqs. (129)-(133), Fig. 23)} should also apply to thin films with "symmetric" surfaces that both favor the same component (say B, cf. Fig. 5) relative to the other. The additional feature, not present in Fig. 22, then is a shift of the critical volume fraction $\bar{\phi}_{crit}(D)$ with thickness. Scaling considerations [216,224,225] predict for this shift

$$\bar{\phi}_{crit}(D) - \bar{\phi}_{crit}(\infty) \propto D^{-\beta/\nu} \approx D^{-0.51} \text{ (Ising case)} \tag{134}$$

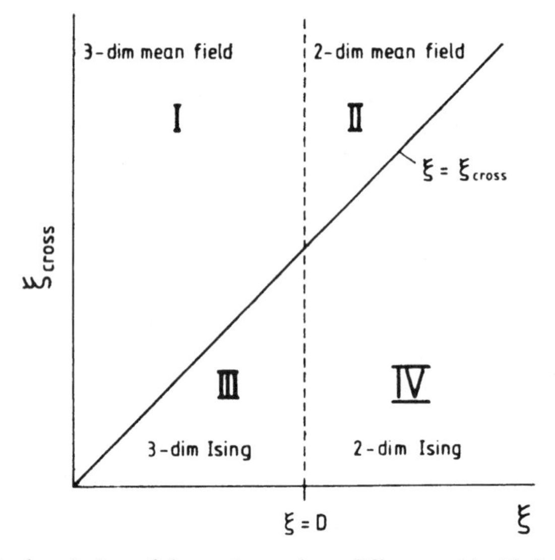

Fig. 23. Schematic description of the regimes where different critical behavior is expected for a polymer mixture in thin film geometry. The correlation length ξ can be varied by choosing suitable temperature distances from the critical points while the crossover correlation length ξ_{cross} can be varied by choosing suitable chain lengths. A crossover from three-dimensional (3-dim) to two-dimensional (2-dim) critical behavior is expected when $\xi \approx D$, while a crossover from mean-field to Ising behavior is expected when $\xi = \xi_{cross}$. In the Ising regime the dimensional crossover (3-dim→2-dim, from regime III to IV) implies a change of critical exponents, while in the mean-field regime the dimensional crossover (from regime I to II) implies only a change of critical amplitude prefactors. Of course, the lines drawn to limit these regimes I,II,III,IV should not be understood as sharp boundaries, because all crossovers are rather smooth. Also the regime close to the origin is not physically meaningful, since both ξ and ξ_{cross} should exceed the chain gyration radius in the critical region. From Rouault et al. [55]

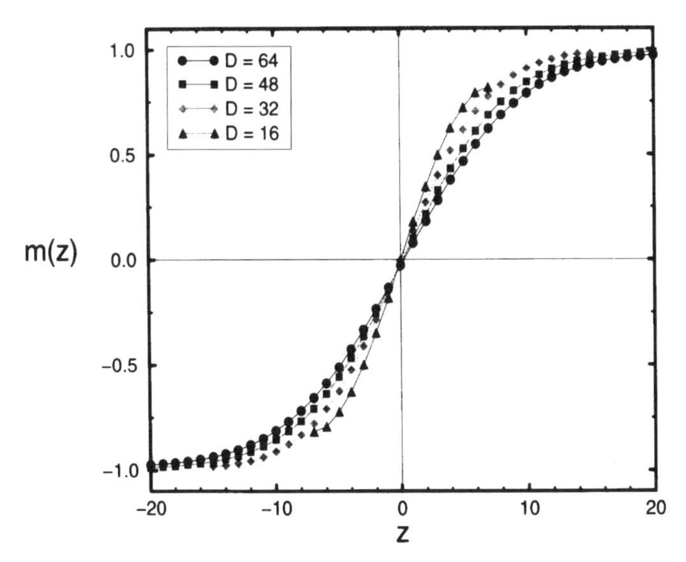

Fig. 24. Order parameter profiles $m(z)=\langle[\rho_A(z)-\rho_B(z)]\rangle/\langle[\rho_A(z)+\rho_B(z)]\rangle$, where $\rho_A(z)$, $\rho_B(z)$ are densities of A-monomers or B-monomers at distance z from the center of the interface, for systems of size 256×256xD, with walls situated at $z=\pm D/2$, for four different film thicknesses as indicated. All lengths are being measured in units of the lattice spacing. All data refer to a symmetrical polymer mixture modelled by the bond fluctuation model, with chain length N=32, choosing $\varepsilon_{AA}=\varepsilon_{BB}=-\varepsilon_{AB}=-\varepsilon=0.03\ k_BT$, i.e. in the strong segregation regime ($T/T_c=0.48$). A wall interaction $\varepsilon_w=0.1$, acting on monomers in the first two layers adjacent to the walls, and favoring species B on the left side and species A on the right side, is chosen. The lines through the data points are fits to $m(z)=m_b \tanh(z/w)$. From Werner et al. [266]

but we are not aware of any test of this relation yet. For the mean-field case, meanfield exponents $\beta=\nu=1/2$ would imply a shift proportional to D^{-1}, but Fig. 9 shows that presumably extremely large film thicknesses D are necessary to verify this relation.

Finally, we return to the case of "antisymmetric" surfaces, i.e. a situation where one surface of the thin film favors the A-rich phase and the other the B-rich phase (Fig. 1d). Simulations were recently carried out [266] in order to test the predictions {Eq. (127)} on the anomalous interfacial broadening (Sect. 2.5). Figure 24 demonstrates that this phenomenon can indeed be readily observed. Using the interfacial tension σ that has been independently measured [215], σ= 0.015, and the correlation length $\xi_b \approx 3.6$ lattice spacings from a direct study of the bulk correlation function, one can evaluate Eq. (127) quantitatively, using $\omega=\left(4\pi\xi_b^2\sigma\right)^{-1}$ and for w_0 the estimate resulting from self-consistent field theory (ω_0^{SCF}=4.65 lattice spacings). Ignoring the last additive constant on the right-hand side of Eq. (127), which presumably is small, the prediction shown in Fig. 25 is obtained, which compares favorably with the Monte Carlo results

[266]. It is remarkable, however, that large deviations do occur for film thicknesses up to $D \approx 20$, i.e. for thicknesses about three times the gyration radius: for $D \leq 16$ the observed width w is even smaller than the "intrinsic width" ω_0^{SCF}, i.e. the confinement effect on the interface is so strong that the "intrinsic" interfacial profile is squeezed! In fact, in these ultrathin films one observes [266] a behavior $w \propto D$ (as long as $w \leq \omega_0^{SCF}$), that is not accounted for by the theory of Sect. 2.5, which only describes the asymptotic behavior for very large D. However, the simulations [266] gave direct evidence for the exponential variation of the correlation length $\xi_{||}$ associated with the interfacial fluctuations, Eq. (114), with film thickness $\{\xi_{||} \propto \exp(\kappa D/4)\}$, and the general concepts described in Sect. 2.5 have been nicely corroborated.

2.7 Computer Simulation of Confined Block Copolymers

This subsection has to be very brief, since only rather preliminary qualitative studies are available [34,35]. The model used are self-avoiding walks with N=16

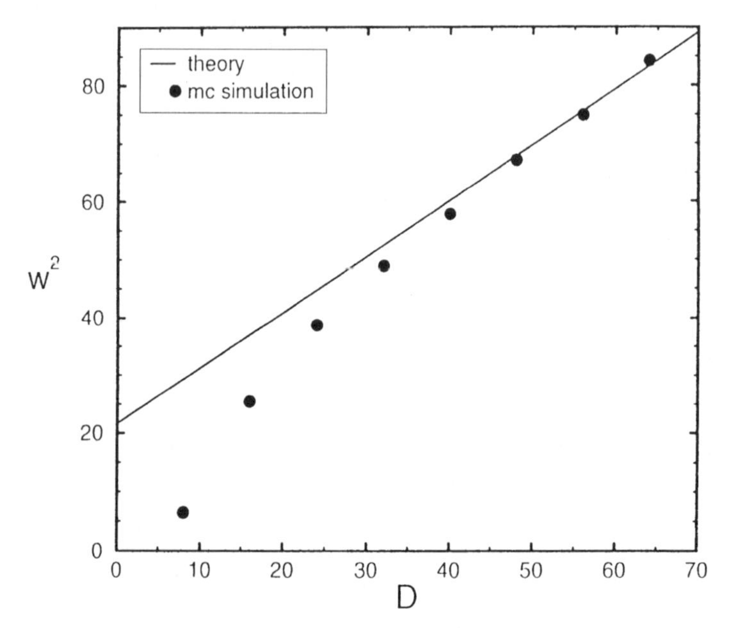

Fig. 25. Squared interfacial width w^2 vs film thickness D, for the model of Fig. 24. The *straight line* is the corresponding theoretical prediction (cf. text). All lengths are measured in units of the lattice spacing. From Werner et al. [266]

effective monomers on the simple cubic lattice, each effective monomer taking one site of the lattice, and one works at a density of ϕ=0.8 occupied sites. Only the symmetric case f=1/2 (i.e. $N_A=N_B=N/2=8$) was studied, choosing a repulsive nearest neighbor interaction $\varepsilon_{AB}>0$ between monomers ($\varepsilon_{AA}=\varepsilon_{BB}=0$), and a monomer-wall interaction that repels A-monomers, $\varepsilon_{AW}=\varepsilon_{AB}/2$. Lattice geometries L x L x D were chosen, with two hard walls of size L x L a distance D apart, D being varied from D=8 to D=32 lattice spacings, the lateral linear dimension L being L=16 or L=24, to check for size effects. Configurations were equilibrated by a combination of "slithering snake" moves [198] and head-tail interchanges of the chains.

Figure 26 shows typical order parameter profiles obtained across the films at various temperatures. Due to the repulsive wall-A interaction the order parameter $\rho_A(z)-\rho_B(z)$ is always depressed towards negative values for z near the walls. From previous studies of the same model in the bulk [190,191] one expects an order-disorder transition to a lamellar phase for T≈2 (note $\varepsilon_{AB}/k_B\equiv1$), and consequently one sees for D=20 and D=30 well-developed lamellae oriented parallel to the walls, evidenced by the periodic variation of the order parameter across the film, with an amplitude close to the saturation value unity already. However, for D=14 the system does not develop this type of order, and for D=24 there seems to be well-developed order near the right surface but not near the left surface, which is unexpected since both surfaces are equivalent.

The snapshot pictures (Fig. 27) show that a nice lamellar arrangement of the type proposed in Fig. 3a indeed occurs for both D=20 and D=30, while the case D=14 is an example for the structure of Fig. 3b, where lamellae run perpendicular to the walls. These results thus confirm qualitatively the theoretical considerations of Sect. 2.3, where the perpendicular arrangement was indeed found to be more stable in certain cases where the parallel arrangement of lamellae was too strongly "frustrated". The case D=24 seems to be an example for the structure proposed in Fig. 3c, but it may well be that this configuration is not the true equilibrium structure, since also Monte Carlo simulations of block copolymer ordering are hampered by very slow relaxation and hysteresis. Also, for the model of Figs. 26, 27 no attempt has been made to estimate the parameters λ_b, Γ and δ that enter the theoretical description of Sect. 2.3, Eqs. (85)-(93). Since no full segregation inside the domains is yet achieved (Fig. 26), no quantitative agreement with the theory of Sect. 2.3 anyhow can be expected. In the disordered phase (1/T=0.3 and 0.4), the order parameter profiles show surface-induced order of the type discussed in Sect. 2.2, see Fig. 12a, but again the model is not enough characterized to allow a more specific comparison.

As observed for block copolymers in the bulk [190–193], there is a detailed interplay between the order in the system and the chain linear dimension: chains like to stretch out somewhat in the direction perpendicular to the lamellae. If one studies the mean square gyration radius component R_{gz}^2 in the z-direction perpendicular to the walls, one hence observes an oscillatory dependence on thickness D for 1/T≥0.5, since only for $n\lambda_b=D$ (n=1,2,3,...) the parallel morphology fits nicely into the film, cf. Sect. 2.3, and the "dumbbell"-like block copoly-

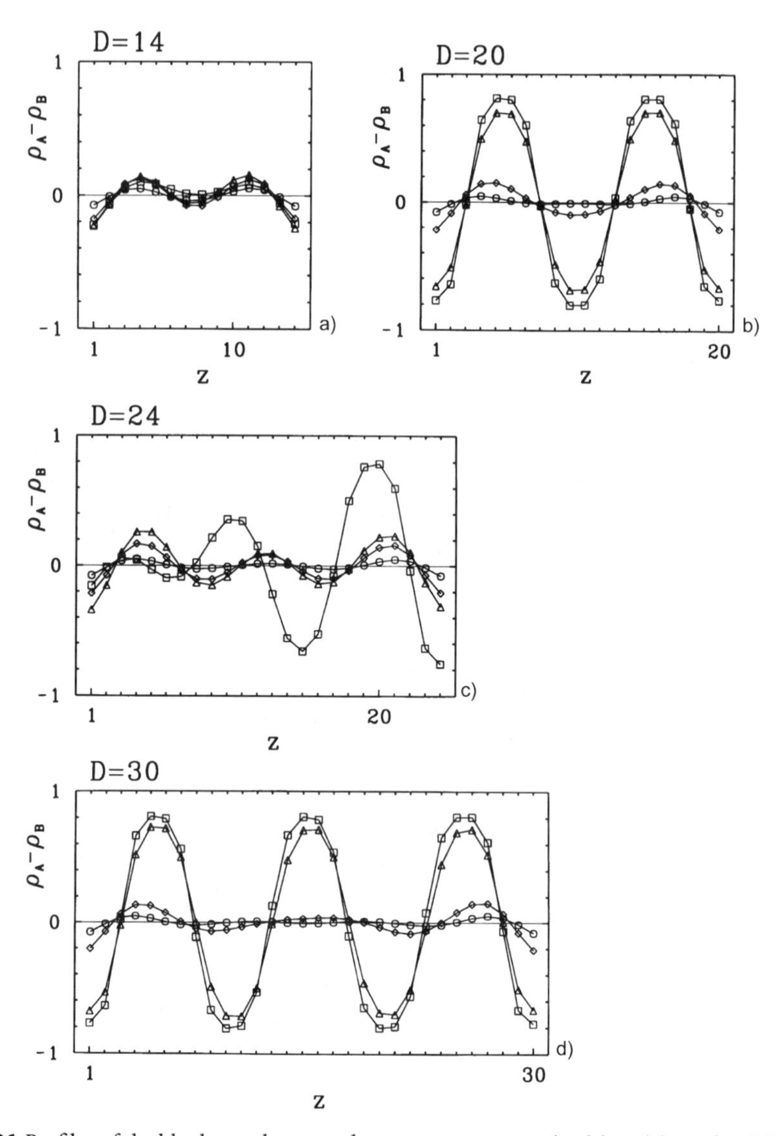

Fig. 26. Profiles of the block copolymer order parameter $\rho_A - \rho_B$ ($\rho_A(z), \rho_B(z)$ are densities of A-monomers or B-monomers at distance z from the left wall), plotted for the thicknesses D=14(**a**), 20(**b**), 24(**c**) and 30(**d**). Temperatures are (in units of ε_{AB}/k_B) 1/T=0.3 (*circles*), 0.4 (*diamonds*), 0.5 (*triangles*) and 0.6 (*squares*). From Kikuchi and Binder [35]

mer coils are then also oriented perpendicular to the walls, while for the perpendicular ordering (as it occurs for D=14, for example) the stretched "dumbbells" are oriented parallel to the walls, and R_{gz}^2 is not enhanced [35].

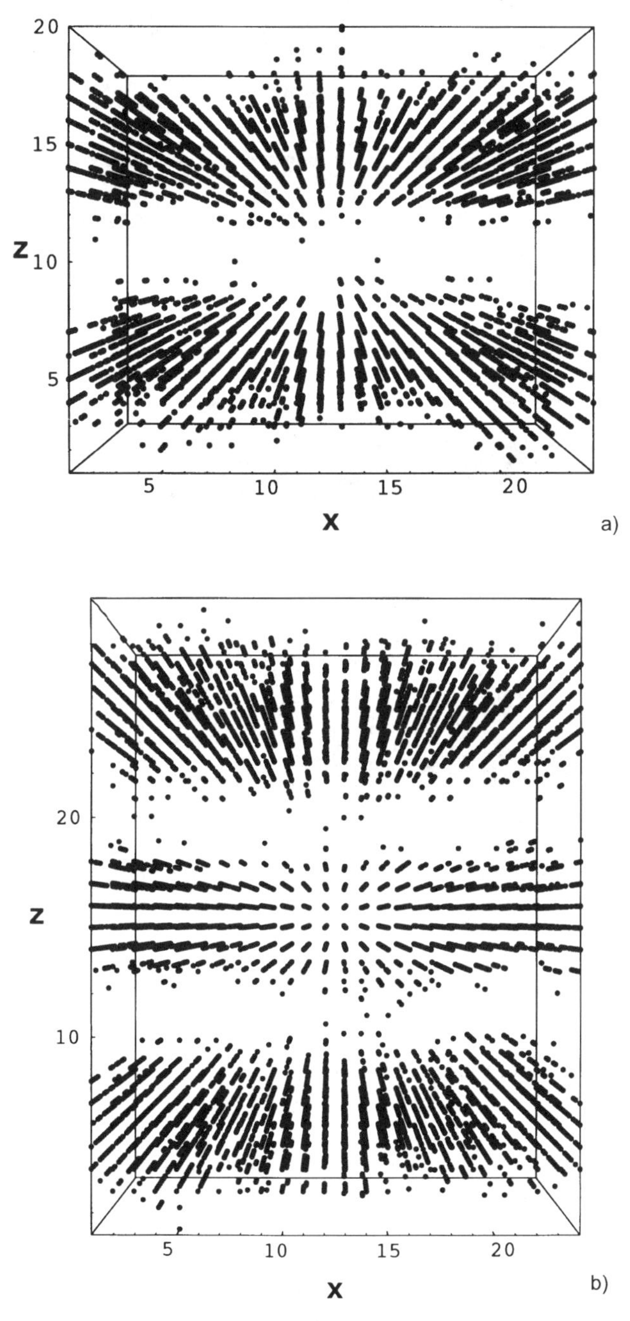

Fig. 27. a ,b Snapshots of the monomer configurations at 1/T=0.6 for D=20(**a**), 30(**b**)Only A-monomers are shown as *dots*. The two walls are at the top and the bottom of the figures. From Kikuchi and Binder [35]

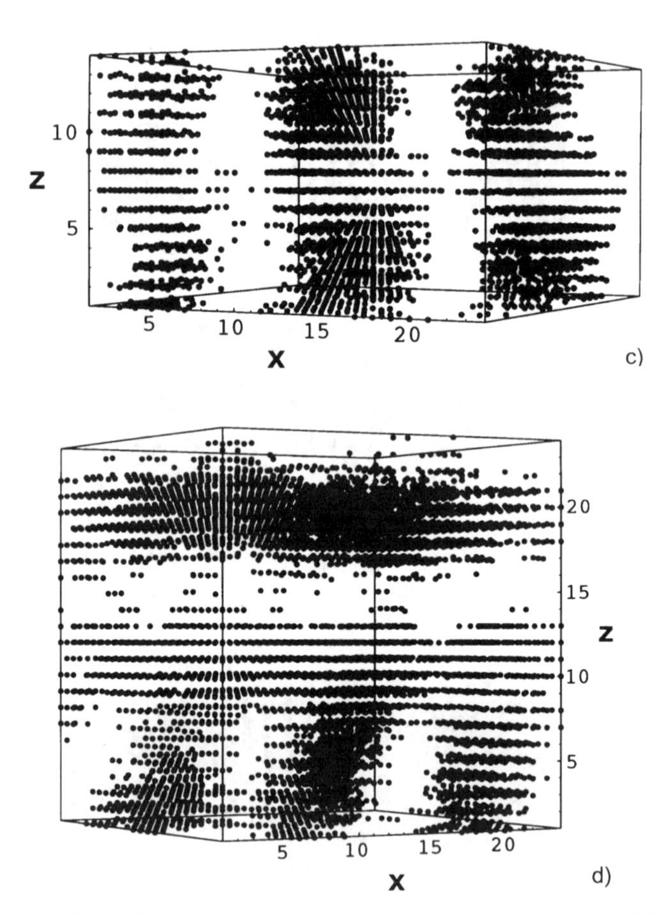

Fig. 27. c,d. Snapshots of the monomer configurations at 1/T=0.6 for D= 14(**c**) and 24(**d**). Only A-monomers are shown as *dots*. The two walls are at the top and the bottom of the figures. From Kikuchi and Binder [35]

2.8 Dynamics of Phase Separation in Films: an Introduction

As we have seen in Sects. 2.1, 2.5, 2.6, polymer blends confined in thin films in equilibrium typically occur in a two-phase morphology, just as polymer blends in the bulk typically are incompatible and hence separated in two coexisting phases [186,199–202,288]. Since polymer coils due to their large molecular weight are very slowly relaxing objects [200], also the dynamics of phase separation in polymer blends is very slow, both in the bulk [186,206,207] and in thin film geometry [5,158]. Thus, these systems have become model systems for the study of the initial stages of phase separation in the bulk [289], and the possibility to control the morphology of the system by quenching intermediate stages of the slowly coarsening two-phase structure from the fluid to the glassy phase is of practical interest [290]. Similarly, polymers in thin film geometry have proven to be model systems for the study of surface effects on spinodal decomposition [5,158]. We give here only a tutorial introduction, since excellent recent reviews of both theory [158] and experiment [5] exist.

By "spinodal decomposition" [186,290–294] one means the spontaneous growth of long wavelength concentration fluctuations in a mixture that has been quenched from the one-phase region to the unstable region inside the spinodal curve. In the bulk, wavenumbers in the range $0<q<q_c$ are expected to grow, with the critical wavenumber q_c being given by [186]

$$q_c = \frac{2\pi}{\lambda_c} = \frac{6}{b}\sqrt{\phi_0(1-\phi_0)}\left[\chi - \chi_s(\phi_0)\right]^{1/2} \tag{135}$$

ϕ_0 being the average volume fraction in the mixture, and $\chi_s(\phi_0)$ is the value of the Flory-Huggins parameter χ at the spinodal curve $\chi=\chi_s(\phi_0)$ {of course, the unstable region in the phase diagram of the mixture is defined by $\chi>\chi_s(\phi_0)$}. Maximum growth (in the weak segregation limit [186]) occurs for $q_m=q_c/\sqrt{2}$.

While in the bulk the phases of the growing concentration waves are random, and also the directions of the wavevectors $\vec{q}$ are controlled by random fluctuations in the inital states, a surface creates a boundary condition, and working out a dynamic extension [45,129,132,133,144,156] of the model in Sect. 2.1 {Eqs. (7)–(10)} one finds that under typical conditions wavevectors oriented perpendicular to the walls occur, with phases such that the maxima of the waves occur at the walls (Fig. 28). In terms of a normalized order parameter $\psi(Z,\vec{R},\tau)$ where τ is a scaled time and Z, $\vec{R}$, are scaled coordinates perpendicular and parallel to the walls, $Z=z/2\xi_b$, $\vec{R}=\vec{q}/2\xi_b$, $\psi=(\phi-\phi_{crit})/(\phi_{coex}-\phi_{crit})$, this dynamic extension is the Cahn-Hilliard equation [291–294]

$$\partial\psi\left(\vec{R},Z,\tau\right)/\partial\tau = -\nabla^2\left\{\psi - \psi^3 + \nabla^2\psi/2\right\} \tag{136}$$

applying boundary conditions such that the z-component of the concentration current vanishes for Z=0 and $Z=\tilde{D}$, and at Z=0, $Z=\tilde{D}$ boundary terms resulting from Eqs. (7) and (10) are added, e.g. at Z=0 one has [129,132,133,144,156]

$$\partial\psi/\partial\tau = h_1 + \tilde{g}\,\psi + \gamma\partial\psi/\partial Z \tag{137}$$

where h_1, $\tilde{g}$ are related to μ_1,g in Eq. (10) with suitable rescalings, and γ is related to the constant of the gradient energy κ, Eq. (8) {see Refs. 129,144 for microscopic derivations of these phenomenological equations, Eqs. (136) and (137)}. In Fig. 28 ψ is averaged over the longitudinal coordinate $\vec{R}$ and this average then is denoted as $\phi_{av}(Z,\tau)$. One can clearly see the damped concentration wave, whose amplitude is maximal at both walls, and whose wavelength can be shown to be consistent with Eq. (135), see Ref. 156.

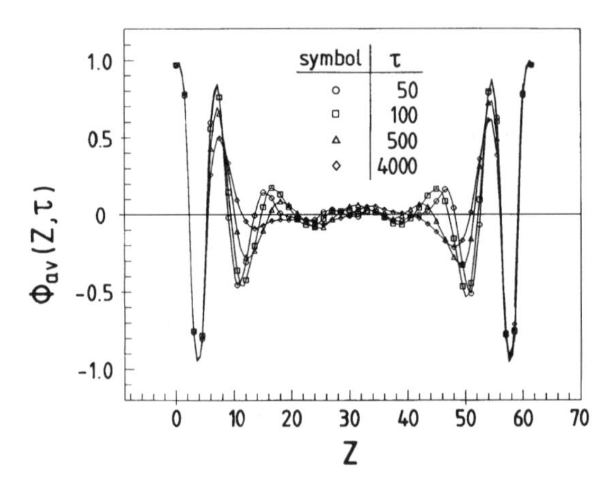

Fig. 28. Averaged order parameter profiles $\phi_{av}(Z,\tau)$ plotted vs the scaled distance $Z=z/2\xi_b$ from the left wall at z=0 for four different scaled times τ after the quench as indicated, for a scaled distance $\tilde{D}=D/2\xi_b=60$. Choosing a rescaled distance $\tilde{L}_{||}/2\xi_b=600$, and a discretization $\Delta X=1.5, \Delta\tau=0.05$, the resulting equations are solved by the cell-dynamics method. The results shown are for parameters $h_1=\gamma=4$, $\tilde{g}=-4$, and averaged over 2000 independent initial conditions, corresponding to random fluctuations in a state with $\int\phi_{av}(Z,0)dZ=0$. The parameters h_1 and $\tilde{g}$ were chosen such that both walls prefer A but one is still in the non-wet region of the equilibrium surface phase diagram of the corresponding semi-infinite system. From Puri and Binder [145]

While the maximum of these concentration waves in this "surface-directed" spinodal decomposition process at the walls forms already during the early stages and then stays constant, as well as the subsequent minimum of this "wave", the second and further maxima and minima are more and more damped out in the course of this phase separation process (Fig. 28). This is as expected, since well-defined concentration waves result from the Cahn-Hilliard equation, Eq. (136), only when one linearizes the $\psi 3$ term around the average order parameter [186,291–294], which is accurate only for the early stages. As a consequence of the non-linear term in Eq. (136), however, coarsening of the initially formed concentration inhomogeneities sets in during the intermediae stages of the process, as is exemplified by the "snapshot pictures" of Fig. 29. During the late stages of phase separation where the characteristic linear dimension $l(\tau)$ of the A-rich and B-rich domains that form in the bulk region of the film is much larger than the bulk correlation length ξ_b in thermal equilibrium, Eq. (136) leads to the growth law of Lifshitz and Slyozov [295], $l(\tau)\propto\tau^{1/3}$. However, Eq. (136) is too simplified to describe phase separation of actual polymer blends in the late stages in the bulk, where hydrodynamic mechanisms [296] predict a faster growth, $l(\tau)\propto\tau$, which is indeed observed for "critical quenches" (i.e., quenches at critical volume fraction) [290]. The extent to which such hydrodynamic mechanisms are also operative in thin film geometry still needs to be investigated, however.

We also emphasize that Figs. 28, 29 are characteristic for the situation where the surface forces due to the walls are strictly short range {this is assumed already in Eqs. (10), (13), (14) and (137)}, and a situation which is on the nonwet side but close to the wetting transition. Therefore, the equilibrium value of the scaled order parameter $\phi_{av}(Z,\tau)$ in Fig. 28 at the surfaces is close to unity. If the equilibrium condition of the corresponding semi-infinite system corresponds to a wet surface, however, then $\phi_{av}(Z,\tau)$ at the surfaces exceeds unity [144,145], and in addition the phase separation in the bulk competes with the growth of wetting layers (or precursors thereof) at the surfaces of the thin film. For short range surface forces, however, the thickness of the wetting layer $l_w(\tau)$ grows only logarithmic in time, $l_w(\tau)\propto \ln\tau$, and snapshot pictures of the growth would still be very similar to Fig. 29, over the regime of scaled times that has been studied. A faster growth, $l_w(\tau)\propto\tau^x$ occurs for long-range surface potentials, however[228]. Figure 30 shows typical order parameter profiles for a surface potential $V(Z)\propto Z^{-3}$ and a semi-infinite geometry (or one surface of a very thick film, respectively) [228]. One recognizes at later times ($\tau\geq 900$) the two-step decay of the profile typical for growing wetting layers: the order parameter $\phi_{av}(Z,\tau)$ starts out at $Z=0$ at a value larger than unity (i.e., a higher value than occurring in the bulk A-rich phase), but decays on a scale of Z of order unity to a plateau where $\phi_{av}(Z,\tau)\approx 0$, and the thickness of this plateau grows with the power law quoted above, the exponent x being $x\approx 0.16$ [228]. While in the initial stages ($\tau\leq 30$) one can still see the damped oscillatory decay characteristic for surface-directed spinodal decomposition, at later stages the growing wetting layer and a subsequent broad depletion zone have removed this structure completely.

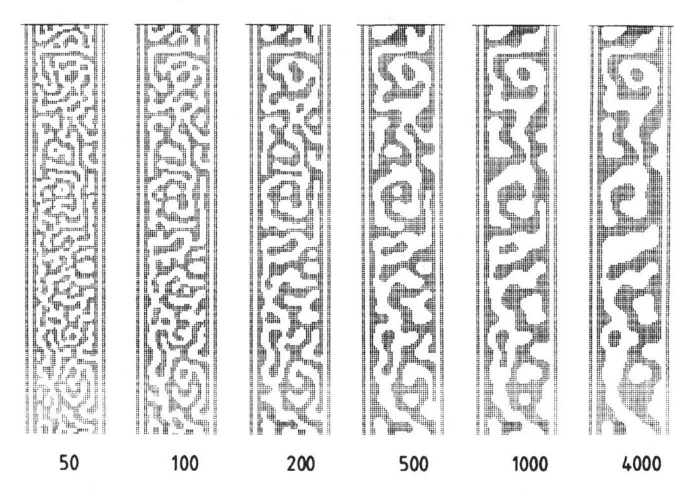

Fig. 29. Snapshot pictures of one particular time evolution of the model described in Fig. 28, showing positive $\psi(Z,R,\tau)$ *dark*, while negative $\psi(Z,R,\tau)$ is left *white*. Note that both Figs. 28,29 refer to a two-dimensional model system, where ρ and R are scalar coordinates. From Puri and Binder [145]

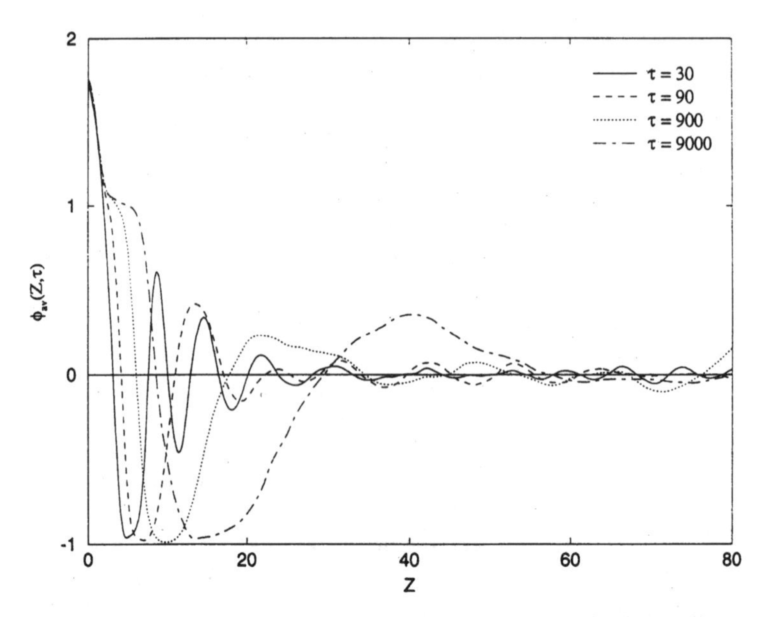

Fig. 30. Laterally averaged order parameter profiles $\phi_{av}(Z,\tau)$ vs Z for four different times, for the same model as in Fig. 28, but choosing $\tilde{L}_{||}=400$, $\tilde{D}=300$, $\Delta X=1.0$, $\Delta\tau=0.03$, and a surface potential $V(Z)=-h_1/Z^3$, with parameters $h_1=\gamma=4$, $\tilde{g}=-4$. Averages were taken over 200 independent runs. From Puri et al. [228]

We emphasize that Eqs. (136) and (137) provide a very coarse-grained description applicable at best in the weak segregation limit – any effects specific for polymers are absorbed in the scale factors for spatial ($Z=z/2\xi_b$) and temporal coordinates $\{\tau=(T_c/T-1)\, t/(8\tau_s\, \xi_b^2)$, t being the physical time and τ_s being the microscopic time constant of the underlying model, i.e. a Rouse time for unentangled polymers}. Eqs. (136) and (137) apply for short range wall forces only, but a comparison of Figs. (28) and (30) shows that long range forces do change the behavior significantly, and in the general case we expect a competition between short range and long range surface forces. Neither has this competition been addressed so far, nor the influence of hydrodynamic effects in thin film geometry, and thus it is clear that more theoretical work is desirable to address these issues.

3
Outlook on Pertinent Experimental Results

As discussed already in the introduction, there is a huge and diverse literature on thin films of polymer blends and block copolymers. The field is rapidly developing, and many phenomena are as yet not fully understood. In this section, we hence cannot attempt to give a fair evaluation of this rich literature, but focus attention only on those subjects which are closely related to the selection of theoretical issues discussed in the previous section.

3.1 Polymer Blends in Thin Film Geometry in Thermal Equilibrium

Many experimental studies on polymer blends in thin film geometry actually are performed in a desire to test theoretical concepts on the surface behavior of polymer blends, which were worked out for a semi-infinite geometry. Recent studies along such lines are due to Genzer et al. [74,86] and Scheffold et al. [81,82,85], for instance. E.g., Genzer et al. [74] analyzed a mixture of protonated and deuterated polystyrene (with chain lengths N_{dPS}=9190, N_{PS}=17300 and varying ϕ_{dPS} in the "bulk" from zero to about 22%, using the data of Zhao et al. [67] at 184°C, where the χ parameter is $\chi=1.5\times10^{-4}$) and studied the relation between the total surface excess ϕ_s {Eq. (5)} and the local concentration ϕ (z=0) of dPS at the surface. They concluded that the long wavelength approximation of Schmidt and Binder [11] describes the observed variation of ϕ_s and $\phi(z=0)$ with ϕ_{dPS} in the bulk only qualitatively while a numerical self-consistent field calculation along the lines of Shull [20] provided a significantly better fit. They showed that under typical conditions the surface enrichment decays on a length scale of 20 nm, while the gyration radius of PS is about 36 nm, i.e. the condition $b(d\phi/dz)^2 << 1/N$ necessary for the validity of the long wavelength approximation [11] did not hold in the experiment. Genzer et al. [74] also extracted a numerical function $df_s^{(bare)}(\phi_0)/d\phi_0$ from this analysis and concluded that it could not be described by the model assumption, Eq. (10). In their more recent work, Genzer and Composto [86] analyze the even more symmetric mixture of deuterated polystyrene (N_{dPS}=1650) and the statistical copolymer poly(styrene-*co*-4-bromostyrene, PBrS, N_{PBrS}=1670), for various temperatures near the critical point (T_c=189 °C, ϕ_{PBrS}^{crit} =0.502) over the full range of PBrS volume fractions. They find that the surface excess has a maximum near ϕ_{PBrS}^{crit} which grows as T approaches T_c, and again calculate $df_s^{(bare)}(\phi_0)/d\phi_0$ as function of ϕ_0 for four temperatures near T_c.

While studies of this type are clearly very pioneering and useful, the author of this review believes that the quantitative results on $df_s^{(bare)}(\phi_0)/d\phi_0$ and their interpretation must be taken as preliminary, since the analysis refers to a semi-infinite geometry throughout, while the actual experiments are done on thin films, spin-coated on silicon wafers. Genzer and Composto [86] quote a film thickness of about 220 nm, corresponding to about 20 times the gyration radius of dPS in the bulk ($R_g \approx 11$ nm). If we could take over the model calculation of Fig. 9 (a thin film between completely symmetric walls) to this less symmetric situation (with one surface being given by the silicon wafer, the other by air) we would predict a shift of T_c of about 4 K, comparable to the smallest temperature distance from T_c studied in [86] (6 K). This argument already shows that the finite film thickness is expected to have important effects. The same conclusion is reached when we estimate the bulk correlation length ξ_b using the mean-field formula [186] $\xi_b=(R_g/\sqrt{3})(T/T_c-1)^{-1/2}$ which yields $\xi_b \approx 24$ nm for T=222 °C but $\xi_b \approx 56$ nm for T=195 °C. Since finite thickness effects are of order $\exp[-D/2\xi_b]$, cf. Eq. (28), they are not negligible here. Thus a re-interpretation of the findings of Genzer and Composto [86] may be called for. This is not straightforward since the two surfaces (silicon,air) are not equivalent.

Particularly illuminating is the neutron reflectometry study of surface segregation of Norton et al. [80] carried out for a nearly symmetric mixture of protonated and deuterated poly(ethylenepropylene), with $N_{PEP}=2360$, $N_{dPEP}=2140$. Film thicknesses of about 150 nm were used, about 10 times the radius of gyration ($R_g = b\sqrt{N/6} \approx 15.5\,\text{nm}$ since $b \approx 8$ Å), using Si as a substrate. While the roughness of this substrate was measured as 5 Å and the roughness of the polymer-air interface about 10Å, the surface enrichment profiles revealed a flat region near the surface of about 50 Å, i.e. one third of the gyration radius. This effect could be observed for a broad range of compositions in the bulk of the thin film, and was pronounced as soon as the volume fraction of dPEP at the surface exceeded about 50%. Norton et al. [80] concluded that neither long range surface forces nor wetting phenomena nor surface roughness can be responsible for this effect, and concluded that it implies a basic breakdown of the mean field theory of surface enrichment. At this point we recall a qualitative similarity of these profiles with the simulation results [56] shown in Fig. 20, which apply for strictly neutral surfaces, and were interpreted assuming that the surface-enriched layer consists of coils of the preferred species oriented with their long axis in parallel to the surface, but having otherwise an essentially undeformed conformation.

Norton et al. [80] also extract $df_s^{bare}(\phi_0)/d\phi_0$ as function of ϕ_0 from their data, using both the mean field theory for semi-infinite systems [11] and a different method based on the "Gibbs adsorption equation" [80]. They find from both analyses that $-df_s^{bare}(\phi_0)/d\phi_0$ stays constant for $\phi_0 \leq 0.6$ and steeply rises when $\phi_0 \to 1$. Again the caveat must be mentioned that possible effects due to the finite thickness of the total film have not been considered in this work.

Budkowski et al. and Bruder et al. [70] have tried to estimate the bulk phase diagram of polymer mixtures, using a geometry where for $T<T_c$ a phase separation with an interface running parallel to the substrate occurs (Fig. 1d). They obtained order parameter profiles across the film and associate the order parameter close to both walls with the order parameters $\phi_{coes}^{(1)}$, $\phi_{coes}^{(2)}$ of the two coexisting phases in the bulk. As shown by the model calculations of Ref. [62], such a procedure also leads to systematic errors due to the finite film thickness D, in particular for T near T_c where ξ_b is large. However, the accuracy with which Budkowski et al. and Bruder et al. [70] estimated $\phi_{coes}^{(1)}$, $\phi_{coes}^{(2)}$ was rather limited, and

thus it is probably not warranted to consider these finite thickness effects on the basis of these rather rough data. A similar method was also used by Scheffold et al. [81], where typically films of thickness 1000 nm were used to study mixtures of deuterated and protonated model polyolefins, consisting of branched ethylethylene units $(EE)_x$ and linear di-ethylene units $(E)_{1-x}$, using a degree of polymerization between N=707 and N=2030. Altogether 13 different blends were studied, but the critical temperature could be estimated only with an uncertainty of at least ±4 K (and sometimes as large as ±10 K). Thus these data probably do not allow any conclusions on the change of the critical behavior due to finite film thickness yet.

In subsequent work [82,85], Scheffold et al. studied surface enrichment phenomena, preparing mixtures d66/h52, d86/h75 and d94/h86 {here d,h refers to deuterated or protonated and the number refers to the content of ethylethylene C_2H_3 (C_2H_5) in percent}. These mixtures were studied in the one phase region below T_c (at $T/T_c \approx 0.95$ in the d66/h52 sample), preparing 8–9 different concentrations between $\phi_{d66}=0$ and $\phi_{d66}=\phi_{coex}^{(1)}$, and the surface excess as function of bulk concentration was measured. Using again an analysis following Ref. 11, appropriate for semi-infinite systems but adding a term $-Y(d\phi/dz)_{z=0}$ to $f_s^{(bare)}$ $(\phi(z=0))$ in Eq. (10) in a somewhat ad-hoc manner (Y is an adjustable constant) [82], Scheffold et al. estimated also the concentration variation of $-df_s^{(bare)}$ $(\phi)/d\phi$ as function of the (local) volume fraction ϕ at the surfaces of their blends. While Genzer and Composto [86] found that $-df_s^{(bare)}$ $(\phi)/d\phi$ is roughly constant for $\phi \leq \phi_{crit}$ and then increases, Scheffold et al. [82] find $-df_s^{(bare)}$ $(\phi)/d\phi$ is roughly constant for $\phi \leq \phi_{crit}$ and then decreases (for those samples which are fairly close to the critical temperature [85]). However, again we must raise the objection that most of the data are derived from films in the thickness range from 300 to 500 nm, the correlation lengths being of the order of 10 nm (or larger, for those samples which were close to criticality [85]), and so finite thickness effects could again play a role. It is interesting to note that for systems at temperatures far below T_c [85] a smooth increase of $-df_s(\phi_0)/d\phi_0$ with surface composition ϕ_0 was found over the entire range of ϕ_0 – in this case the film thickness is two orders of magnitude larger than the correlation length and therefore a better approximation of semi-infinite behavior should be obtained. We also recall from Fig. 16a that the profiles of the surface enrichment are sensitive to the precise interaction range, to surface roughness of a free surface of a polymer blend, etc., and these effects are all most pronounced when the volume fraction ϕ at the surface differs appreciably from the bulk. Thus the quantitative understanding of these experiments may be rather difficult.

A particularly striking observation was reported by Bruder and Brenn [72] for films of asymmetric blends of deuterated polystyrene (dPS) and poly(styrene-4-co-bromostyrene), PBr_xS, with $N_{dPS}=5714$, $N_{PB_xS}=1660$, using films of

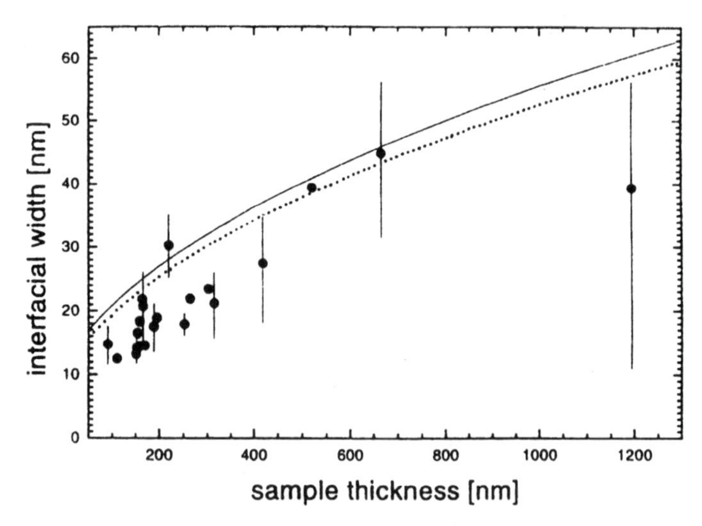

Fig. 31. Plot of interfacial width w vs D for a blend of olefinic copolymers d75/h66 (cf.text) at T_0=356 K, extracted from nuclear reaction depth profiling experiments, based on the reaction $^3He+^2H \rightarrow ^4He+^1H+18.35$ MeV and backward angle detection of ^1H. Note that the spatial resolution is optimal near the air surface (4 nm) but quickly deteriorates for large distances from the air surface. Therefore, the error bar on w/2 increases strongly with increasing D. Full and dotted curves represent the approximate asymptotic formula $w^2 = w_0^2 + \xi_b D/4$ (a factor $\pi\omega/(1+\omega/2)$ in Eq. (127) being approximated as unity), choosing $w_0=\xi_b$, and ξ_b=11.8 nm or ξ_b=10.6 nm, respectively. From Kerle et al. [84]

thickness 103 nm on Si substrates. Their data seem to indicate a first-order surface phase transition, i.e. the surface excess as function of bulk composition first increases smoothly and then exhibits a jump before the coexistence curve is approached. But unlike the prewetting transition shown in Fig. 6b, the jump goes from a larger surface excess towards a smaller one, so ϕ_s vs ϕ_b varies non-monotonically.

Choosing conditions where the equilibrium of the thin film is of the type shown in Fig. 1d, experimental study of the concentration profile across a thin film yields information on the interfacial profile between coexisting phases [32,84,125,297,298]. While sometimes [32,298] it has been tried to account for a broadening of interfacial profiles by capillary wave fluctuations, using Eq. (117) and relating q_{max} to the lateral resolution length of the experiment, other researchers claim [125,297] that one observes directly the "intrinsic profile" (comparable to theory [242–253]), provided the initial roughness before the annealing process of the sample is subtracted. Recently [84,298] it has been shown that the interfacial thickness as extracted from such measurements in thin film geometry does depend on the thickness D of the thin film, in agreement with the theoretical considerations of Sect. 2.5. Kerle et al. [84] study a mixture of the same type as used by Scheffold et al. [82,85], namely d75/h66, with chain lengths

N_{d75}=1625, N_{h66}=2030, at T=356 K (since T_{cb}=374 K this means $T/T_{cb} \approx 0.95$). At the coexistence curve at this temperature, the bulk correlation length $\xi_b \approx$ 11 nm, and choosing film thicknesses from D=100 nm up to D=1200 nm an increase of the interfacial halfwidth w/2 from about $w/2 \approx 12$ nm at small D to about $w/2 \approx 40$ nm at large D is observed (Fig. 31). This increase is interpreted in terms of the behavior $w \propto \sqrt{D}$, appropriate for short range forces acting from the surfaces of the film on the interface {Eq. (127)}. In contrast, Sferrazza et al. [298] study the interface between deuterated polystyrene and polymethylmethacrylate for film thickness between 5 and 500 nm and find in this regime an increase $w \propto \ln D$ only, as appropriate for long range forces {see Eq. (128)}. Since this latter study uses chemically more dissimilar polymers than used by Kerle et al. [84], it is likely that in the system of [298] long range forces are indeed more important. At this point, we also draw attention to a recent simulation study [299] where it was argued in a study of the wetting transition at the surface of polymer blends that long range interactions change the order of the wetting transition from second to first order (as they do for small molecule mixtures [220,300]) only for small molecular weight. It was suggested that the effect of the surface is essentially described by a short range model if the "effective range" σ of the van der Waals interaction, $V_{vdw}(z)/k_BT=(\sigma/z)^3$, is much smaller than the polymer coil size.

While the critical behavior of polymer blends in d=3 dimensions has already been studied very thoroughly (e.g. [286,287]), we are not aware of any counterpart in d=2 dimensions, which would be very interesting particularly for ultrathin films (with thicknesses comparable to the gyration radius). We also are not aware of any study of the shift of T_c as function of thickness. Thus there is no data to which the results of Sect. 2.6 could be compared with.

3.2 Ordering in Thin Films of Block Copolymers

In this section we shall not deal with all the work treating films on substrates with a free surface against air (Fig. 4b-d), but rather focus attention only on the problem where the block copolymer melt is confined to a film with uniform given thickness D (Fig. 3, Fig. 4a), assuming that there is a hard wall both on top and on the bottom of the film. This situation is a natural boundary condition for theoretical work (Sects. 2.2–2.4,2.7), but hard to realize experimentally. Lambooy et al. [112] created such a system for the first time for films of symmetric diblock copolymers of perdeuterated polystyrene (dPS) and poly(methylmethacrylate), PMMA, with a molecular weight of 80 000. In the bulk, this system (f_{dPS}=0.51) orders in the lamellar phase with layers of wavelength λ_0=30 nm. On a suitable substrate (polished Si passivated by a 15–20 Å oxide layer at the surface) films of thicknesses in the range from about $6\lambda_0$ to $9\lambda_0$ were spin cast, then covered with a 100 Å thick buffer layer of high molecular weight PMMA, and then a 1.2 µm thick film of $SiO_{1.4}$ was evaporated onto the PMMA. The samples were annealed at 175°C and then quenched to room temperature, i.e. below the

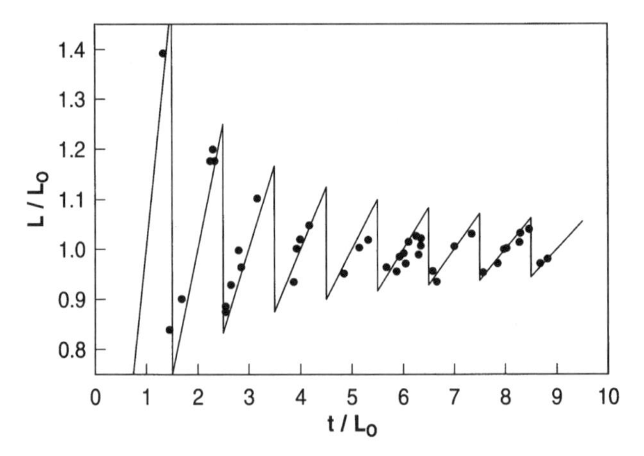

Fig. 32. Period λ of dPS-PMMA symmetric diblock copolymers relative to the period λ_0 in the bulk, plotted as a function of the film thickness D in units of λ_0. The solid line in the figure shows what would be expected if the period λ varied linearly with D/λ_0 between $(n-1/2)\lambda_0$ and $(n+1/2)\lambda_0$, n =1,2,... From Lambooy et al. [112]

glass transition temperature of both blocks before neutron reflectivity measurements were made. These study revealed lamellae oriented parallel to the confining walls, and measuring the period λ in units of λ_0 as function of thickness D/λ_0 one finds oscillatory behavior, Fig. 32 [112,301], qualitatively similar to corresponding theoretical predictions (Fig. 15). One notable distinction between this experiment and the corresponding theory (Sects. 2.3, 2.4, 2.7) is the lack of any evidence for perpendicular alignment of the lamellae for this system. The reason why the parallel orientation prevails here is that the surface energy preferring PMMA at both surfaces is very strong.

In more recent work, Kellogg et al. [119] pointed out that one can vary the strength of block-wall interactions by coating the SiO_x walls with thin layers of random copolymers composed of the two block components. Studying sample thicknesses in the range $D=1.4\lambda_0$ to $D=3.2\lambda_0$, compelling evidence for perpendicular orientation of the lamellae for $D/\lambda_0=2.52$ was obtained [119]. As expected theoretically, for D/λ_0 close to integer values the parallel orientation is still the stable structure, but it is clear that for $D/\lambda_0=2.52$ either a situation with n=3 rather compressed lamellae or with n=2 strongly expanded lamellae is unfavorable. The existence of the perpendicular orientation, which thus avoids this "frustration effect" on the parallel orientation, could be directly visualized by electron micrograph techniques. A complicating feature, however, is that these perpendicular lamellae exhibit short range order only; it appears that there exist a lot of defects in the structure. Thus the qualitative pictures, Figs. 3a-c, on which the theoretical analysis [34–36,63] (Sects. 2.3,2.4,2.7) is based, clearly are somewhat to idealized. At this point, we recall that the boundary condition provided experimentally by the random copolymer layers should perhaps be treated not just in terms of a homogeneous surface perturbation, such as written in Eqs. (10,48,87,89), but there could also be the need to add a kind of "random surface field", describing the fluctuations in the local interactions acting on the block copolymer layer. This randomness could be responsible for the lack of long range order observed experimentally [119]. While it was shown that "random fields" coupling linearly to the order parameter of block copolymers in the bulk do disrupt their long-range order [302], we are not aware of any corresponding studies of random surface effects on block copolymers yet.

A second system of block copolymer films confined between hard walls is the system poly (ethylenepropylene)-poly(ethylethylene), PEP-PEE, with molecular weight M_n=106 000 and composition f_{PEP}=0.54 [114]. Here the confinement was achieved by two similar hard walls of solid amorphous polystyrene (PS) and by two dissimilar hard walls (Si and PS, respectively). Film thicknesses in the range from $D/\lambda_0 \approx 2.8$ to $D/\lambda_0 \approx 4$ were investigated, and it was shown by neutron reflectometry techniques that the parallel arrangement of lamellae was stable throughout. The wavelength λ/λ_0 varied linearly from about $\lambda/\lambda_0 \approx 0.95$ for $D/\lambda_0 \approx 2.8$ to $\lambda/\lambda_0 \approx 1.12$ for $D/\lambda_0 \approx 3.5$, where then a discontinuous jump from n=3 to n=4 lamellae did occur, with a second region of linear variation for D/λ_0 in the range from $\lambda/\lambda_0 \geq 3.7$. Both sets of data (PS-PS confinement and PS-Si confinement) fall on the same lines in this plot, consistent with the theoretical expectations.

A third system, diblock copolymers formed from deuterated polystyrene (dPS) and poly(2-vinylpyridine), PVP, was studied by Koneripalli et al. [118], confining the copolymer melt between a silicon substrate on one side and a glassy polymer, poly(2methylvinyl-cyclohexane) on the other side. It was shown that under these boundary conditions (which lack any particular symmetry) there exists a range of film thicknesses D where the complex morphology of Fig. 3c (or Fig. 27c, respectively) occurs. Koneripalli et al. [118] suggest that in their case due to the lack of symmetry this is a truly stable configuration, unlike the simulation [35] where the boundaries were strictly symmetric and one expects this phase with both lateral and perpendicular structure (Fig. 27c) to be only metastable [63]. Thus, all the structures shown qualitatively in Fig. 3 have indeed been observed experimentally! The possibility of forming patterns of block copolymer films with lateral structure on the scale of tens of nanometers is of potential interest in fabricating optoelectronic devices [8,303], and hence this structure formation in block copolymer films (Figs. 3, 27) is perhaps of more than academic interest. Finally, we note that even for the case of "pseudo-confined" block copolymer films (i.e., films on a substrate with the upper surface "free" against air, Fig. 4b,c) the island formation which usually occurs [5] may be suppressed for very thin films [111]. Studying films of the symmetric diblock copolymer poly(styrene-b-paramethylstyrene) by nuclear reaction analysis ($^{15}N+^{1}H \rightarrow ^{4}He+^{12}C+\gamma$) for film thicknesses in the range of D=12 nm to D=92 nm near the bulk order-disorder transition, order parameter profiles could be measured in a range from $D/\lambda_0 \approx 3/8$ to $D/\lambda_0 \approx 2$. It was found that for $D/\lambda_0 < 1$ the profile was rather asymmetric, and only weak segregation did occur, in contrast to the other studies reported here [112,114,119], which all referred to strongly segregated films. The effective wavelength λ did increase linearly from about $\lambda \approx 15$ nm for D=18 nm to $\lambda = \lambda_0 \approx 45$ nm for $D = \lambda_0$, and then stayed constant up to $D \approx 2\lambda_0$. Remarkably, parallel alignment is not suppressed for $D = 3\lambda_0/2$. This fact can be attributed to the strong affinity of paramethylstyrene to the free surface [111]. Russell et al. [304] studied the evolution of ordering in thin films of symmetric diblock copolymers and found that first a bicontinuous network formed which later transformed into the multilayered structure.

Finally, we mention that interesting behavior can also be inferred from measurements with very thick films [113]. Mutter and Stühn [113] studied films of symmetric diblocks of polystyrene-b-polyisoprene with M_w=15 700, for film thicknesses of about 1 µm, while the lamellar wavelength only is 15.3 nm, so for perfect long-range order a stacking of about 70 lamellae parallel to the substrate is formed. Studying the kinetics of ordering in time-dependent neutron reflectivity measurements, it was shown that the lamellae order first in a random orientation and later get oriented parallel to the substrate and air surface on a time scale that increases strongly with decreasing temperature just below the order-disorder transition temperature.

3.3 Dynamics of Phase Separation

Since the first discovery [128] of surface-induced spinodal decomposition in thin films of polymer blends, there has been much further experimental activity [5,130,131,134–137,141, 147,148,150–155,157]. We summarize only some key features here. Figure 33 shows that similar to the simulation (Fig. 28) a surface enrichment layer forms, followed by a depletion region. The scale of these waves coarsen with time, and there are a number of studies (e.g. [136]) that report a growth of the charateristic length scale $\ell(t)$ with time t after the quench as $\ell(t) \propto t^{1/3}$. This behavior is well-known from the coarsening of solid binary mixtures [295] in the bulk, as well as from strongly off-critical fluid binary mixtures [292–294] (coarsening due to diffusion and coagulation of droplets [305]). For (near-)critical fluid binary mixtures in the bulk, a faster growth law $\{\ell(t) \propto t\}$ is predicted [296] and observed [292–294].

It turns out that the behavior of thin polymer blend films confined between plates, on which systems this review focuses, is much more complicated and there is a subtle interplay with wetting phenomena and the nature of hydrodynamic effects also changes under confinement. Thus rather conflicting observations are reported in the literature. For example, Pan and Composto [152] study blends of polystyrene (PS)-poly(vinylmethylether) (PVME) confined between glass plates at thicknesses D from $D \approx 200$ nm to $D \approx 700$ nm. They find that $\ell(t)$ increases more rapidly than linear with time up to sizes of the order of 100 µm (i.e., much larger than D), and then the growth suddenly stops. Because of this confinement, the shape of the domains is very anisotropic, flattened in the directions parallel to the walls. Pan and Composto [152] suggest that the fast growth

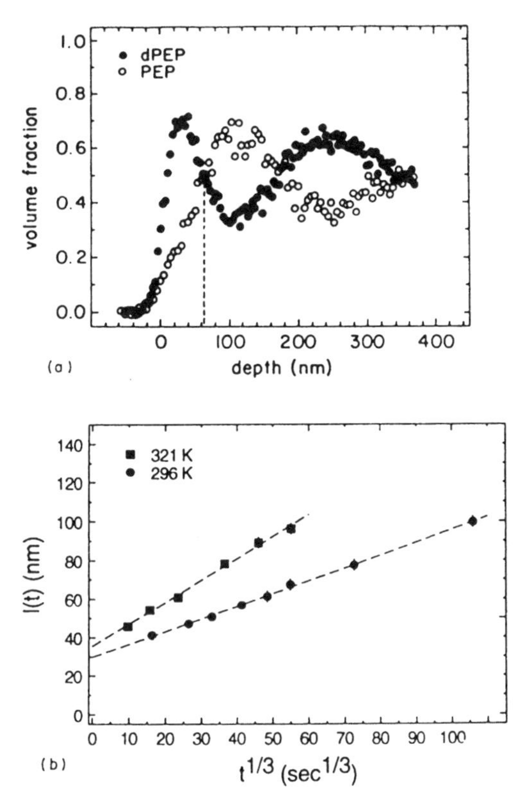

Fig. 33. a Volume fraction of deuterated poly(ethylenepropylene), dPEP (*full dots*) and protonated PEP (*open circles*) versus depth, for a degree of polymerization N≈2300 for both constituents, after a 4 h quench to T=294 K (T_{cb}≈365 K). Profiles are obtained with the time of flight forward recoil spectrometry (TOF-FRES). The dashed line indicates the surface domain thickness l(t). **b** Plot showing the growth of the surface domain thickness $\ell(t)$ vs $t^{1/3}$. From Krausch et al. [136]

is due to wetting phenomena, but this clearly needs further study. On the other hand, Sung et al. [157] studying the phase separation of thin films of polystyrene and polybutadiene on a silicon substrate (with a free film boundary) with 20 nm and 100 nm thicknesses, reveal a rather different behavior: for the "thick" film (D=100 nm), a crossover from $\ell(t) \propto t^{1/3}$ to $\ell(t) \propto t^1$ is observed, similarly as in the bulk, although at late time also a "pinning" of this lateral linear dimension occurs {at a size $\ell(t \to \infty) \approx 20$ μm}. In contrast, for the thinner film (D=20 nm), one observes a behavior $\ell(t) \propto t^{0.44}$, before the pinning occurs {$\ell(t \to \infty) \approx$ 15 μm}. Sung et al. [157] attribute this different power law to a two-dimensional version of a hydrodynamic mechanism of coarsening. It seems that the reason for the pinning of the structure at later stages, which is at variance with the theory (Sect. 2.8), is not at all understood. For the rather thin films used by Sung et al. [157], a study of the equilibrium phase diagram would be valuable (measure-

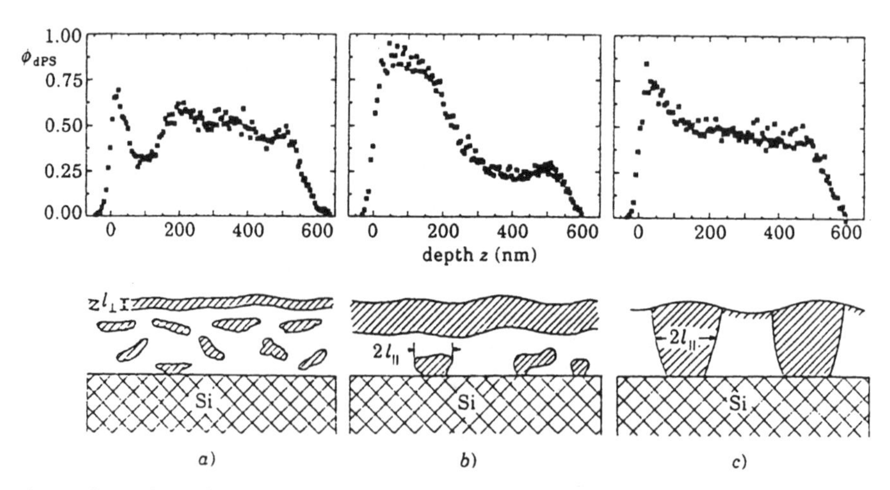

Fig. 34. dPS volume fraction ϕ_{dPS} vs depth z for a 50/50 dPS/PBr$_x$S mixture after annealing for different times at T=180 °C as revealed by TOF-FRES **a** t_A=10 min, **b** t_A=160 min, **c** t_A=6175 min). In the bottom part, a sketch of the domain structure is shown as revealed from the ion beam and scanning near field optical microscopy (SNOM) results (shaded areas: dPS-rich domains; white areas: dPS-poor domains). From Straub et al. [153]

ments were done at T=25 °C at critical volume fraction ϕ_{PSd}=0.7, the bulk critical temperature being $T_{cb}\approx 51$ °C), since due to the formation of surface enrichment layers there could be an appreciable shift of critical composition in the thin film (Fig. 5), and thus the change of coarsening behavior with thickness could be related with a crossover from critical to off-critical morphology.

Particularly interesting observations, however, are reported by Straub et al. [153] who combine ion-beam analysis and scanning near-field optical microscopy to characterize the domain structure of their mixture (dPS, N_{dPS}=1700, and partially brominated polystyrene, N_{PB_xS}) in a fully three-dimensional manner, and thus are able to characterize the transient wetting and quasi-two-dimensional spinodal decomposition. They use a nearly critical mixture on silicon wafers with a free air surface, for films of 600 nm thickness. In the initial stages, the volume fraction of dPS has a surface enrichment layer {of thickness $\ell_\perp(t)$} followed by a depletion layer and further oscillations around ϕ_{crit}=0.5, qualitatively similar to surface directed spinodal decomposition. This surface enrichment layer then grows in thickness, and the volume fraction profile is monotonic. The optical microscopy reveals, however, that the other phase (near the Si substrate) is inhomogeneous, there occur large regions of dPS-rich phase of linear dimension $\ell_{||}(t)$. It turns out that both $\ell_{||}(t)$ and $\ell_\perp(t)$ grow according to $t^{1/3}$ laws. At still later times, the growing dPS-rich droplets in the interior coalesce with the dPS rich surface enrichment layer, which then breaks up and a morphology with domain walls perpendicular to the thin film (Fig. 5b) remains. The lateral phase separation is also clearly visible from the time of flight-forward recoil spetroscopy (TOF-FRES) volume fraction profiles, which settle down near $\phi_{dPS} \approx \phi_{crit}$ in the bulk of the film, as expected from the discussion presented in Sects. 1,2.1 (Fig. 34). The situation is somewhat complicated since at the free surface the lateral phase separation is accompanied with a "roughening" of the film (i.e., the dPS-rich domains stick a bit out of the film, as shown schematically in Fig. 34).

While the above system is an example where two-dimensional phase separation in the sense of Fig. 1a,b (or Fig. 5) occurs, there exist also good examples where no lateral phase separation exists in equilibrium, and the system forms a single interface parallel to the surfaces (Fig. 1d). However, if one chooses the initial state such that the phase preferred by air is close to the substrate and the phase preferred by the substrate is next to the air surface [77], the system is unstable and surface phase inversion takes place. A laterally inhomogeneous state then occurs only as a transient phenomenon necessary to trigger the inversion kinetics, but not as an equilibrium state [77].

4 Discussion and Conclusions

This review has discussed the phase behavior of polymer blends and symmetric block copolymer melts in thin film geometry, considering mostly films confined between two symmetrical hard walls. Occasionally, also an "antisymmetric" boundary condition (i.e. one wall prefers component A while the other wall prefers component B) is studied. These boundary conditions sometimes approximate the physically most relevant case, namely a polymeric film on a solid substrate exposed to air or vacuum with a free, flat surface (Fig. 1). The case where the film as a whole breaks up into droplets (Fig. 2) due to dewetting phenomena is not considered, however, nor did we deal with the formation of islands or holes or terraces in the case of ordered block copolymer films (Fig. 4b-d).

The main interest of this article was the inhomogeneity in composition {described e.g. by the local volume fraction $\phi(\vec{\rho},z)$ of species A in the film} that develops in the film, and to relate this inhomogeneity to the corresponding phase transitions in the bulk: unmixing of a (symmetrical) binary polymer mixture (A,B), or lamellar ordering of (symmetrical) block copolymers. While qualitatively the behavior of asymmetrical polymer blends would not entail many new features, for block copolymers asymmetrical composition would yield other ordering phenomena (e.g. bicontinuous ordered phases or hexagonal or cubic structures in the bulk competing with lamellar ordering at the walls, the latter due to surface enrichment of the preferred component) which are beyond the scope of the present article. Also we have not paid much attention to the interesting interplay between the composition inhomogeneity and molecular architecture of the polymer chains (flexible vs stiff chains, etc.). Only flexible linear polymers were treated.

The common theme of this article was hence the formation of surface enrichment layers in thin films and their effect on the phase transitions mentioned above, with emphasis on the accompanying interface formation: i.e., the lamellar phase of block copolymers can be viewed as a parallel arrangement of AB-interfaces (where in the strong segregation limit the junction points are localized). Phase separation in a blend, at fixed overall volume fraction $\bar{\phi}$ inside the two-phase coexistence curve, also implies interface formation.

It is an intriguing question, and has been discussed at length in the present article, to what extent the phase transitions occurring in the thin film remain sharp or become rounded due to the finite thickness D of the thin film. For strictly symmetrical block copolymers in the framework of the Leibler theory predicting a second-order transition in the bulk, the transition would be rounded in the thin film, since the surface enrichment causes a damped oscillatory composition profile in the z-direction already above the bulk transition temperature, and this profile develops gradually towards a more strongly ordered situation. Taking fluctuation effects (and compositional asymmetry) into account, one would expect that for thick films there is still a first-order transition possible, and wetting-type phenomena ("surface-induced ordering" causes the formation of strongly ordered surface layers at the walls, separated from the disordered bulk of the film by well developed interfaces) occur.

An intriguing point also is that for block copolymers with lamellar ordering parallel to the walls in general there will be a misfit between the wavelength that the system likes to form in the bulk (λ_0) and the film thickness D (i.e., $D \neq n\lambda_0$).

As a result, the wavelength λ in units of λ_0 exhibits a "sawtooth"-profile as function of the film thickness (Figs. 15,32): for a parallel arrangement of the lamellae between symmetric walls (Fig. 3a, Fig. 26b,d, Fig. 27b,d) the domains are compressed or expanded to enforce the condition $D=n\lambda$ in the whole range from $D=\left(n-\frac{1}{2}\right)\lambda_0$ to $D=\left(n+\frac{1}{2}\right)\lambda_0$, while at the boundaries of this regime first-order transitions in the number of layers ($n-1 \rightarrow n$ or $n \rightarrow n+1$, respectively) occur. This parallel orientation, however, occurs only for sufficiently strong preference of the walls for the B-rich phase: simulations (Fig. 27a) and experiment support the theoretical prediction, that perpendicular domain orientation(Fig. 3b) occurs for weak enough surface forces. Also the inhomogeneous asymmetric domain pattern of Fig. 3d, first seen in simulations (Figs. 26c, 27c) and later shown to be stable for asymmetric boundary conditions at the walls, has been found experimentally. Thus, one might expect that in thin film geometry also thermally driven first-order transitions between parallel and perpendicular arrangements of lamellae could occur, since λ_0 is not a constant but temperature dependent, but this phenomenon remains to be investigated. This "frustration effect" due to the misfit between thin film thickness D and preferred wavelength λ should also lead to oscillatory dependence of various properties on film thickness already in the weak segregation limit of the system.

For the polymer blend with symmetric walls, approaching the bulk critical temperature from the disordered region there occurs a completely smooth and gradual formation of surface enrichment layers on both walls (Fig. 1c), without any phase transition. The unmixing phase transition involves formation of concentration inhomogeneity in lateral directions (Fig. 1a,b), with interfaces oriented perpendicular to the walls. This is true irrespective whether or not in semi-infinite geometry a wetting transition can occur – this wetting transition anyway is rounded off in a thin film, and thus there is no sharp transition between the situation in Fig. 1a and Fig. 1b, since the thickness of the surface enrichment layer in Fig. 1b is rather limited (only prewetting transitions, Fig. 6b, still could occur). It is easy to see that vertical phase separation (Fig. 1c) cannot be the thermodynamically stable situation for a thick film ($D\to\infty$) for symmetric walls: the total interfacial energy then would be (for a LxLxD geometry with $L>>D$) $2L^2\sigma_{AB}+2L^2\sigma_{BW}$, with σ_{AB} and σ_{BW} the interfacial tensions for AB and B-wall interfaces, respectively. For a fraction x of the A-rich phase, we have for the situation of Fig. 1b a total interfacial energy $2L^2\sigma_{BW}+LD\sigma_{AB}+2xL^2\sigma_{AB}$. It is seen that the structure in Fig. 1c exceeds the interfacial free energy cost of the structure of Fig. 1b by $2(1-x)L^2\sigma_{AB}-LD\sigma_{AB}$, which for $L/D\to\infty$ is positive for any $x>0$. As a consequence, structures as shown in Fig. 1c are stable only in the one-phase region of the blend, where the thickness of the B-rich surface enrichment layers are microscopic, and do not take a finite fraction of the film thickness D as $D\to\infty$. For very small D, however, there is an appreciable depression of $T_c(D)$ in comparison to the bulk, and structures of the type of Fig. 1c occur under conditions where in the bulk phase separation already would occur, e.g. for $T_c(D)<T<T_{cb}$. A similar reasoning applies to the non-wet situation, where Fig. 1a) is preferable over Fig. 1c), since the total interfacial energy is $(1-x)2L^2\sigma_{BW}+2L^2\sigma_{AW}+LD\sigma_{AB}$, and hence the structure of Fig. 1c exceeds the free energy cost of Fig. 1a by $2L^2(\sigma_{AB}+x\sigma_{BW}-x\sigma_{AW})-LD\sigma_{AB}=2L^2(1-x)\sigma_{AB}-LD\sigma_{AB}+2L^2 x(\sigma_{AB}+\sigma_{BW}-\sigma_{AW})$. For a non-wet (partially wet) situation we have $\sigma_{AB}>\sigma_{AW}-\sigma_{BW}$, and hence again the free energy cost for $L/D\to\infty$ is positive (for the wet case, $\sigma_{AB}=\sigma_{AW}-\sigma_{BW}$, we recover the free energy cost found above, of course).

Similar arguments apply for the case of "anti-symmetric" boundary conditions (one surface preferring the A-rich phase, the other one the B-rich phase, Figs. 1d,e): again the transition of the thin film from the disordered phase to the structure with vertical phase separation is smooth, only the extent of the rounding of the transition near T_{cb} decreases gradually as the film thickness D increases. However, a true phase transition involving lateral phase separation (Fig. 1e) can occur also in this case, although we are not aware of any experimental evidence for this transition yet. While the geometry of Fig. 1d) is very popular for a study of interfacial profiles, and the geometry of Fig. 1c) is used for studying the surface enrichment profiles in order to extract the pure surface free energy experimentally, we have argued that in both cases the finite thickness D affects such analyses substantially and needs to be carefully considered.

While the static equilibrium behavior of polymer blends in thin film geometry thus is rather well understood, at least in principle, the kinetic behavior (Sects. 2.8,3.3) is much less well understood, since there is a delicate interplay between surface-directed spinodal decomposition, thickness-limited growth of wetting layers, and the hydrodynamic mechanisms of coarsening in this constrained geometry still needs investigation.

Finally, we draw attention to a topic somewhat neglected in this review, namely the interplay between concentration inhomogeneities $\phi\ \vec{\rho},z$ near the surfaces and interfaces and the local configurational properties of the polymer coils (enrichment of chain ends, orientation and possibly distortion of polymer coils, etc.). The reason for this omission was that not so much general features are known about these questions. Clearly, the subject of phase transitions of polymer blends and block copolymer melts in thin film geometry will remain a challenge in the future.

Acknowledgements. The author is greatly indebted to J. Baschnagel, B. Dünweg, T. Flebbe, G.H. Fredrickson, H.L. Frisch, T. Kerle, M. Kikuchi, J. Klein, M. Müller, S. Puri, Y. Rouault, F. Schmid, S. Stepanow and A. Werner for their fruitful collaboration on various research problems that have been reviewed here. The author thanks T.P. Russell for granting permission to use Fig. 15 (from Ref. 36) and Fig. 32 (from Ref. 112), M.W. Matsen for Fig. 17 (from Ref. 63), F. Schmid for allowing him to reproduce Fig. 16 (from Ref. 57), S.K. Kumar for Fig. 19 (from Ref. 39) and G. Krausch for Fig. 33 (from Ref. 136) and Fig. 34 (from Ref. 153). The author would like to thank all the colleagues mentioned above for stimulating discussions and many insights. Partial support of this work by the Deutsche Forschungsgemeinschaft (DFG, grant Nos. SFB262/D2, Bi314/3), the Bundesministerium für Bildung, Wissenschaft, Forschung und Technologie (BMBF, grant No 03N8008C) and the Materialwissenschaftliches Forschungszentrum Mainz (MWFZ) is acknowledged.

5
References

1. Charvolin J, Joanny J-F, Zinn-Justin J (eds)(1990) Liquids at Interfaces. North-Holland, Amsterdam
2. Sanchez IC (ed) (1992) Physics of Polymer Surfaces and Interfaces. Butterworth-Heinemann, Boston
3. Tirrell M, Parsonage EE (1993) In: Thomas EL (ed) Materials Science and Technology, Vol 12: Structure and Properties of Polymers. VCH Weinheim, p. 653
4. Fleer GJ, Cohen Stuart MA, Scheutjens JMH, Cosgrove T, Vincent B (1993) Polymers at Interfaces. Chapman & Hall, London
5. Krausch G (1995) Mat Sci Eng Rep R14:1
6. Binder K (1995) Acta Polymer 46: 204
7. Tirrell M (ed) (1996) Annu Rev Mater Sci, Vol 26
8. Russell TP (1996) Current Opinion in Colloid & Interface Sci 1:107
9. Mayes AM, Kumar SK (1997) MRS Bulletin 22:43
10. Nakanishi H, Pincus P (1983) J Chem Phys 79:997
11. Schmidt I, Binder K (1985) J Phys (France) 46:1631
12. Fredrickson GH (1987) Macromolecules 20:2535
13. Carmesin I, Noolandi J (1989) Macromolecules 22:1689
14. Cohen SM, Muthukumar M (1989) J Chem Phys 90:5749
15. Hariharan A, Kumar SK, Russell TP (1990) Macromolecules 23:3584
16. Chen ZY, Noolandi J, Izzo D (1991) Phys Rev Lett 66:727
17. Wang JS, Binder K (1991) J Chem Phys 94:8537; Pereira GG, Wang JS (1996) J Chem Phys 104:5296
18. Wang JS, Binder K (1992) Macromol Theory Simul 1:49
19. Fredrickson GH, Donley JP (1992) J Chem Phys 97:8941
20. Shull KR (1992) Macromolecules 25:2122

21. Turner MS, Joanny J-F (1992) Macromolecules 25:6681
22. Eggleton CD (1996) Phys Lett A223:394
23. Tang H, Freed KF (1992) J Chem Phys 97:4496
24. Turner MS (1992) Phys Rev Lett 69:1788
25. Cifra P, Karasz FE, MacKnight WJ (1992) Macromolecules 25:4895
26. Jerry RA, Naumann EB (1992) Phys Lett A 167:198
27. Jerry RA, Naumann EB (1992) J Colloid Interface Sci 154:122
28. Jerry RA, Naumann EB (1993) Phys Rev E 48:1583
29. Hariharan A, Kumar SK, Russell TP (1993) J Chem Phys 98:4163; (1991) Macromolecules 24:4909
30. Ausserre D, Raghuntan VA, Maaloum M (1993) J Phys II (France) 3:1485
31. Jones RAL (1993) Phys Rev E 47:1437
32. Shull KR, Mayes AM, Russell TP (1993) Macromolecules 26:3929
33. Cifra P, Bruder F, Brenn R (1993) J Chem Phys 99:4121
34. Kikuchi M, Binder K (1993) Europhys Lett 21:427
35. Kikuchi M, Binder K (1994) J Chem Phys 101:3367
36. Walton DG, Kellogg GJ, Mayes AM, Lambooy P, Russell TP (1994) Macromolecules 27:6225
37. Yethiraj A, Kumar SK, Hariharan A, Schweizer KS (1994) J Chem Phys 100:4691
38. Tang H, Szleifer I, Kumar SK (1994) J Chem Phys 100:5367
39. Kumar SK, Tang H, Szleifer I (1994) Mol Phys 81:867
40. Ding EJ (1994) J Phys: Condensed Matter 6:3641
41. Hariharan A, Harris JG (1994) J Chem Phys 101:3353
42. Wong KY, Trachi M, McMullen WE (1994) J Chem Phys 101:5372
43. Brazhnik PK, Freed KF, Tang H (1994) J Chem Phys 101:9143
44. Turner MS, Rubinstein M, Marques CM (1994) Macromolecules 27:4986
45. Brown G, Chakrabarti A (1994) J Chem Phys 101:3310
46. Brown G, Chakrabarti A (1995) J Chem Phys 102:1440
47. Williams DRM (1995) Phys Rev Lett 75:453
48. Turner MS, Johner A, Joanny JF (1995) J Phys I (France) 5:917
49. Turner MS (1995) Phys Rev Lett 75:977
50. Singh N, Bates FS, Tirrell (1995) Phys Rev Lett 75:976
51. Donley JP, Fredrickson GH (1995) J Polym Sci B Polym Phys Ed 33:1345
52. Carignano MA, Szleifer I (1995) Europhys Lett 30:525
53. Kumar SK, Yethiraj A, Schweizer KS, Leermakers FAM (1995) J Chem Phys 103:10332
54. Yethiraj A (1995) Phys Rev Lett 74:2018
55. Rouault Y, Baschnagel J, Binder K (1995) J Stat Phys 80:1009
56. Rouault Y, Dünweg B, Baschnagel J, Binder K (1996) Polymer 37:297
57. Schmid F (1996) J Chem Phys 104:9191
58. Flebbe T, Dünweg B, Binder K (1996) J Phys II (France) 6:667
59. Wu DT, Fredrickson GH, Carton J-P (1996) J Chem Phys 104:6387; Donley JP, Wu DT, Fredrickson GH (1997) Macromolecules 30:2167
60. Milner ST, Morse C (1996) Phys Rev E54:3797
61. Binder K, Frisch HL, Stepanow S (1997) J Phys II (Paris) 7:1353
62. Binder K, Nielaba P, Pereyra V (1997) Z Phys B 104:81
63. Matsen MW (1997) J Chem Phys 106:7781
64. Reich S, Cohen Y (1981) J Polym Sci B, Polym Phys Ed 19:1255
65. Jones RAL, Kramer EJ, Rafailovich MH, Sokolov J, Schwarz SA (1989) Phy Rev Lett 62:280
66. Jones RAL, Kramer EJ (1990) Philos Mag B62:129
67. Zhao X, Zhao W, Schwarz SA, Wilkens BJ, Rafailovich MH, Sokolov J, Jones RAL (1991) Macromolecules 24:5991
68. Clark MB, Burkhardt JA, Gardella JA (1991) Macromolecules 24:799
69. Steiner U, Eiser E, Klein J, Budkowski A, Fetters LJ (1992) Science 258:1126

70. Budkowski A, Steiner U, Klein J, Schatz G (1992) Europhys Lett 18:705; Bruder F, Brenn R (1991) Macromolecules 24:5552; see also Bruder F, Brenn R, Stühn B, Strobl GR (1989) Macromolecules 22:4434
71. Budkowski A, Steiner U, Klein J (1992) J Chem Phys 97:5229
72. Bruder F, Brenn R (1993) Europhys Lett 22:707
73. Cowie JMG, Devlin BG, McEwen IJ (1993) Macromolecules 26:5628
74. Genzer J, Faldi A, Composto RJ (1994) Phys Rev E 50:2373
75. Guckenbiehl B, Stamm M, Springer T (1994) Colloids Surf 86:311
76. Steiner U, Klein J, Eiser E, Budkowski A, Fetters LJ (1994) Ber Bunsenges Phys Chem 93:366
77. Steiner U, Klein J, Fetters LJ (1994) Phys Rev Lett 72:1498
78. Hong PP, Boerio FJ, Smith SD (1994) Macromolecules 27:596
79. Steiner U, Eiser E, Budkowski A, Fetters LJ, Klein J (1994) In: Teramoto A (ed) Ordering in Macromolecular Systems. Springer, Berlin
80. Norton LJ, Kramer EJ, Bates FS, Gehlsen MD, Jones RAL, Karim A, Felcher GP, Kleb R (1995) Macromolecules 28:8621
81. Scheffold F, Eiser E, Budkowski A, Steiner U, Klein J, Fetters LJ (1996) J Chem Phys 104:8786
82. Scheffold F, Eiser E, Budkowski A, Steiner U, Klein J, Fetters LJ (1996) J Chem Phys 104:8795
83. Steiner U, Klein J (1996) Phys Rev Lett 77:2526
84. Kerle T, Klein J, Binder K (1996) Phys Rev Lett 77:1318
85. Budkowski A, Scheffold F, Klein J, Fetters LJ (1997) J Chem Phys 106:719
86. Genzer J, Composto RJ (1997) Europhys Lett 38:171
87. Coulon G, Russell TP, Deline VR, Green PF (1989) Macromolecules 22:2581
88. Russell TP, Coulon G, Deline VR, Green PF (1989) Macromolecules 22:4600
89. Anastasiadis SH, Russell TP, Satija SK, Majkrzak CF (1989) Phys Rev Lett 62:1852
90. Anastasiadis SH, Russell TP, Satija SK, Majkrzak CF (1990) J Chem Phys 92:5677
91. Coulon G, Ausserre D, Russell TP (1990) J Phys (France) 51:777
92. Coulon G, Collin B, Ausserre D, Chatenay D, Russell TP (1990) J Phys (France) 51:2801
93. Green PF, Christensen TM, Russell TP, Jérôme J (1990) J Chem Phys 92:1478
94. Russell TP, Menelle A, Anastasiadis SH, Satija SK, Majkrzak CF (1991) Macromolecules 24:6263
95. Collin B, Chatenay D, Coulon G, Ausserre D, Gallot Y (1992) Macromolecules 25:1621
96. Menelle A, Russell TP, Anastasiadis SH, Satija SK, Majkrzak CF (1992) Phys Rev Lett 68:67
97. Foster MD, Sikka M, Singh N, Bates FS, Satija SK, Majkrzak CF (1992) J Chem Phys 96:8605
98. Maaloum M, Ausserre D, Chatenay D, Coulon G, Gallot Y (1992) Phys Rev Lett 68:1575
99. Sikka M, Singh N, Karim A, Bates FS, Satija SK, Majkrzak CF (1993) Phys Rev Lett 70:307
100. Mayes AM, Johnson RD, Russell TP, Smith SD, Satija SK, Majkrzak CF (1993) Macromolecules 26:1047
101. Coulon G, Collin B, Chatenay D, Gallot Y (1993) J Phys II (France) 3:697
102. Coulon G, Daillant J, Collin B, Benattar JJ, Gallot Y (1993) Macromolecules 26:1582
103. de Jeu WH, Lambooy P, Vaknin D (1993) Macromolecules 26:4973
104. Ehlich D, Takenaka M, Hashimoto T (1993) Macromolecules 26:492
105. Bassereau P, Brodbeck D, Russell TP, Brown HR, Shull KR (1993) Phys Rev Lett 71:1716
106. de Jeu WH, Lambooy P, Hamley IW, Vaknin D, Pederson S, Kjaer K, Seyger R, van Huten P, Hadziioannou G (1993) J Phys II (France) 3:139
107. Morkved TL, Wiltzius P, Jaeger HM, Grier DG, Witten TA (1994) Appl Phys Lett 64:422
108. Sikka M, Singh N, Bates FS, Karim A, Satija SK, Majkrzak CF (1994) J Phys II (France) 4:2231
109. Carvalho BL, Thomas EL (1994) Phys Rev Lett 73:3321

110. Karim A, Singh N, Sikka M, Bates FS, Dozier WD, Felcher GP (1994) J Chem Phys 100:1620
111. Gießler K-H, Rauch F, Stamm M (1994) Europhys Lett 27:605
112. Lambooy P, Russell TP, Kellogg GJ, Mayes AM, Gallagher PD, Satija SK (1994) Phys Rev Lett 72:2899
113. Mutter R, Stühn B (1995) Colloid Polymer Sci 273:653
114. Koneripalli M, Singh N, Levicky R, Bates FS, Gallagher PD, Satija SK (1995) Macromol ecules 28:2897
115. Grim PCM, Nyrkova IA, Semenov AN, ten Brinke G, Hadziioannou G (1995) Macromolecules 28: 7501
116. Mansky P, Chaikin P, Thomas EL (1995) J Mater Sci 30:1987
117. Pickett GD (1996) J Chem Phys 104:1657
118. Koneripalli M, Levicky R, Bates FS, Ankner J, Kaiser H, Satija SK (1996) Langmuir 26:6681
119. Kellogg GJ, Walton DG, Mayes AM, Lambooy P, Russell TP, Gallagher PD, Satija SK (1996) Phys Rev Lett 76:2503
120. Morkved TL, Lu M, Urbas AM, Ehrichs EE, Jaeger HM, Mansky P, Russell TP (1996) Science 273:931
121. Morkved TL, Jaeger HM (1997) preprint
122. Wong GCL, Commandeur J, Fischer H, de Jeu WH (1997) preprint
123. Wu S (1982) Polymer Interfaces and Adhesion. Dekker, New York
124. Koberstein JT (1987) In: Mark H, Overberger CG, Bikales NM, Menges G (eds) Encyclopedia of Polymer Science and Engineering, Vol 8. Wiley, New York, p. 237
125. Stamm M, Schubert DW (1995) Annu Rev Mater Sci 25:326
126. Genzer J, Faldi A, Oslanec R, Composto RJ (1996) Macromolecules 29:5438
127. Ball RC, Essery RLH (1990) J Phys: Condensed Matter 2:10303
128. Jones RAL, Norton LJ, Kramer EJ, Bates FS, Wiltzius P (1991) Phys Rev Lett 66:1326
129. Binder K, Frisch HL (1991) Z Phys B 84:403
130. Wiltzius P, Cumming A (1991) Phys Rev Lett 66:3000
131. Cumming A, Wiltzius P, Bates SF, Rosedale JH (1992) Phys Rev A45:885
132. Puri S, Binder K (1992) Phys Rev A46:R4487
133. Brown G, Chakrabarti A (1992) Phys Rev A46:4829
134. Bruder F, Brenn R (1992) Phys Rev Lett 69:1326
135. Tanaka H (1993) Phys Rev Lett 70:53; ibid 70:2770; Europhys Lett 24:665
136. Krausch G, Dai C-A, Kramer EJ, Bates FS (1993) Phys Rev Lett 71:3669
137. Krausch G, Dai C-A, Kramer EJ, Marko JF, Bates FS (1993) Macromolecules 26:5566
138. Marko JF (1993) Phys Rev E 48:2861
139. Puri S, Frisch HL (1993) J Chem Phys 99:5560
140. Troian S (1993) Phys Rev Lett 71:1399; (1994) Phys Rev Lett 72:3739
141. Shi BQ, Harrison C, Cumming A (1993) Phys Rev Lett 70:206
142. Sagui C, Somozo AM, Roland C, Desai RC (1993) J Phys A: Math Gen 26:L1163
143. Keblinski P, Ma WJ, Maritan A, Koplik J, Banavar JR (1994) Phys Rev Lett 72:3738
144. Puri S, Binder K (1994) Phys Rev E 49:5359
145. Puri S, Binder K (1994) J Stat Phys 77:145
146. Vaksman MA, McMullen WE (1994) Phys Rev E49:4724
147. Krausch G, Mlynek J, Straub W, Brenn R, Marko JF (1994) Europhys Lett 28:323
148. Krausch G, Kramer EJ, Bates FS, Marko JF, Brown G, Chakrabarti A (1994) Macromolecules 27:6768
149. Brown G, Chakrabarti A, Marko JF (1994) Phys Rev E 50:1674
150. Kim E, Krausch G, Kramer EJ, Osby OJ (1994) Macromolecules 27:5927
151. Krausch G, Kramer EJ, Rafailovich MH, Sokolov J (1994) Appl Phys Lett 64:2655
152. Pan Q, Composto RJ (1995) Mat Res Soc Symp Proc 366:27
153. Straub W, Bruder F, Brenn R, Krausch G, Bielefeldt A, Kirsch A, Marti O, Mlynek JL, Marko JF (1995) Europhys Lett 29:353

154. Geoghegan M, Jones RAL, Clough AS (1995) J Chem Phys 103:2719
155. Harrison C, Rippard W, Cumming A (1995) Phys Rev E 52:723
156. Frisch HL, Nielaba P, Binder K (1995) Phys Rev E 52:2848
157. Sung L, Karim A, Douglas JF, Han CC (1996) Phys Rev Lett 76:4368
158. Puri S, Frisch HL (1997) J Phys: Condensed Matter 9:2109
159. Irvine DJ, Gersappe D, Balazs AC (1995) Langmuir 11:3848
160. Alexander S (1977) J Phys (Paris) 38:983
161. De Gennes PG (1980) Macromolecules 13:1069
162. Halperin A, Tirrell M, Lodge TP (1991) Adv Pol Sci 100:31
163. Milner ST (1991) Science 251:905
164. Halperin A (1994) In: Rabin Y, Bruinsma R (Eds) Soft Order in Physical Systems. Plenum Press, New York, p. 33
165. Grest GS, Murat M (1995) In: Binder K (ed) Monte Carlo and Molecular Dynamics Simulations in Polymer Science. Oxford Univ Press, New York, p. 476
166. Szleifer I, Carignano MA (1996) In: Prigogine I, Rice SA (eds) Advances in Chemical Physics, Vol XCIV. J. Wiley & Sons, New York, p. 165
167. Szleifer I (1996) Current Opinion in Colloid & Interface Sci 1:416
168. Leibler L, Mourran A (1997) MRS Bulletin 22:33
169. Grest GS, Lacasse MD, Murat M (1997) MRS Bulletin 22:27
170. Balazs AC, Singh S, Zhulina E, Gersappe D, Pickett G (1997) MRS Bulletin 22:16
171. Marko JE, Witten TA (1991) Phys Rev Lett 66:1541; (1992) Macromolecules 25:296
172. Lai P-Y (1994) J Chem Phys 100:3351
173. Baumgärtner A, Heermann DW (1986) Polymer 27:1777
174. Chakrabarti A, Toral R, Gunton JD, Muthukumar M (1989) Phys Rev Lett 63:2072
175. Forrest BM, Heermann DW (1991) J Phys II (France) 1:909
176. Redon C, Brochard-Wyart F, Rondelez F (1991) Phys Rev Lett 66:715
177. Reiter G (1992) Phys Rev Lett 68:75
178. Reiter G (1993) Langmuir 9:1344
179. Redon C, Broska JB, Brochard-Wyart F (1994) Macromolecules 27:468
180. Brochard-Wyart F, de Gennes PG, Hervet H, Redon C (1994) Langmuir 10:1566
181. Milchev A, Binder K (1997) J Chem Phys 106:1978
182. Keblinski P, Kumar SK, Maritan A, Koplik J, Banavar JR (1996) Phys Rev Lett 76:1106
183. Jandt KD, Kramer EJ, Heier J (1996) Langmuir 12:3716
184. Krausch G, Straub W (1995) Europhys Lett 29:353
185. Bates SF, Fredrickson GH (1990) Annu Rev Phys Chem 41:525
186. Binder K (1994) Adv Polymer Sci 112:181
187. Matsen MW, Schick M (1996) Current Opinion in Colloid & Interface Sci 1:329
188. Minchau B, Dünweg B, Binder K (1990) Polymer Commun 31:348
189. Bates FS, Rosedale JH, Stepanek P, Lodge TP, Wilkins P, Fredrickson GH, Hjelm P Jr (1990) Phys Rev Lett 65:863
190. Fried H, Binder K (1991) J Chem Phys 94:8349
191. Fried H, Binder K (1991) Europhys Lett 16:237
192. Barrat JL, Fredrickson GH (1991) J Chem Phys 95:1281
193. Binder K, Fried H (1993) Macromolecules 26:6878
194. Weyersberg A, Vilgis TA (1993) Phys Rev E 48:377
195. Weyersberg A, Vilgis TA (1994) Phys Rev E 49:3097
196. Micka U, Binder K (1995) Macromol Theory Simul 4:419
197. Leibler L (1980) Macromolecules 13:1602
198. Binder K (1995) In: Binder K (ed) Monte Carlo and Molecular Dynamics Simulations in Polymer Science. Oxford Univ Press, New York, p. 356
199. Flory PJ (1953) Principles of polymer chemistry. Cornell University Press, Ithaca
200. De Gennes PG (1979) Scaling concepts in polymer physics. Cornell University Press, Ithaca
201. Koningsveld R (1968) Adv Colloid Interface Sci 2:151

202. Koningsveld R, Kleintjens LA, Nies E (1987) Croat Chim Acta 60:53
203. Huggins ML (1941) J Chem Phys 9:440
204. Flory PJ (1941) J Chem Phys 9:660
205. Staverman AJ (1941) Recl Trav Chim 60:640
206. De Gennes PG (1980) J Chem Phys 72:4756
207. Binder K (1983) J Chem Phys 79:6387
208. Cahn JW, Hilliard JE (1958) J Chem Phys 28:258
209. Vrij A (1968) J Polym Sci, Part A 2, 6:1919
210. Fredrickson GH, Helfand E (1987) J Chem Phys 87:697
211. Stepanow S (1995) Macromolecules 28:8233
212. Matsen MW, Schick M (1994) Phys Rev Lett 72:2660
213. Matsen MW, Bates FS (1996) Macromolecules 29:1091
214. Yoon DY, Vacatello M, Smith GD (1995) In: Binder K (ed) Monte Carlo and Molecular Dynamics Simulations in Polymer Science. Oxford University Press, New York, p. 433
215. Müller M, Binder K, Oed W (1995) J Chem Soc: Faraday Trans 91:2369
216. Binder K (1983) In: Domb C, Lebowitz JL (eds) Phase Transitions and Critical Phenomena, Vol 8. Academic Press, New York, p. 1
217. De Gennes PG (1985) Rev Mod Phys 57:825
218. Sullivan DE, Telo da Gama MM (1986) In: Croxton CA (ed) Fluid Interfacial Phenom ena. Wiley, New York, p. 45
219. Fisher ME (1986) J Chem Soc, Faraday Trans 82:1569
220. Dietrich S (1988) In: Domb C, Lebowitz JL (eds) Phase Transitions and Critical Phenomena, Vol 12. Academic Press, New York, p. 1
221. Schick M (1990) In: Charvolin J, Joanny JF, Zinn-Justin J (eds) Liquids at Interfaces. North Holland, Amsterdam, p. 415
222. Blokhuis EM, Widom B (1996) Current Opinion in Colloid & Interface Sci 1:424
223. Parry AO (1996) J Phys: Condens Matter 8:10761
224. Fisher ME, Nakanishi H (1981) J Chem Phys 75:5857
225. Nakanishi H, Fisher ME (1983) J Chem Phys 78:3279
226. Cahn JW (1977) J Chem Phys 66:3667
227. Binder K, Hohenberg PC (1972) Phys Rev B 6:3461
228. Puri S, Binder K, Frisch HL (1997) Phys Rev E 56:6991
229. Fisher ME (1974) Rev Mod Phys 46:587
230. Le Guillou JC, Zinn-Justin J (1980) Phys Rev B 21:3976
231. Bates FS, Schulz MF, Khandpur AK, Förster S, Rosedale JH (1994) Faraday Discuss 98:7
232. Gompper G, Schick M (1989) Phys Rev Lett 62:1647
233. Schmid F, Schick M (1993) Phys Rev E 48:1882
234. Fredrickson GH, Binder K (1989) J Chem Phys 91:7265
235. Lipowsky R (1984) J Appl Phys 55:2485; (1987) Ferroelectrics 73:69
236. Fisher ME (1971) In: Green MS (ed) Critical Phenomena, Proc 1970. E. Fermi Int. School Phys. Academic Press, London, p. 1
237. Barber MN (1973) In: Domb C, Lebowitz JL (eds) Phase Transitions and Critical Phenomena, Vol 8. Academic Press, London, p. 145
238. Binder K (1987) Ferroelectrics 73:43; (1992) In: Gausterer H, Lang CB (eds) Computational Methods in Field Theory. Springer, Berlin, p. 59
239. Privman V (ed) (1990) Finite Size Scaling and Numerical Simulation of Statistical Systems. World Scientific, Singapore
240. Helfand E (1975) Macromolecules 8:552; Helfand E, Wasserman ZR (1976) Macromolecules 9:879
241. Ohta T, Kawasaki K (1986) Macromolecules 19:2621
242. Helfand E, Tagami Y (1971) J Chem Phys 56:3592; J Polym Sci B 9:741
243. Helfand E, Tagami Y (1972) J Chem Phys 57:1812
244. Helfand E, Sapse AM (1975) J Chem Phys 62:1327
245. Helfand E (1975) J Chem Phys 62:999

246. Hong KM, Noolandi J (1981) Macromolecules 14:727
247. Noolandi J, Hong KM (1982) Macromolecules 15:482
248. Hong KM, Noolandi J (1983) Macromolecules 16:1083
249. Noolandi J, Hong KM (1984) Macromolecules 17:1531
250. Shull KR, Kramer EJ (1990) Macromolecules 23:4769
251. Shull KR (1991) J Chem Phys 94:5723
252. Shull KR (1993) Macromolecules 26:2346
253. Schmid F, Müller M (1995) Macromolecules 28:8639; Müller M, Werner A (1997) J Chem Phys 107:10764
254. Werner A, Schmid F, Binder K, Müller M (1996) Macromolecules 29:8241
255. Ypma GJA, Citra P, Nies E, van Bergen ARD (1996) Macromolecules 29:1252
256. Semenov AN (1985) Sov Phys JETP 61:733
257. Vavasour JD, Whitmore MD (1993) Macromolecules 26:7070
258. Matsen MW, Schick M (1994) Macromolecules 27:4014
259. Vavasour JD, Whitmore MD (1996) Macromolecules 29:5244
260. Scheutjens JMHM, Fleer GJ (1979) J Chem Phys 83:1619
261. Dickmann R, Hall CK (1986) J Chem Phys 85:4108
262. Deutsch H-P, Dickman R (1990) J Chem Phys 93:8983
263. Carmesin I, Kremer K (1988) Macromolecules 21:2819
264. Deutsch H-P, Binder K (1991) J Chem Phys 94:2294
265. Paul W, Binder K, Heermann DW, Kremer K (1991) J Phys II (France) 1:37
266. Werner A, Schmid F, Müller M, Binder K (1997) J Chem Phys 107:8175
267. Albano EV, Binder K, Heermann DW, Paul W (1989) Surface Sci 223:151
268. Parry AO, Evans R (1990) Phys Rev Lett 64:439
269. Parry AO, Evans R (1992) Physica A 181:250
270. Binder K, Landau DP and Ferrenberg AM (1995) Phys Rev Lett 74:298; Phys Rev E 51:2823
271. Binder K, Evans R, Landau DP and Ferrenberg AM (1996) Phys Rev E 53:5023
272. Rowlinson JS, Widom B (1982) Molecular Theory of Capillarity. Clarendon, Oxford
273. Jasnow D (1984) Rep Progr Phys 47:1059
274. Semenov AN (1993) Macromolecules 26:6617; (1994) Macromolecules 27:2732
275. Boulter CJ, Parry AO (1995) Phys Rev Lett 74:3403; Physica A 218:109
276. Parry AO, Boulter CJ (1995) Physica A 218:77
277. Brezin E, Halperin BI, Leibler S (1983) Phys Rev Lett 50:1387
278. Lipowsky R, Kroll DM, Zia RKP (1983) Phys Rev B 27:4499
279. Fisher ME, Wen H (1992) Phys Rev Lett 68:3654
280. Deutsch H-P, Binder K (1992) Macromolecules 25:6214
281. Eisenriegler E, Kremer K, Binder K (1982) J Chem Phys 77:6296
282. Binder K (1974) Thin Solid Films 20:367
283. De Gennes PG (1977) J Phys Lett (Paris) 38:L 441
284. Joanny JF (1978) J Phys A 11:L 117
285. Binder K (1984) Phys Rev A 29:341
286. Meier G, Schwahn D, Mortensen K, Janssen S (1993) Europhys Lett 22:577
287. Schwahn D, Meier G, Mortensen K, Janssen S (1994) J Phys II (Paris) 4:837
288. Solc K (1980) Polymer compatibility and incompatibility – principles and practices. Harwood Acad Publ., Chur
289. Bates FS, Wiltzius P (1989) J Chem Phys 91:3258
290. Hashimoto T (1993) In: Cahn RW, Haasen P, Kramer EJ (eds) Materials Science and Technology, Vol 12: Structure and Properties of Polymers. VCH, Weinheim, p. 251
291. Cahn JW (1961) Acta Metall 9:795
292. Gunton JD, San Miguel M, Sahni PS (1983) In: Domb C, Lebowitz JL (eds) Phase Transitions and Critical Phenomena Vol 8. Academic, London, p. 267
293. Komura S, Furukawa H (eds) (1988) Dynamics of Ordering Processes in Condensed Matter. Plenum Press, New York

294. Binder K (1991) In: Cahn RW, Haasen P, Kramer EJ (eds) Materials Science and Technology, Vol 5: Phase Transformations in Materials. VCH, Weinheim, p. 405
295. Lifshitz IM, Slyozov VV (1961) J Chem Phys Solids 19:35
296. Siggia E (1979) Phys Rev A 20:595
297. Schubert DW, Stamm M (1996) Europhys Lett 35:419
298. Sferrazza M, Xiao C, Jones RAL, Bucknall DG, Webster J, Penfold J (1997) Phys Rev Lett 78:3693
299. Pereira GG, Wang J-S (1996) Phys Rev E 54:3040
300. Dietrich S, Schick M, (1985) Phys Rev B 31:4718
301. Lambooy P et al. (1997) to be published
302. Stepanow S, Dobrynin AV, Vilgis TA, Binder K (1996) J Phys I (Paris) 6:837
303. Pickett GT, Balazs AC (1997) Macromolecules 30:3097
304. Russell TP, Mayes AM, Kunz MS (1994) In: Teramoto A, Kobayashi M, Norisuye T (eds) Ordering in Macromolecular Systems p. 217. Springer, Berlin; Mayes AM, Russell TP, Bassereau P, Baker SM, Smith GS (1994) Macromolecules 27:749
305. Binder K, Stauffer D (1974) Phys Rev Lett 33:1006

Received: January 1998

Flexible Polymers in Nanopores

Pierre-Gilles de Gennes

Collège de France, 11 place Marcelin Berthelot75231 Paris Cedex 05, France
e-mail: pierre-gilles.degennes@espci.fr

The abstract corresponds to the first part of the first chapter.

Keywords: branched polymers, nanopores, confined polymers, polymer characterisation, polymer rheology

1 The need for New Methods of Polymer Characterisation 92

1.1 Branched Structures 92
1.2 Probing Methods 92
1.3 Forced Permeation 93

2 A Reminder on Linear Chains in Pores 94

2.1 Competition Between Squeezing and Aspiration 94
2.2 Permeation Without Flow 95
2.3 Forced Permeation 95
9Z2.4 Another Hydrodynamic Picture 96

3 Theoretical Predictions for Stars in Pores 96

3.1 Statics : Optimal Shape Without Flow 96
3.2 Dynamics : Optimal Shape with Flow 98
3.3 Problems of Presentation. 99

4 Theoretical Predictions for Statistically Branched Polymers 99

4.1 Maximum Stretching and the Ariadne Length. 99
4.2 The Equivalent Semi-Dilute Solution. 100
100Z4.3 Critical Currents. 102

5 Perspectives 103

5.1 Characterisation. 103
5.2 Extension to Other Semi -Compact Structures 103
5.3 Problems of Chain Rupture 104

6 References 105

Advances in Polymer Science, Vol.138

1 The Need for New Methods of Polymer Characterisation

1.1 Branched Structures

A major breakthrough in polymer production occurred with the discovery of metallocene catalysts [1]. We are now able to make polyolefins with a controlled level of branching (and tacticity). The simplest object is a statistically branched polymer, with a certain overall degree of polymerisation X, and a certain distance (monomer units) between successive branch points, which we shall call b. The basic goal of characterisation is to measure X and b from a minimum number of experiments in dilute solutions.

Let us first recall some basic principles on the conformation of these branched objects. Our main assumption is that the molecule is flexible and tree-like: although it is branched it does not contain any closed cycle.

In good solvents, we know from theory [2,3] and from experiments [4] that the radius of gyration R_g follows a very strange law

$$R_g(X,b) \sim aX^{1/2}b^{1/10} \tag{1}$$

where a is something like a monomer size. The exponents can be derived from a calculation of the Flory type [5], including the correct value of the ideal branched object size R_0 ,which was first derived long ago by Zimm and Stockmayer [6,7]

$$R_0 = a(bX)^{1/4} \tag{2}$$

A partial check on Eq.(1) is that if we make b=X (i.e.: if we return to linear chains) we recover the standard Flory formula $R_g \sim aX^{3/5}$.

1.2 Probing Methods

What information can we obtain about these branched systems by standard-measurements ?

a) The hydrodynamic radius (proportional to R_g) can be derived from viscosities in dilute solutions, from measurements of a diffusion constant,and from gel permeation chromatography.
b) The molecular weight is in principle accessible from light scattering intensities – preferably via quasi elastic scattering to eliminate the rôle of dust, etc. For small values of X, the recent forms of mass spectroscopy can also be used.
c) A classical approach is based on rheology. In dilute solutions these compact coils are not easy to distort. In melts, we can achieve strong distortions; but the mechanical response of these complicated structures, both entangled and branched, is hard to predict in a reliable way.

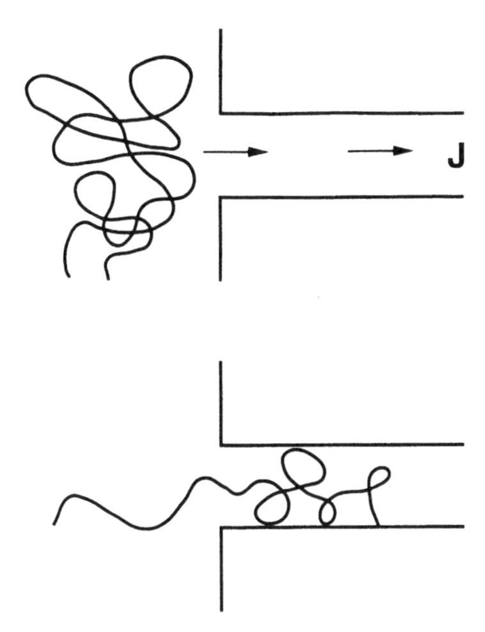

Fig. 1. A linear chain facing a pore

All the above led to a different approach where a) the coils would be kept dilute, but would be forced to distort quad and b) linear chains would behave in a trivial way.

1.3
Forced Permeation

One possible approach is based on the *aspiration* of the dilute polymer inside a nanopore (Fig. 1). The nanopores are produced easily by particle-track etching [8], and can be bought commercially. Three requirements must be imposed:

a) the polymer must *not adsorb* on the tube surface;
b) the tube diameter D must be smaller than the polymer size R_g;
c) the tube cross section must be constant (no rugosity).

In the1980s, Léger and coworkers [9,10] studied the aspiration of *linear* polystyrene in pores of polycarbonates. A theoretical picture was also constructed for this [11] and is summarised in Fig. 2. The main result is that linear chains are sucked in when the pore current J (the solvent volume flowing per unit time) is beyond a certain threshold value J_c.This J_c turns out to be independent of tube diameter and of the molecular weight of the linear polymer

$$J_{cl} = \text{const.} \frac{kT}{\eta} \tag{3}$$

η being the solvent viscosity. If we translate this into a pressure drop p, over a length L of the tube, the scaling law, derived from Poiseuille flow is

$$p \cong \frac{kT}{D^4} L \qquad (4)$$

These p values are sizeable, but not incompatible with the strength of standard pore materials with suitable(ceramic) supports.

A more detailed explanation of Eqs 3 and 4 shall be given in section 2. For our present purposes, the main conclusion is that J_c is not dependent on X for linear chains. But, without any calculation, we can expect that a branched object will enter much more painfully in a narrow tube. Thus permeation studies may provide an interesting assessment of branching. In section 3, we discuss briefly the case of star polymers [11] which is relatively easy to visualise. Then in section 4, we study the morerealistic problem of branched objects [12]. All our discussion is very crude; restricted to polymers in very good (athermal) solvents; and restricted to the level of scaling laws.

2 A Reminder on Linear Chains in Pores

2.1 Competition Between Squeezing and Aspiration

Figure 2 summarises the situation: the chain has partially entered into the pore (over a length L >D). It can be pictured as a sequence of blobs, each of size D and of monomer number $g_D=(D/a)^{5/3}$ (theFlory law). The confinement energy per blob is of the order kT [13]and the overall confinement energy is thus kT ℓ/D (ℓ/D being the number of squeezed blobs. The force tending to pull the chain out of the pore is thus kT/D.

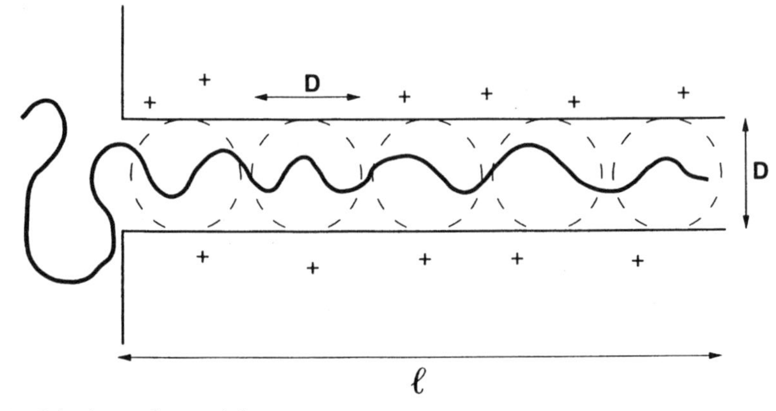

Fig. 2. Blob picture for partial entry

2.2
Permeation Without Flow

The first (classical) application of these ideas concerns chains which move in and out of the tube without any flow. In order to insert a full chain into the tube, where it exhibits a length L, we need an energy

$$E = kT\frac{L}{D} = kT\frac{X}{g_D}$$
$$E = kTX\left(\frac{a}{D}\right)^{5/3} = kT\left(\frac{R_g}{D}\right)^{5/3} \tag{5}$$

The permeation coefficient of a membrane based on these pores is thus proportional to

$$\exp(-E/kT) = \exp-\left[(const).\left(\frac{R_g}{D}\right)^{5/3}\right] \tag{6}$$

This allows, in principle, for a selection according to molecular weight. However it acts rather like a cut-off: chains of size R_g larger than D do not permeate. Equation (6) is, in fact, the basis of gel permeation chromatography.

2.3
Forced Permeation

We now want to force the chains in by strong flows. If we impose a flow velocity $V=J/D^2$ we must have a Stokes force $\sim \eta VD$ on each blob, and thus an overall hydrodynamic force $(\ell/D)\eta VD=\eta V\ell$. The smallest hydrodynamic force available corresponds to ℓ=D. If the first blob goes in, the other ones will follow more easily. Thus the threshold corresponds to ℓ=D, or

$$\frac{kT}{D} = \eta VD = \eta\frac{J_c}{D} \tag{6a}$$

Thus the critical current is $J_c=kT/\eta$ (Eq. 3). Numerically, at room temperature, with viscosities η comparable to the viscosity of water, $J_c \sim 5.10^{-12}cm^3/s$. Then in a pore of diameter D=50Å, the velocity is $V = J_c/D^2 \sim 20 cm/s$.

Experimentally, the threshold is not very sharp, but the cross over from zero permeation to strong permeation is quite visible for polystyrene in good solvents inside polycarbonate nanotubes [10]. At the time of these experiments, the fact that J_c was independent of molecular weight was considered as rather discouraging. But for our present purposes this may in fact be a bonus.

2.4
Another Hydrodynamic Picture

Our discussion on the aspiration of linear chain has been basedon the behavior inside the tube. It is also helpful to consider what happens outside (in the entry region). Here, in the most naive picture, we have a convergent flow of velocity $v(r)=J/r^2$, where r is the distance to the center of the entry disc. The velocity gradient is $\dot{\gamma} = \nabla v \sim J/r^3$. At a certain radius r_c, this is such that $\dot{\gamma} T_z = 1$, where T_z is the relaxation time of a coil $T_z = \eta R_g^3 / kT$. At distances $r<r_c$, the deformation is fast, the chain cannot relax and deforms affinely: the final lateral dimension of the distorted coil is $r_f \sim D/r_c R_g$. When r_f becomes as small as D, the chain can get in. This corresponds to $r_c \sim R_g$ and $J \sim J_c$, where J_c is again given by Eq.(3). Thus this different picture leads to the same conclusion.

3
Theoretical Predictions for Stars in Pores

3.1
Statics: Optimal Shape Without Flow

Let us think of a star, with f arms, each of N monomers ($X=fN$). We assume long arms ($N >> f^{1/2}$). The opposite case would yield a compact nodule. In dilute solutions, these stars build up a structure of radius

$$R = aN^{3/5} f^{1/5} \tag{7}$$

This result can be derived from a simple extension of the Flory calculation; more detailed pictures of the swollen star can be constructed cite[13]and lead to the same form. (An attentive reader will note that if we put $N=f^{1/2}$ in eq. (7), we end up with $R=aX^{1/3}$, i.e. a compact nodule as announced).

Consider now the star inserted in a tube of diameter $D<R$, and assume that the arms extend over lengths much larger than D. Then each arm isessentially uniformly stretched.

As we shall see later, the conformation which gives the lowest energy is 'symmetric' (Fig. 3a) with a number $\tilde{f} = f/2$ of arms forward, and an equal number $\tilde{f}$ backwards. Each arm is confined into a small tube of cross section $\xi^2 = D^2/\tilde{f}$. The stretched length of the arms is thus

$$L = \xi \frac{N}{g_\xi} \tag{8}$$

where $g_\xi = (\xi/a)^{5/3}$ is the number of monomers per subunit of size ξ. Equation 8 can then be rewritten as

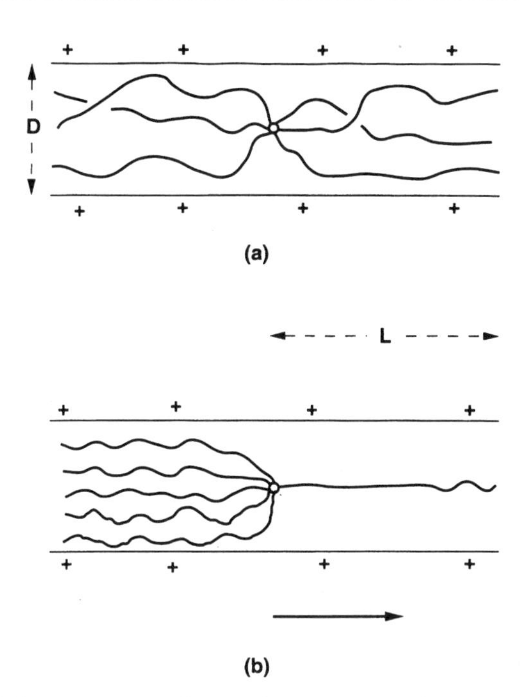

Fig. 3. A star trapped in a nanopore. The case showncorresponds to f=6. **a** Symmetric mode. **b** One arm forward

$$\frac{L}{D} = \left(\frac{R}{D}\right)^{5/3} \tag{9}$$

where R is the star radius of Eq. (7). The minimal diameter allowing passage is $D=D_m$ and is such that ξ=a (no solvent left). Thus

$$D_m \simeq a f^{1/2} \left(\sim a f^{1/2}\right) \tag{10}$$

Then the arms are completely stretched and L=Na. The energy F of the confined star is again kT per submit, giving a total

$$\begin{aligned} F &= kTf\frac{N}{g\xi} \\ F &= kTN\left(\frac{a}{D}\right)^{5/3} f^{11/6} \end{aligned} \tag{11}$$

The power law for $F(\tilde{f})$ involves an exotic exponent p=11/6. The crucial feature here is that p>1, or equivalently, that theenergy is a convex function of the number of arms inserted. From this, we can see that the lowest energy corresponds to the symmetric case. If we had chosen a non symmetric conformation, with f_1 arms forward, and a different number $f-f_1$ of arms backward, we would find an energy

$$F(f_1)+F(f-f_1)>2F\left(\frac{f}{2}\right) \tag{11a}$$

Thus, when we discuss permeation without aspiration, we can safely omit then on symmetric modes. The energy barrier for entry is well described by Eq. (11).

3.2 Dynamics: Optimal Shape with Flow

The major surprise (encountered in ref. 11) is that the symmetric mode described above is not necessarily the major mode of aspiration.

To understand this, let us now assume that we suck in the solvent, with acurrent J, slightly larger than the threshold J_c of Eq. (3): then our star will be able to insert one arm into the tube, as shown on Fig. 3b.

Of course, the rest of the star is plugged in at the entry, and offers a large resistance. But this resistance is finite: and if the arm length N is long enough, the hydrodynamic force over N/g_D blobs, $\eta DV\ N/g_D$, can be made very large: at some point, it will be able to overcome the resistance from the $(f-1)$ other arms.

This remark (by F. Brochard) prompted us to investigate the hydrodynamic thresholds for non symmetric conformations: the calculation is described in [11]. There is a certain regime $(N<(D/a)f^{3/2})$, wherethe optimal number of forward arms is

$$f_1 = f^{1/2}\left(\frac{D}{Na}\right)^{1/3} \tag{12}$$

giving a critical current

$$J_c = J_{c1} f\left(\frac{D}{Na}\right)^{2/3} \tag{13}$$

which is larger than J_{c1}, and dependent on f (and N). On the other hand, if N is larger $(N>(D/a)\ f^{3/2})$, we expect $J_c=J_{c1}$, and the experiment would bring in no useful information.

This permeation by asymmetric modes is a surprise. In most physical phenomena (e.g. in chemical reactions) the static energy barrier is not very different from the dynamic barrier. Here, they can be deeply different.

3.3 Problems of Presentation

The actual scenario for entry of stars under aspiration may be quite complex: upon arrival at the tube opening, the star will usually not show up the optimal number of arms. If, for instance, one arm went in, while the optimal number f_1 was higher, it may take a long time for the star to bring a second arm in the right position. Fortunately, these complications, which would be very serious for stars, are probably much less important for statistically branched polymers.

4 Theoretical Predictions for Statistically Branched Polymers

4.1 Maximum Stretching and the Ariadne Length

What are the conformations of a statistically branched object when it is squeezed into a tube?

This question was solved some years ago by Vilgis and coworkers [13, 14]. The main result can be understood simply in terms of a Flory calculation, with a free energy F of the form

$$\frac{F}{kT} = \frac{L^2}{X^{1/2}a^2b^{1/2}} + \frac{X^2a^3}{LD^2} \tag{14}$$

The first term is an elastic energy, for a molecule of extension L along the tube, with an unperturbed radius R_0 given by Eq. (2). The second term describes monomer/monomer repulsions (in an athermal solvent). Optimising Eq. (14), we arrive at

$$L = DX^{5/6}\left(\frac{a}{D}\right)^{5/3} b^{1/6} = D\left(\frac{R}{D}\right)^{5/3} \tag{15}$$

The internal concentration is

$$\phi = \frac{Xa^3}{LD^2} = \left(\frac{D_m}{D}\right)^{4/3} \tag{16}$$

where we have introduced (following Vilgis) a minimum tube diameter

$$D_m = a\left(\frac{X}{b}\right)^{1/8} \tag{17}$$

Whatever the hydrodynamic force, the polymer will not be able to enter in a tube thinner than D_m. Note that if we return to linear chains (b=X) we recover the natural result $D_m = a$.

Of particular interest is the length L_A associated with the diameter D_m.

$$L_A = aX^{3/4}b^{1/4} \tag{18}$$

We call L_A the *Ariadne length*. Ariadne helped Theseus through the Minoan labyrinth, by giving him a reel of thread, which kept a track of his march. L_A represents the length of the shortest path (from the starting point) to the monster: in our context, the thread distance between two arbitrary points in the cluster. When we squeeze the molecule at its utmost, we arrive at a length of order L_A.

The Ariadne length is related to what is called the *spectral dimension* of a cluster in statistical physics. Readers interested in this aspect will find more information in refs [12–14].

4.2 The Equivalent Semi-Dilute Solution

When our polymer is confined, with a certain interval concentration f (Eq. 16) we have to understand first what is the *mesh size* (or correlation length) ξ which this 'solution' of branched objects builds up. For solutions of linear polymers, the relation between ξ and ϕ is classical [15]

$$\xi = a\phi^{-3/4} \tag{19}$$

This will still hold here, provided that branching is negligible at the scale ξ: i.e. that the size of a submit of b monomers is larger than ξ. Or

$$b > g_\xi = \left(\xi / a\right)^{5/3} = \phi^{-5/4} \tag{20}$$

This corresponds to ϕ values which are relatively large: we call this regime *strong confinement*.

In the opposite limit ($\phi < b^{-4/5}$) we talk of weak confinement.Here, the mesh size ξ is associated with a branched submit (the 'pearl'), with a number g_p of monomers, by a transposition of Eq.(1)

$$\xi = a g_p^{1/2} b^{1/10} \qquad \left(\phi < b^{-4/5}\right) \tag{21}$$

The available space is densely filled by pearls, and this imposes

$$\phi = g_p a^3 / \xi^3 \tag{22}$$

Eliminating g_p between Eqs (20 and 21) we arrive at

$$\xi = ab^{-1/5}\phi^{-1} \qquad \left(\phi < b^{-4/5}\right) \tag{23}$$

The crossover between weak and strong confinement occurs when $\phi=b^{-4/5}$: i.e. when the correlation length is equal to the size of a linear chain, made of b monomer, in good solvent conditions $\xi = \xi_b = ab^{3/5}$. This corresponds to a crossover diameter for the tube

$$D^* = D_{min} b^{3/5} \tag{24}$$

The two confinement regimes are qualitatively shown on Fig. 4.

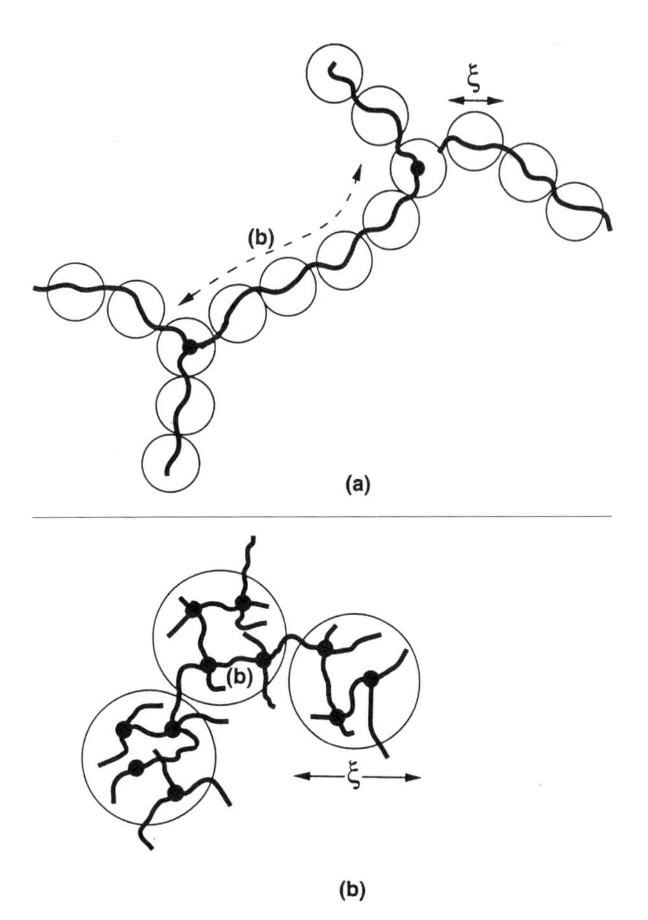

Fig. 4. Two forms of organisation for the squeezed polymer **a** strong confinement: locally the structure resembles a solution oflinear chains **b** weak confinement : the mesh unit ('pearl') contains many branch points

4.3 Critical Currents

Once we know the mesh size, we know the osmotic pressure of the confined solution

$$\Pi = \frac{kT}{\xi^3} \tag{25}$$

The energy required to confine a length ℓ, of the compacted polymer is then

$$E_c = \Pi D^2 \ell = kT \frac{D^2 \ell}{\xi^3} \tag{26}$$

An attentive reader will note that this E_c is (slightly) different from the optimal energy derived from the Flory approach (Eq. 14). This is not a surprise: the Flory (mean field) calculation is good to compute sizes, but less good to compute energies [15].

The hydrodynamic force acting on one submit is $\eta \xi V$. The number of submits is $D^2 \ell / \xi^3$. Thus, the resulting energy is

$$E_h = -\frac{1}{2} \frac{\ell^2}{\xi^2} \eta J \tag{27}$$

Optimising the sum $E_c + E_h$, we arrive at

$$\frac{\ell}{D} = \frac{J_{c1}}{J} \frac{D}{\xi} \tag{28}$$

and a barrier energy

$$E_b = kT \left(\frac{D}{\xi} \right)^4 \frac{J_{c1}}{J} \tag{29}$$

The critical current J_c corresponds to $E_b \sim kT$ and thus scales according to

$$\frac{J_c}{J_{c1}} = \left(\frac{D}{\xi} \right)^4 \tag{30}$$

In this formula, we insert the relevant value for ξ: given by Eq.(19) for $D<D^*$ and by Eq.(23) for $D>D^*$. The results are

$$\frac{J_c}{J_{c1}} = X^{2/3} (a/D)^{4/3} b^{2/15} \qquad \left(D > D^* \right) \tag{31}$$

$$\frac{J_c}{J_{c1}} = \left(\frac{X}{b}\right)^{1/2} \qquad \left(D < D^*\right) \tag{32}$$

Equations 31and 32 are the basic predictions of the model[1].

5 Perspectives

5.1 Characterisation

We have seen in Section 4, that a measurement of the critical current J_c, performed at *two* diameters ($D < D^*$ and $D > D^*$) should allow for the determination of *two* unknowns: the molecular weight (proportional to X) and the distance between branch points.

This conclusion is promising, but all our discussion is still very far from real life. The following is a list of expected complications:

a) of course, we are interested in polydisperse systems. The fluctuations of b may be somewhat averaged out inside one single molecule. But thedispersion of X values brings in serious problems. However it is not hopeless to come back from experimental plots (giving the total weight of polymer permeated versus the permeation current J) to distributionfunctions;

b) the possibility of *clogging* by the largest molecules is always present. Our group is planning experiments on fluorescent DNAs to observe these effects, and all the other entry problems, at least on linear polymers;

c) in some cases, the hydrodynamic forces may be large enough to *break* the polymer. This is a significant worry for star polymers [11] and is discussed in Section 5.3 below;

d) another problem is that for high J_c, the mechanical strength of the filter must be quite high: see Eq.(4) for the pressure drop.

All these questions are serious. But, as far as we can see, none of them is obviously catastrophic. We need a thorough reflection on the chemical engineering aspects. Should we work with one plate, or with a sequence of plates of different diameters? Should we work in *d.c.* hydrodynamic flow, or in modulated flows? Is there any interest in tapered tubes? To what extent could we replace the tubes by conventional (non adsorbing) porous media... etc.

5.2 Extension to Other Semi-Compact Structures

Our theoretical discussion can easily be transposed to other molecules: comb polymers being a typical example. Another amusing example is a swollen,

---

1 Equation 31 has a misprint in the original ref [12]

spherical gel particle of original size R, somewhat larger than D [12]. A third category of interest would be based on wormlike micelles, which are known to be flexible. In suitably controlled conditions, these micelles systems can branch. But a major difference is that all these micellar structures are labile: they can break and reform.

Going further in this direction, certain block copolymers, with successive sequences which are soluble/insoluble in the solvent, can associate their insoluble parts and form branched structures (ultimately terminating in a gel above certain thresholds). Thus, very generally, the permeation properties of labile aggregates may be instructive, and possibly useful.

5.3 Problems of Chain Rupture

All our discussion has assumed that our branched structures are robust. But they are in fact subjected to significant hydrodynamic tensions. Consider a star polymer under a current J largerthan J_{c1}. Then we shall meet situations where one (or a few) arm(s) has entered the tube, while the central nodule is still stuck at the entry as in Fig. 3b. If ℓ is the length of tube which has been invaded by the arm(as in Sect. 2.3), the hydrodynamic force $\eta v \ell$, may be large, and can possibly reach the threshold force τ_m for rupture near the nodule. This leads to a critical value ℓ_m:

$$\ell_m = \frac{D^2 \tau_m}{kT} \frac{J_{c1}}{J} \tag{33}$$

What is the scaling structure for τ_m? Here, we shall consider two distinct examples.

a) Chemical rupture, with a bond energy U_b, will be associated with a force $\tau_m \sim U_b/a$.
b) As mentioned in Sect. 5.2, another 'nodule' of interest is obtained with block copolymers: for instance (with water as the solvent) a long hydrophilic arm, terminated at one end by a short hydrophobic piece, will end up forming micelles or stars with a hydrophobic core. Here the relevantenergy U_h is the transfer energy, required to pull out one hydrophobic monomer, from an environment of the same monomer, into water. U_h is much smaller than U_b. Ultimately, we can write for both cases :

$$\ell_m \sim \frac{U}{kT} \frac{D^2}{a} \frac{J_{c1}}{J} \tag{34}$$

The corresponding polymerisation index for the arm, is:

$$N_m = \frac{\ell_m}{D} \left(\frac{D}{a}\right)^{5/3} = \frac{U}{kT} \left(\frac{D}{a}\right)^{8/3} \frac{J_{c1}}{J} \tag{35}$$

Even in mild conditions (i.e. when J is not much more larger than J_{c1}) this N_m is not huge. For instance, in case (b), if we take U/kT~1, D=10 nm a=0.3 nm, and choose $J=3J_{c1}$, we arrive at $\ell_m>100$ nm, and $N_m>3000$.

Similar ideas can be used for statistically branched polymers; let us, for instance, consider the case of large distances b between branch points, i.e. 'strong confinement' as defined by the inequality (20). Then the tube (of cross section D^2) can be subdivided into $(D/\xi)^2$ 'subtubes', and the chain, in each of them, is subjected to the tension $\eta v\ell$. When the tension equals the threshold force, this leads again to Eq.(34). Here, the current threshold J_c is much larger than J_{c1}, and rupture may often occur. To avoid this, we need to operate in situations of mild compression, i.e. with D/R_g not too small.

Thus, for the characterisation problem, even with strongly bonded branch-points, we must be careful. On the other hand, the rupture process may be interesting in itself: for instance, with diblock copolymers making micellar stars, the rupture current could be measured, and would provide an estimate of the bonding energy U_h.

Acknowledgements. This work was initiated by discussions with T.Mc Leish and S. Milner. We are greateful to F. Brochard-Wyart, C. Gay and E. Raphaël for their ideas und comments. Written exchanges with T. Vilgis and S. Havlin were also very helpful.

6 References

1. Raminsky W, Arndt M (1997) Adv Polymer Sci 127:144
2. Isaacson J, Lubensky T (1980) J Physique Paris 41:L469
3. Daoud M, Joanny JF (1981) J Physique Paris 42:1359
4. Adam M, Delsanti M, Munch JP, Durand D (1987) J PhysiqueParis 48:1809
5. Flory P Principles of polymer chemistry, Cornell University Press
6. Zimm B, Stockmayer W (1949) J Chem Phys 17:1301
7. de Gennes PG (1968) ([6]) for a more compact derivation ofsome results in; Biopolymers 6:715
8. Fleischer R, Price P, Walker RM (1975) Nuclear track in solids, University of California Press
9. Guillot G, Läger L, Rondelez F (1985) Macromolecules 18:2531
10. Guillot G (1985) Thesis, Paris
11. Brochard F, de Gennes PG (1996) C R Acad Sci Paris 323 II:473
12. Gay C, de Gennes PG, Raphaâl E, Brochard F (1996)Macromolecules 29:8379
13. Vilgis TA (1992) J Physique II 2:2097
14. Vilgis TA, Haronska P, Benhamou M. (1994) J Physique II 4:2187
15. See for instance de Gennes PG (1985) Scaling concepts in polymer physics, Cornell University Press, 2nd printing

Received: January 1998

Polymer-Silicate Nanocomposites: Model Systems for Confined Polymers and Polymer Brushes

E. P. Giannelis[1,3], R. Krishnamoorti[2,4], E. Manias[1,5]

[1] Department of Materials Science & Engineering, Cornell University, Ithaca, NY 14853, USA
[2] Department of Chemical Engineering, University of Houston, Houston, TX 77204, USA
[3] *emails: epg2@cornell.edu*
[4] *ramanan@bayou.uh.edu*
[5] *manias@msc.cornell.edu*

The static and dynamic properties of polymer-layered silicate nanocomposites are discussed, in the context of polymers in confined spaces and polymer brushes. A wide range of experimental techniques as applied to these systems are reviewed, and the salient results from these are compared with a mean field thermodynamic model and non-equilibrium molecular dynamics simulations.
Despite the topological constraints imposed by the host lattice, mass transport of the polymer, when entering the galleries defined by adjacent silicate layers, is quite rapid and the polymer chains exhibit mobilities similar to or faster than polymer self-diffusion. However, both the local and global dynamics of the polymer in these nanoscopically confined galleries are dramatically different from those in the bulk. On a local scale, intercalated polymers exhibit simultaneously a fast and a slow mode of relaxation for a wide range of temperatures, with a marked suppression (or even absence) of cooperative dynamics typically associated with the glass transition. On a global scale, relaxation of polymer chains either tethered to or in close proximity (<1nm as in intercalated hybrids) to the host surface are also dramatically altered. In the case of the tethered polymer nanocomposites, similarities are drawn to the dynamics of other intrinsically anisotropic fluids such as ordered block copolymers and smectic liquid crystals. Further, new non-linear viscoelastic phenomena associated with melt-brushes are reported and provide complementary information to those obtained for solution-brushes studied using the Surface Forces Apparatus.

1 Introduction 108

1.1 Layered-Silicate Structure 110

2 The Structure of Nanocomposites 113

2.1 Morphologies of Polymer/Silicate Nanocomposites 113
2.2 Thermodynamics of Nanocomposite Formation 115

3 Dynamics of Nanoscopically Confined Polymers 118

3.1 Kinetics of Polymer Melt Intercalation 118
3.2 Structural Evolution During Intercalation 121
3.3 Local Dynamics 122
3.3.1 On the Relaxation of nm-Thick Polymer Films Between Walls . . . 122
3.3.2 NMR Measurements 123
3.3.3 Computer Simulations 125
3.3.4 Cooperative Motion – TSC and DSC Measuraments 128

Advances in Polymer Science, Vol.138

4 **The Rheology of Polymer/Silicate Nanocomposites** 131

4.1 Linear Viscoelasticity 131
4.2 Alignment of Nanocomposites 139
4.3 Strain - Hardening of Polymer Brushes 140

5 **Conclusions** 142

6 **References** 143

1 Introduction

Layered-silicate based polymer nanocomposites have become an attractive set of organic–inorganic materials not only for their obvious potential as technological materials, but also for providing a convenient macroscopic system to study fundamental scientific issues concerning confined and tethered polymers. Studying the formation, structure and dynamics of these nanocomposites can lead to a better understanding of organic–inorganic hybrids, polymers in a confined environment or at a solid interface and polymer brushes.

One promising way to synthesize polymer nanocomposites is by intercalating polymers in layered inorganic hosts [1–4]. Graphite, transition metal chalcogenides, metal phosphates, complex oxides, oxychlorides and mica-type layered silicates are some examples of layered solids capable of intercalation. The structure and properties of the resulting nanostructure can be conveniently mediated by controlling subtle guest–host interactions. Beyond the conventional **phase separated** polymer/silicate composites, for which the polymer and the inorganic host remain immiscible, two types of hybrids are possible (Fig. 1): **intercalated** in which a single, extended polymer chain is intercalated between the host layers resulting in a well ordered multilayer with alternating polymer/inorganic layers and a repeat distance of a few nanometers (Fig. 2a), and **exfoliated** or **delaminated**, in which the silicate layers (1 nm thick) are exfoliated and dispersed in a continuous polymer matrix [5, 6] (Fig. 2b).

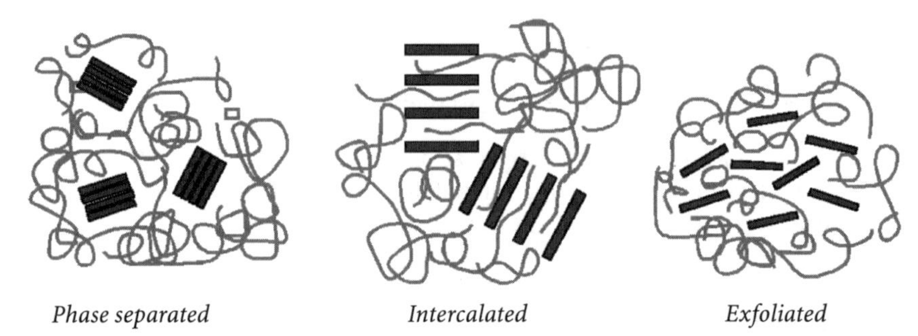

Fig. 1. Schematic representation of different polymer/silicate hybrid structures.

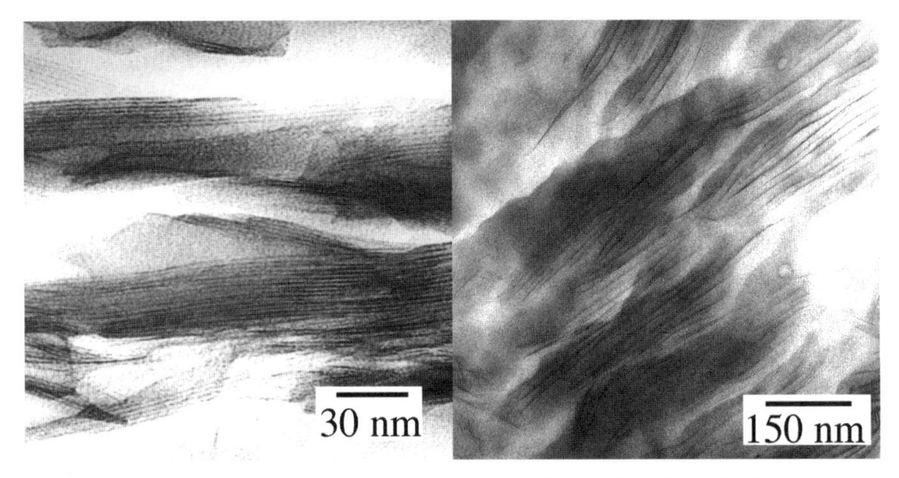

Fig. 2. TEM images of an intercalated (left) and an exfoliated (right) hybrid [5]

In the case of mica-type layered silicates it has been recently demonstrated that nanocomposites (both intercalated and delaminated) can be synthesized by direct melt intercalation even with high molecular weight polymers [7–18]. This synthetic method is quite general and is broadly applicable to a range of commodity polymers from essentially non-polar polystyrene, to weakly polar poly(ethylene terephthalate), to strongly polar nylon. Nanocomposites can, therefore, be processed using currently available techniques such as extrusion, thus lowering the barrier towards commercialization.

The unprecedented mechanical properties of polymer layered silicate (PLS) nanocomposites were first demonstrated by a group at the Toyota research center in Japan using nylon nanocomposites [19–21]. They showed that a doubling of the tensile modulus and strength is achieved for nylon-layered silicate nanocomposites containing as little as 2 vol. % inorganic. More importantly the heat distortion temperature of the nanocomposites increases by up to 100 °C extending the use of the composite to higher temperature environments, such as automotive under-the-hood parts.

PLS nanocomposites have several advantages [6] including: (a) they are lighter in weight compared to conventionally filled polymers because high degrees of stiffness and strength are realized with far less high density inorganic material; (b) their mechanical properties are potentially superior to fiber reinforced polymers because reinforcement from the inorganic layers will occur in two rather than in one dimension without special efforts to laminate the composites; and (c) they exhibit outstanding diffusional barrier properties without requiring a multipolymer layered design, allowing for recycling.

In addition to their potential applications, PLS nanocomposites are unique model systems to study the statics and dynamics of polymers in confined environments. Using both delaminated and intercalated hybrids, the statics and dynamics of polymers confined over distances ranging from the radius of gyration of the polymer to the statistical segment length of the chains can be studied. Even simple notions regarding the conformations of polymers confined to two dimensions are yet to be understood. In three dimensions it is known that for long chain polymers there is significant overlap between molecules. In two-dimensions it has been suggested that different chains should only overlap slightly [22]. Therefore the local and global conformations of the polymers within the host galleries are expected to be dramatically different from those observed in the bulk not only due to the confinement of the polymer chains but also due to specific polymer-surface interactions, normally not observed in the bulk. It is also expected that the local and chain dynamics would be greatly affected by the confinement as well as the polymer-surface interactions.

The behavior of polymer liquids under confinement is in general very different from that in the bulk, especially when the confining dimensions become comparable or smaller than the polymer coil size. Traditional notions such as reptation dynamics governing the relaxation of a long polymer chain are improbable in highly confined intercalated systems (confinement distances comparable to the statistical segment length of the polymer) as it is impossible to imagine a topological entanglement in two dimensions.

1.1 Layered-Silicate Structure

The layered silicates used in the nanocomposites belong to the same structural family as the better known minerals talc and mica [10] (i.e. 2:1 phyllosilicates). Their crystal lattice consists of a two-dimensional, 1 nm thick layers which are made up of two tetrahedral sheets of silica fused to an edge-shaped octahedral sheet of alumina or magnesia. The lateral dimensions of these layers vary from 300 Å to several microns depending on the particular silicate. Stacking of the layers leads to a regular van der Walls gap between them called the *interlayer* or *gallery*. Isomorphic substitution within the layer generates negative charges that are normally counterbalanced by hydrated alkali or alkaline earth cations residing in the interlayer. Because of the relatively weak forces between the layers (due to the layered structure), intercalation of various molecules, even polymers, between the layers is facile.

Pristine mica-type layered silicates usually contain hydrated Na^+ or K^+ ions [16]. Ion exchange reactions with cationic surfactants including primary, tertiary and quaternary ammonium or phosphonium ions render the normally hydrophilic silicate surface organophilic, which makes possible intercalation of many engineering polymers. The role of alkyl ammonium cations in the organosilicates is to lower the surface energy of the inorganic host and improve the wetting characteristics with the polymer. Additionally, the alkyl ammonium cations could provide functional groups that can react with the polymer or initiate polymerization of monomers to improve the strength of the interface between the inorganic and the polymer [17–19].

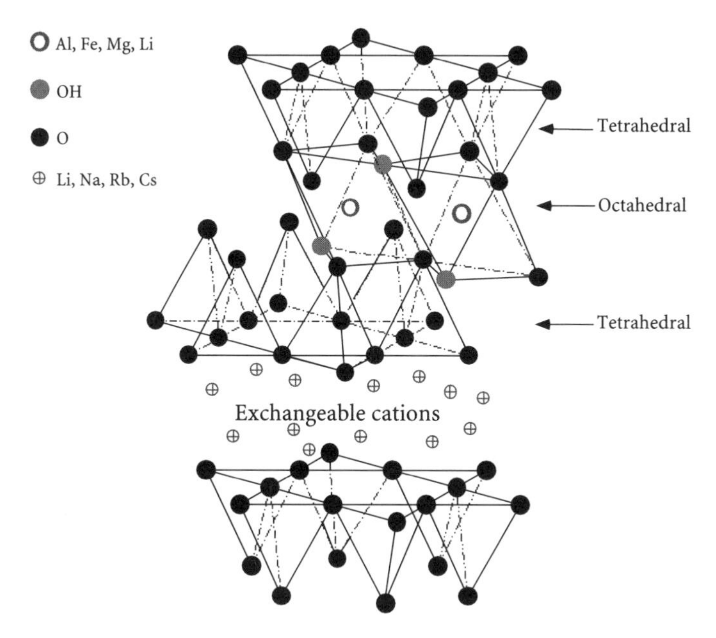

Schematic 1. The structure of 2:1 layered silicates. M is a monovalent charge compensating cation in the interlayer and x is thedegree of isomorphous substitution, which for the silicates of interest is between 0.5 and 1.3. The degree of isomorphous substitution is also expressed as a cation exchange capacity (CEC) and is measured in milli-equivalents/g.

The most commonly used layered silicates are montmorillonite, hectorite and saponite. Details on the structure and chemistry for these layered silicates are provided in Schematic 1 and Table 1. All of these silicates are characterized by a large active surface area (700 – 800 m^2/g in the case of montmorillonite), a moderate negative surface charge (cation exchange capacity) (CEC) and layer morphology, and are regarded as hydrophobic colloids of the constant-charge type. The layer charge indicated by the chemical formula is only to be regarded as an average over the whole crystal because the charge varies from layer to layer (within certain bounds). Only a small proportion of the charge balancing cations are located at

Table 1. Structure and chemistry of Mica-type layered silicates

Silicate	Location of isomorphous substitution	Formula
Montmorillonite	Octahedral	$M_x[Al_{4-x}Mg_x](Si_8)O_{20}(OH)_4$
Hectorite	Octahedral	$M_x[Mg_{6-x}Li_x](Si_8)O_{20}(OH)_4$
Saponite	Tetrahedral	$M_x[Mg_6](Si_{8-x}Al_x)O_{20}(OH)_4$

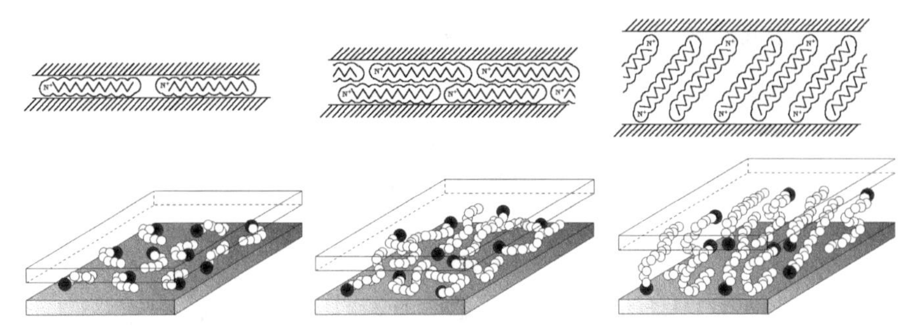

Fig. 3. (top) Idealized structures of organically modified silicates; adopted from Weiss A. (1963) Angew Chem 2: 134 and Lagaly G (1976) Angew Chem 15: 575. (bottom) The structure of organically modified silicates as derived from FTIR experiments (adopted from Vaia et al. (1994) Chem Mater 6: 1017) and molecular dynamics computer simulations of Hackett et al. (1998) J. Chem. Phys. 108: 7410.

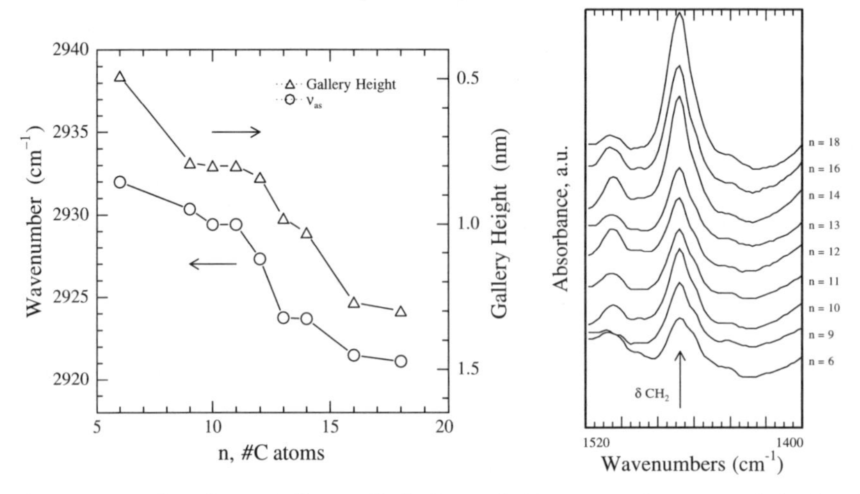

Fig. 4. FTIR studies of organically modified silicates. (left) Asymmetric CH_2 stretch ($\nu_{as}(CH_2)$) and gallery height as a function of surfactant chain length measured at room temperature for a series of organically modified fluorohectorites. The large variation in ν_{as} and gallery height suggests a wide range of molecular configurations and not an all-trans surfactant arrangement [9]. (right) Variation in the CH_2 bending (scissoring) vibration, $\delta(CH_2)$, for a series of organically modified fluorohectorites at room temperature. The observed absoption at 1468 cm^{-1} is characteristic of a partially ordered phase where the chains are mobile while maintaining some orientational order [9].

the external crystal surface with the majority being present in the interlayer space. The cations are exchangeable for others in solution.

Upon replacing the hydrated metal cation from the interlayers in the pristine layered silicates with organic cations such as an alkylammonium or an alkylphosphonium, the layered silicate attains a hydrophobic/organophillic character and typically results in a larger interlayer spacing. Because the negative charge originates in the silicate layer, the cationic head group of the alkylammonium molecule preferentially resides at the layer surface and the aliphatic tail will radiate away from the surface. The equilibrium layer spacing for an organically modified layered silicate depends both on the cation exchange capacity of the layered silicate, as well as on the chain length of the organic cation. Traditional structural characterization to determine the orientation and arrangement of the alkyl chain involved primarily the use of X-ray diffraction. Depending on the packing density, temperature and chain length, the chains were thought to lie either parallel to the silicate layers forming mono or bilayers, or radiate away from the surface forming mono or bimolecular tilted arrangements (Fig. 3). These idealized structures have been shown to be not-realistic by Vaia and coworkers using FTIR experiments [9]. They have shown that the alkyl chains can vary from liquid-like to solid-like, with the liquid-like structure dominating as the interlayer density or chain length decreases (Fig. 4), or as the temperature increases. This can be understood because of the relatively small energy differences between the trans and gauche conformers; the idealized models described earlier assume all trans conformations. In addition, for the longer chain length surfactants, the surfactants in the layered silicate can show thermal transitions akin to melting or liquid–crystalline to liquid-like transitions upon heating.

2 The Structure of Nanocomposites

2.1 Morphologies of Polymer/Silicate Nanocomposites

The structure of the PLS nanocomposites has traditionally been elucidated using X-ray diffraction (XRD) and transmission electron microscopy (TEM) [1,2]. Due to the periodic arrangement of the silicate layers both in the pristine and the intercalated states, with periodicity of 1–4 nm and the presence of high atomic number species in the layers, the choice of X-ray diffraction in determining the interlayer spacing is obvious. However, in the absence of registry, as in an exfoliated or a delaminated nanocomposite, as well as in a disordered nanocomposite, XRD does not provide definite information regarding the structure of the nanocomposite (Fig. 5). In order to provide quantitative information in XRD 'silent' nanocomposites, TEM has proven to be an extremely useful technique. In addition to a description of the spatial correlations of the layered silicates, TEM also provides a means to discern the homogeneity of the mixing process. A bright field TEM image of an organically modified layered silicate intercalated with polystyrene is shown in Fig. 2a. The periodic alternating dark and light bands represent the layers of silicate and the interlayers respectively, with a spacing of ~ 3 nm between the silicate layers. The TEM also reveals the presence of individual crystallites consisting of several tens of such silicate layers, with bulk polymer filling the space between crystallites. The pristine organically modified

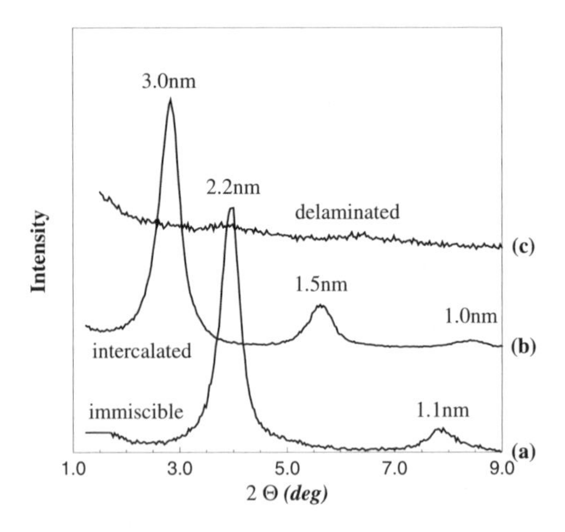

Fig. 5. Typical XRD patterns from polymer/silicate hybrids. (a) XRD obtained from an "immiscible" system (here polyethylene/C18FH), and is identical with the XRD of the neat organo-silicate (C18FH). For intercalated hybrids the d-spacing shifts to a higher value (b) as the gallery expands to accommodate the intercalating polymer (here polystyrene/ C18FH); second and third order reflections – as shown here – are very common and some times intercalated hybrids can have up to 13 order reflections [7], manifesting a remarkable long-range registry. (c) Typical XRD of an exfoliated/delaminated structure or a disordered system (here a siloxane/C18FH delaminated hybrid).

layered silicate exhibits the same microstructure as that observed in Fig. 2a, with the exception of a smaller interlayer spacing. A TEM of an exfoliated hybrid of an organically modified layered silicate dispersed in epoxy is shown in Fig. 2b. Such a hybrid, while being XRD silent, does in fact exhibit some relative layer order with spacing of a few tens of nm between layers. This is attributed to the intrinsic anisotropic dimensions of the layers with the aspect ratios of the individual layers being in the range of 100– 1000. In addition, flexibility of the nanometer thick silicate layers is evident in this micrograph. Similar micrographs have also been obtained by Lan and coworkers [4] and the Toyota group using monomer intercalation followed by polymerization. Micrographs along with small angle X-ray scattering studies of oriented nanocomposites by the Toyota group [19–21] clearly demonstrate the difference observed in short length scale correlations in unaligned exfoliated nanocomposites (as present in Fig. 2b) and the long range correlations that are observed in aligned (using external shear and elongational flows) nanocomposites. These features will be discussed later (see Section 4.2).

Recently, several attempts have been made to understand the underlying structural aspects of the silicate layers as well as the conformations of polymers in layered silicate based nanocomposites, using techniques such as small angle neutron scattering (SANS) and dynamic light scattering. Jinnai and coworkers [23] have studied mixtures of an organically modified vermiculite (modified with n-butylammonium) with poly(vinyl methyl ether) in the presence of n-butylammonium chloride and heavy water, using neutron scattering in order to elucidate the degree of chain interpenetration trapped in lamellar systems. They concluded that the introduction of polymer causes the silicate layers to become more strongly aligned with more regular, but generally decreasing, interlayer spacing in the gel phase. However, the addition of polymer had no effect on the phase transition temperature between the tactoid and gel phases of the layered silicate.

However, the conformation and location of the polymer chains in these mixtures were not unequivocally determined. Carrado et al. [24] have recently used SANS to monitor the change in the structure of the layered silicate (synthetic hectorites) upon hydrothermal crystallization with direct incorporation of poly(vinyl alcohol) (PVA). The results of these experiments established that the PVA appears to coat the small initially formed silicate particles, hindering their further growth. However, upon removal of the polymer no change has been observed in the extended inorganic network. Muzny and coworkers [25] have applied dynamic light scattering to monitor the dispersion of layered silicates in a polymer matrix. Specifically the dispersion of synthetic hectorite clay platelets suitably organically modified in a matrix of polyacrylamide was studied. These studies have clearly shown that a homogeneous ('single layer dispersion') was achievable only when a large excess (equivalent to five times the CEC of the silicate or higher) of the organic cationic surfactant was used.

2.2 Thermodynamics of Nanocomposite Formation

The formation and equilibrium structure of polymer layered silicate nanocomposites, in particular with organically modified layered silicates, has been shown to be a strong function of the nature of the polymer (polar or apolar), the charge carrying capacity of the layered silicate, as well as the chain length and structure of the cationic surfactant. However, both the polymer/silicate compatibility and hybrid equilibrium structure for these nanocomposites are observed to be independent of polymer molecular weight. The experimental results have been summarized by Vaia et al. and a lattice based mean field theory has been developed to explain these results [26].

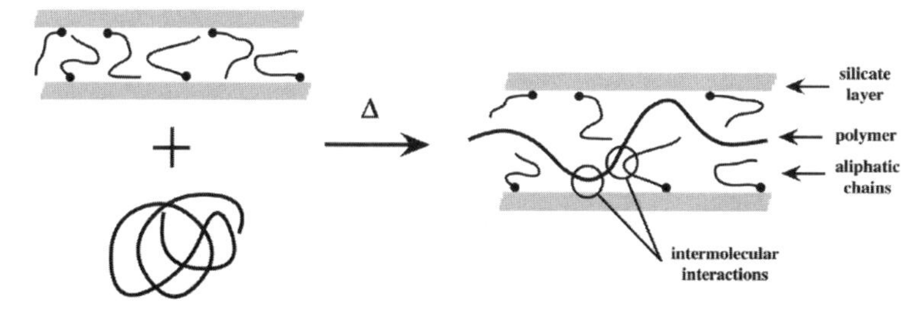

Schematic 2. Schematic representation of the system components before and after the intercalation takes place. The changes in entropy and free energy as a function of the change in gallery height are shown in Fig. 6 and 7.

In general, an interplay of entropic and enthalpic factors determines the outcome of polymer intercalation. Confinement of the polymer inside the interlayers results in a decrease in the overall entropy of the polymer chains. However, the entropic penalty of polymer confinement may be compensated by the increased conformational freedom of the tethered surfactant chains in a less confined environment, as the layers separate (Fig. 6). Since for small increases in gallery height the total entropy change is small, modest changes in the system's total enthalpy will determine if intercalation is thermodynamically possible. Complete layer separation, though, depends on the establishment of very favorable polymer-surface interactions to overcome the penalty of polymer confinement (Fig. 7). The enthalpy of mixing can broadly be classified into two components – apolar, which is generally unfavorable, and polar which originates from the Lewis acid/Lewis base character of the layered silicates, and which could be rendered favorable. A favorable enthalpy change is accentuated by maximizing the magnitude and number of favorable polymer-surface interactions while minimizing the magnitude and number of unfavorable apolar interactions between the polymer and the functionalizing aliphatic chains.

Although simple in comparison to contemporary thermodynamic descriptions of polymer phenomena, the greatest advantage of the current model is the ability to analytically determine the effect of various aspects of the polymer and OLS on hybrid formation. The variation of the free energy of mixing on the gallery spacing and its dependence on enthalpic and entropic factors, based on this model, suggest three possible equilibrium states – immiscible, intercalated and exfoliated. The model has been successful in addressing some of the fundamen-

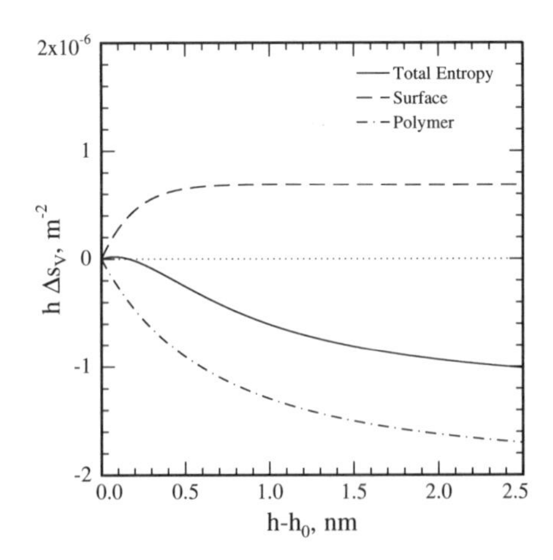

Fig. 6. The change of entropy per area versus the change in gallery height, for the polymer and the surfactant (octadecylammonium) functionalized surface based on the thermodynamic model presented in [26].

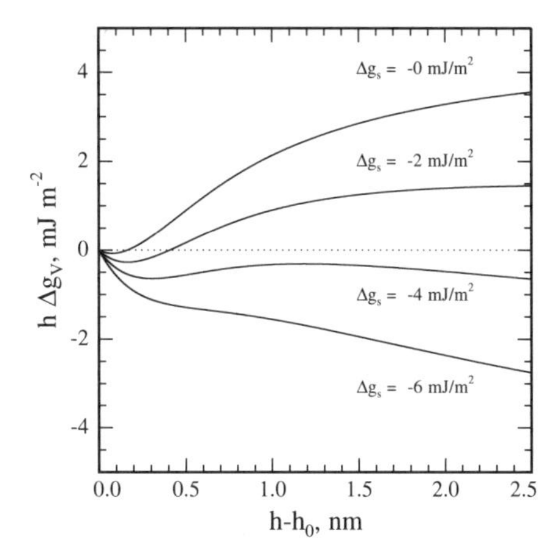

Fig. 7. The change of free energy per area versus the change in gallery height based on the thermodynamic model presented in [26], for various surface-polymer affinities: Δg_s=0, –2, –4 and –6 mJ/m^2. Both figures adopted from [26]

tal and qualitative thermodynamic issues associated with hybrid formation. However, assumptions such as the separation of configurational terms and intermolecular interactions and the further separation of the entropic behavior of the constituents, somewhat limit the usefulness of the model. Additionally, this model is not applicable to situations where the OLS's interlayer is not completely occupied by tethered-chain segments, which is the case for many silicates with low charge densities or modified by short aliphatic chains.

Some of the limitations and gross approximations of the current mean-field, lattice-based model, such as the decoupling of the tethered chain and polymer conformations in the interlayer, may be addressed by developing more sophisticated models of hybrid formation based on theories of Scheutjens and Fleer [27], and Szleifer [28]. Additionally, Monte Carlo simulations of hybrid formation or intercalation of small-molecules in layered materials could lead to additional insights into the mechanisms and important factors associated with intercalation [14]. One very well studied problem is the hydration (water intercalation) in 2:1 silicates which was approached both by Monte Carlo [29, a-c] and molecular dynamic simulations [29, d]. In Fig. 8 the water uptake as a function of gallery height as provided by Grand Canonical MC simulations is shown. From the simulation the thermodynamic properties of the systems – such as disjoining pressure and free energy – can be obtained, and thus the stable d-spacings can be located. The extension of such simulation schemes to the study of polymer/organo-silicate systems is currently underway, and will obviously provide a more detailed picture of the thermodynamics of these systems [29e].

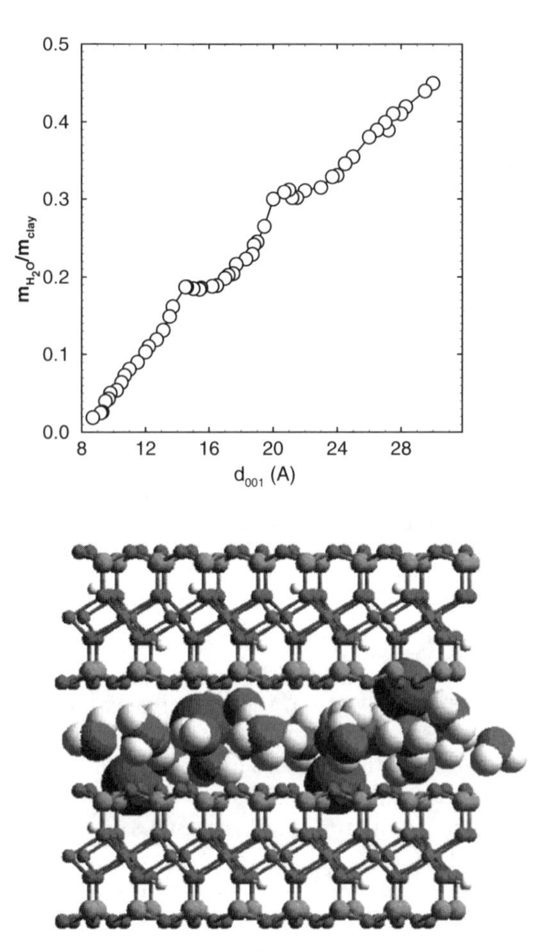

Fig. 8. Simulated water uptake from Monte Carlo simulations, adopted from [29 c]. A typical configuration of a hydrated Na^+ montmorillonite in the d_{001}=12Å stable hydrated state revealed by Monte Carlo simulations [29].

3 Dynamics of Nanoscopically Confined Polymers

3.1 Kinetics of Polymer Melt Intercalation

As outlined in the introduction, polymer melts can intercalate layered inorganic compounds unassisted by shear or solvents. This is a rather surprising result as it implies that polymer chains can undergo large center of mass displacement in almost two dimensional interstices as the distances between the confining surfaces are substantially smaller than the unperturbed radius of gyration of the polymer and are comparable to the monomer size. The reduction in free energy by the intercalate formation (Section 2.2) and the concentration gradient during the intercalation process, give rise to an 'enthalpic force' which 'drives' the polymer coils into the interlayer galleries. On the other hand, the conformational energy cost of stretching the chains, in addition to the topographical constrains and the adsorption on the surfaces are expected to impose severe limitations on diffusion of chains diffusing in a pseudo-two dimensional slit.

Vaia et al. [12] have observed that the kinetics of intercalation even under quiescent conditions (absence of external shear) are quite rapid. Using in-situ XRD (which monitors the angular shift and integrated intensity of the silicate reflections, Fig. 9) they studied the intercalation kinetics of model polymers (monodisperse polystyrene) in organically modified fluorohectorite.

Figure 9 shows a typical temporal series XRD patterns, for a polystyrene M_w= 30,000 (PS30)/octadecyl-ammonium modified fluorohectorite (C18FH) mixture annealed in-situ at 160 °C in vacuum. Details regarding the data collection and analysis are presented in reference [12]. The width of the original unintercalated peak and the final intercalated peak appear to be similar, suggesting that the polystyrene melt intercalation does not drastically alter the coherence length or disrupt the layer structure of the silicate crystallites.

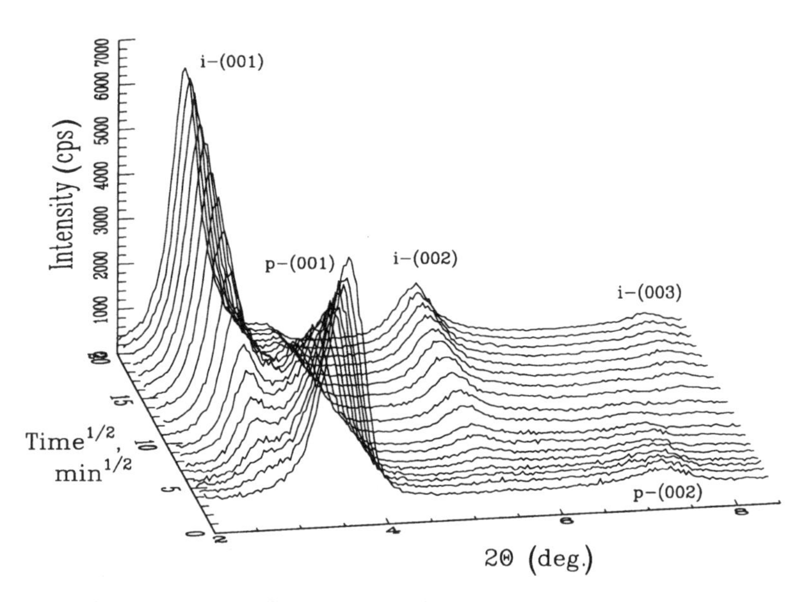

Fig. 9. Typical temporal series of XRD patterns, for a polystyrene (M_w=30,000)/C18FH mixture annealed in-situ at 160 °C in vacuum. Initially (time=0), the basal reflection of the organo-silicate were observed at 2θ=4.15° (d=2.13nm). During the annealing this reflection is progressively reduced in intensity, whereas a new basal reflection develops at 2θ=2.82° (d=3.13nm) that corresponds to the intercalated hybrid.

Moreover the evolution of the XRD during annealing can be modeled to determine the apparent diffusivity, D, of the polymers within the silicate gallery. Namely, the ratio of the amount of intercalated polymer at time t, $Q(t)$, to that at equilibrium $Q(\infty)$ is [30]:

$$\frac{Q(t)}{Q(\infty)} = 1 - \sum_{m=1}^{\infty} \frac{4}{a_m^2} \exp\left(-\frac{D}{\bar{a}^2} a_m t \right) \tag{1.1}$$

where, $D/\bar{a}^2$ is the effective diffusional coefficient, $\bar{a}$ is the mean size of the impermeable surface and a_m are the roots of the zeroth order Bessel function ($J_0(a)=0$). The fraction of intercalated polymer is given by the corresponding fraction of the XRD integrated intensities χ (Fig. 10) and the only adjustable parameter is $D/\bar{a}^2$; the mean silicate size $\bar{a}$ can be measured directly by TEM [12] and in that case was approximately 5μm.

The apparent diffusivity for the intercalation of PS30 in C18FH is the same order of magnitude (10^{-11} cm²/s at 170 °C) [12] as the self diffusion coefficient of polystyrene determined at comparable temperatures and molecular weights [31]. Furthermore, the activation energy of melt intercalation is 166 ± 12 kJ/mol which is comparable to the activation energy measured for self diffusion of polystyrene (167 kJ/mol [31]). Although these first results seem to suggest that the process of intercalation is dictated by the transport of polymer to the silicate agglomerates and not by any process by which the polymer is moved inside the galleries of the silicates [12], more recent investigations – where the effect of the polymer-surface interactions on the intercalation kinetics were explored [32] – have proved that

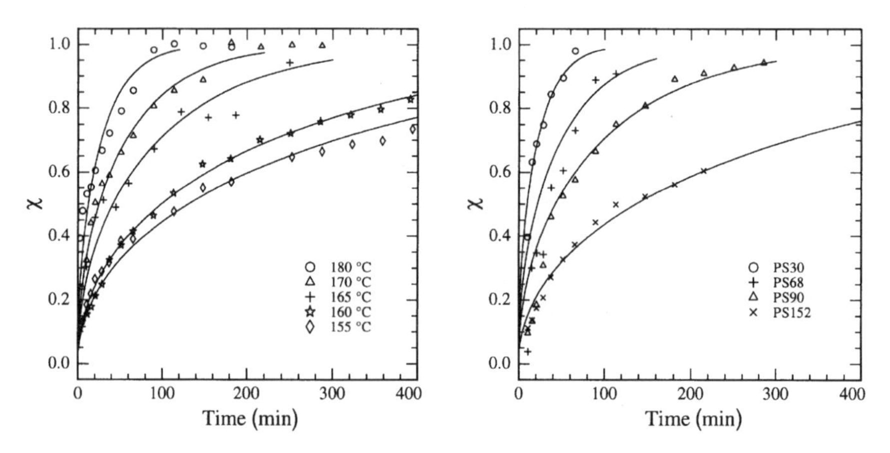

Fig. 10. The fraction of polystyrene intercalated C18FH for various annealing temperatures (PS 30, M_w=30,000) (left) and for various polymer molecular weights at 180 °C (right). The lines are the best fits to the data using equation 1.1. The molecular weights of polystyrene used were M_w=30,000 (PS30), M_w=68,000 (PS68) M_w=90,000 (PS90) and M_w =152,000 (PS152); adopted from reference [12].

this is not always the case. In these more recent studies, it was found that the effective diffusion coefficient depends markedly on the surfactant used, for the same polymer and annealing temperature. Because the surfactant can only affect the polymer motion inside the galleries, one concludes that this motion of the intercalated polymer is the process which dictates the intercalation kinetics [33]. Furthermore, for some systems the intercalating polymers were found to possess a mobility that was much faster than the self-diffusion coefficient of the corresponding polymer in the bulk [31], or in a thin film [34]. This is not unexpected since intercalation is a process where polymers are moving down a concentration gradient, whereas in the other two cases the polymer motion is entropic in origin [35].

3.2
Structural Evolution During Intercalation

The microstructural evolution during the polymer melt intercalation process (leading to well ordered intercalated compounds, disordered intercalated composites or exfoliated nanocomposites) has been studied recently using a combination of XRD, conventional TEM, and high resolution TEM. For well ordered intercalates, as shown in Fig. 2a, the microstructure appears to be very similar to that of the unintercalated layered silicates. The individual crystallites of the layered silicate are tightly packed and oriented along the major axis of the primary particle, with slits between crystallites of comparable dimensions to the gallery height between the silicate layers within a crystallite (Fig. 11). Micrographs ob-

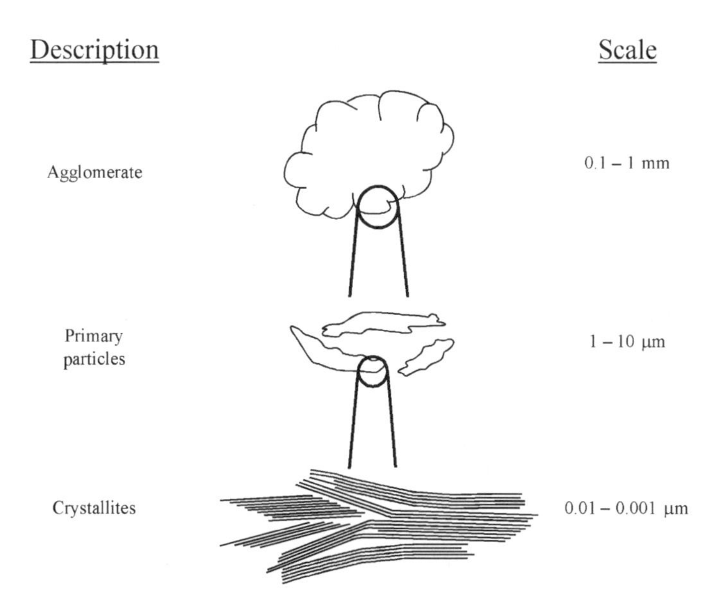

Fig. 11. Hierarchy of structures in a silicate, from reference [12].

tained on partially intercalated nanocomposites revealed the presence of unintercalated crystallites towards the interior of the primary particle (away from the primary particle – bulk polymer interface, where the intercalated crystallites are found), with no crystallites containing both intercalated and unintercalated layers. Based on these observations, it was concluded that polymer intercalation occurs as a front which penetrates the primary silicate particle from the exterior.

In the case of disordered intercalated and exfoliated nanocomposites (examples being polystyrene with a dodecyl-ammonium modified fluorohectorite (C12FH) and poly(3-bromostyrene) (PS3Br) with a C12FH respectively), XRD does not produce well defined basal reflections, and in fact, produces a featureless pattern for the exfoliated nanocomposite. TEM of the PS/C12FH nanocomposite revealed the microstructure of the primary silicate particle to be heterogeneous, with regions of coherent order more prevalent toward the interior than near the bulk polymer – primary particle interface. Layer spacings were found to range from 2.1 to 6.0 nm, with a mean layer spacing of ~ 2.8 nm, comparable to that obtained from the diffuse peak observed in the XRD. In the case of the PS3Br/C12FH nanocomposites, the micrographs indicated that while individual layers were found dispersed in the matrix of the polymer, due to the extremely large aspect ratios of the layers, preferential alignment of the layers even at large separations (> 10nm) was observed. Extensive polymer penetration of the interlayer galleries was observed along with enhanced inter-crystallite gaps within the primary particle. While the layers appear well separated in both the interior and boundaries of the primary particle, the relative order between layers is much better preserved in the interior portions, suggesting that the delamination process occurs by different mechanisms near the bulk polymer – primary particle boundaries, as opposed to the interior of the interlayers.

3.3 Local Dynamics

3.3.1 *On the Relaxation of nm-Thick Polymer Films Between Walls*

Intercalated compounds offer a unique avenue for studying the static and dynamic properties of small molecules and macromolecules in a confined environment. More specifically, layered nanocomposites are ideal model systems to study small molecule and polymer dynamics in restrictive environments with conventional analytical techniques, such as thermal analysis, NMR, dielectric spectroscopy and inelastic neutron scattering. Understanding the changes in the dynamics due to this extreme confinement (layer spacing $\ll R_g$ and comparable to the statistical segment length of the polymer) would provide complementary information to those obtained from traditional Surface-Force Apparatus (SFA) measurements on confined polymers (confinement distances comparable to R_g) [36].

3.3.2
NMR Measurements

The local dynamics of segments of polymer chains can be very conveniently probed using solid-state NMR. The local dynamics of intercalated chains of polyethylene oxide (PEO) in a pristine layered silicate were studied using variable temperature solid state NMR [13]. Figure 13 compares the ^{2}H line shapes for bulk d-PEO and intercalated d-PEO (in Li-fluorohectorite), wherein the d-PEO chains (M_w = 180,000, M_w / M_n =1.2, Polysciences) are confined to a 0.8 nm gap between the silicate layers. For the bulk polymer, at the lowest temperatures where the local segmental motion of the polymer is quiescent, the NMR signal possesses well-formed singularities and sharp step-edges characteristic of the solid/glassy dynamics. With increasing temperature the 'sharp' spectral features gradually broaden with a build-up of the intensity at the center of the powder pattern, indicative of the onset of polymer dynamics. The central peak results from increased segmental motion which causes temporal averaging of the signal. At the highest temperatures i.e., in the melt state of the polymer, where the dynamics of the polymer is expected to be fairly rapid, the anisotropic pattern disappears and a single line spectrum is obtained. In contrast, the intercalated polymer chains even at the lowest experimental temperature of 220 K exhibit small amplitude dynamics as evidenced by the loss of spectral definition. With increasing temperature the powder pattern shows some narrowing and a progressive increase in intensity at the center, consistent with progressively increasing rates of C-D bond re-orientations. However, unlike the bulk sample, the intercalated sample at 400 K (well above any bulk dynamic transition) displays significant spectral features other than the central peak, indicative of significant preferential orientation of the trapped polymer segments with respect to the silicate structure confining the polymer. The temperature dependence of the line shapes indicates that the intercalated chains have more 'freedom' to sample a

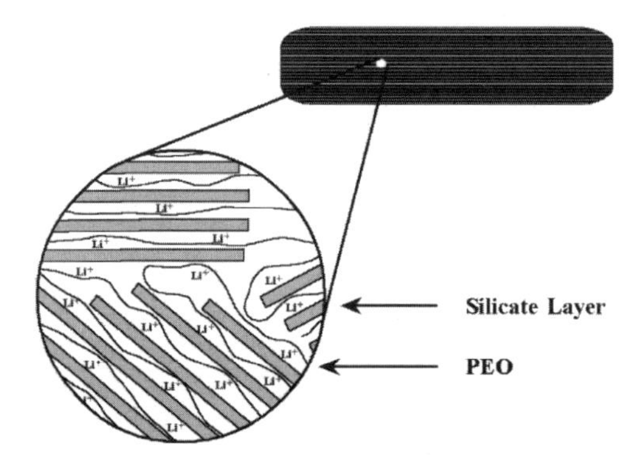

Fig. 12. Schematic representation of the PEO/Li$^+$ fluorohectorite intercalated hybrid.

distribution of local chain configurations, resulting in increased signal averaging. However, at the highest temperatures, the intercalated d-PEO still exhibits a broad base structure reminiscent of the powder pattern whereas the bulk d-PEO shows complete motional narrowing of the signal. This indicates that even though the local segmental motion of the intercalated d-PEO appears more dynamic at lower temperatures, the silicate layers still restrict motion such that some local configurations of the chain are not accessible and thus complete signal averaging is not possible.

Complementary spin-lattice relaxation measurements corroborate the observations made using the ^{2}H line-shape measurements. Based on these measurements the low temperature relaxation times are dramatically shorter in the intercalated sample as compared to the bulk, indicating enhanced polymer re-orientation dynamics in the intercalated samples. Furthermore, the temperature dependence of the relaxation time in the bulk and intercalated sample show dramatic differences. While the relaxation time for the intercalated sample passes smoothly from low to high temperatures, the bulk sample shows a break between the crystalline state and melt state, with the melt state relaxation times at least one order of magnitude faster than those observed in the intercalated sample at the same temperature.

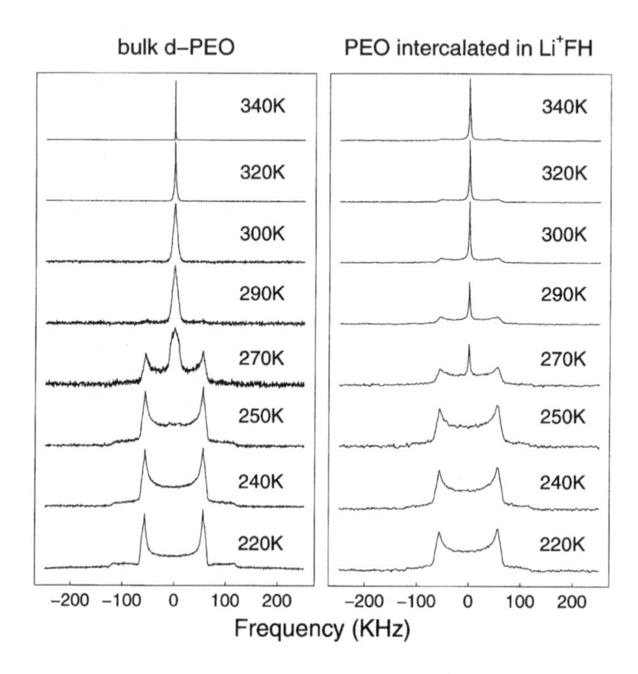

Fig. 13. Temperature dependence of ^{2}H NMR spectra, for bulk PEO and for PEO intercalated in Li^+ fluorohectorite. Adopted from [13].

The enhanced local chain dynamics of the intercalated d-PEO reflect the absence of chain entanglements and the presence of excess free volume associated with the packing constraints of intercalated chains. On average, the constraints to local chain dynamics on the length scale of a few monomers are less than those of the complementary chain in the bulk. However, constraints to local chain motion are not wholly absent, as indicated by the residual powder pattern at elevated temperatures. Based on geometrical considerations of the extremely narrow 0.8 nm interlayer, a large fraction of the intercalated PEO units are in direct contact with the surface of the silicate. Topologically, the surfaces of the opposing silicate layers will restrict the motions of the d-PEO chains. In addition, interactions of the relatively polar PEO segments with the polar silicate surfaces as well as with the interlayer cations will bias the local motion of the d-PEO chain to certain energetically favorable orientations. Because of the extended pseudo-2D nature of the chain and the absence of chain entanglements, the local dynamics of the intercalated chains may approach that of a single chain with no excluded volume effects.

Similar results of strong hindering of polymer mobility by surfaces have also been found by Brik et al. [37]. They studied the motional heterogeneity of PEO grafted between silica particles using electron paramagnetic resonance as well as NMR. They found that the organic phase exhibits marked variations in dynamic behavior with the liquid-like mobility of the polymer strongly hindered at the silica grafting nodes. Computer simulations probing the dynamics of confined oligomers and short polymers have indicated a strong influence of the confining surface on the local mobility of the molecules. Two modes are observed in these simulations – the presence of a fast mode corresponding to the 'tail' segments far away from the wall and a slow mode (whose slowing down depends on the surface interaction) corresponding to the 'train' segments close to the confining surface, we further discuss these simulation findings in the next section.

3.3.3
Computer Simulations

Computer simulations of nanoscopic confined fluids have revealed many details of the dynamics under confinement. The nature of the confined fluids – especially in the immediate vicinity of attractive surface – has been shown to be strongly altered by the confining surfaces, and this is manifested by a behavior dramatically different from the bulk fluids in the local relaxation [38a], the mobility [38c] and rheological properties [39] of molecules near adsorbing surfaces. For monomeric systems many computer simulation studies [40] provide a clear enough picture for the dynamics of confined films of small spherical molecules. On the other hand, for confined oligomers and polymers less has been done, especially towards the understanding of the dynamics of nanoscopic films [41].

Some of the pioneering work has focused on the behavior of abstract Lennard-Jones oligomers in nanoscopic confinements and consists of studies of the relaxation times of different 'modes' [38a,b,d], of the transport properties like the self-diffusion coefficients and the mean square displacements [38c,d], and of the rheological properties and the viscosity [39]. Although the oligomers used in these simulations are too short to exhibit genuine polymeric behavior they are characterized by very high flexibility. This renders the Rouse modes of short oligomers to be well defined near confining walls, when they are far apart (~10 monomer diameters) to create two well separated and non-interacting fluid-wall interfaces [38a], as well as in thiner films (3–6 monomer diameters) [38d] – geometries which are comparable with the confining environment of the intercalated polymer. Moreover, the length of this oligomer is comparable to the average train size of high molecular weight polymers [42] and thus the dynamics of the adsorbed oligomers will mirror the dynamics of the adsorbed parts (trains) of much longer polymer coils in confinement [38a]. In these systems the end-to-end relaxation (first Rouse mode) is studied through the time correlation function of the end-to-end vector and its relaxation time is the rotational relaxation time of the chains. The time correlation function of the end-to-end vector is defined [41–44]:

$$
\begin{aligned}
<\boldsymbol{P}(t)\cdot\boldsymbol{P}(0)> &= <(\boldsymbol{R}_N(t)-\boldsymbol{R}_0(t))\cdot(\boldsymbol{R}_N(0)-\boldsymbol{R}_0(0))> \\
&\sim \exp\left(-\frac{t}{\tau_r}\right), \qquad t\geq\tau_r
\end{aligned}
\tag{1.2}
$$

where $\boldsymbol{P}$ is the end-to-end vector, N the number of beads in the chain, $\boldsymbol{R}_i(t)$ is the position of the i bead of the chain at time t and τ_r is the first Rouse mode relaxation time.

Using Eq. 1.2 the time correlation function of the end-to-end vector was calculated for the hexamer systems in slits of the same thickness as the interlayer gallery of PEO/fluorohectorite intercalated hybrids [38d: Section 2.2], for the whole film (Fig. 14a) and for the adsorbed chains (Fig. 14b). The film as a whole exhibits a multimodal relaxation, including **fast** relaxing and **much slower** relaxing species. The fast, bulk-like, relaxation is due to the free chains and the slower part is due to the superposition of the slower modes of the adsorbed chains and the relaxation times involved for each mode depend on the number of adsorbed segments (Fig. 15).

The most crucial parameter that determines quantitatively the relaxation of the confined chains is proved to be the wall affinity (ε_w). Physically, ε_w is the excess attractive energy per segment of the wall–polymer interaction, compared to the polymer–polymer interaction on a segmental basis. For neutral walls (ε_w=0) the term 'adsorbed chain' is rather ambiguous, since the interaction between the solid particles and the fluid segments has no attractive part, and has only the meaning of chains with segments located inside the first fluid layer. In this sense, the chains touching the neutral walls (ε_w=0) are in an environment analogous to the one felt by free chains in the ε_w=2 or ε_w=3 slits or by the free pentamers inside the second layer of a very wide pore. The effect of the neutral wall is a weak slowdown ($\tau_{conf.}\sim\tau_{bulk}$) of the chain relaxations and the segment mobilities [38]. The main simulation findings [38] are as outlined below.

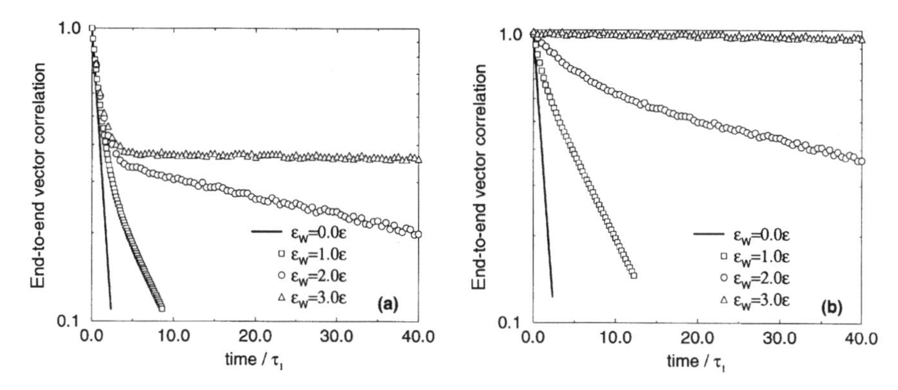

Fig. 14. End-to-end vector time correlation functions for various wall-polymer affinities (**a**) for the whole film (approximately same width as the intercalated PEO interlayer gap), (**b**) for the adsorbed chains independently of the number of contacts. Adopted from reference [38d].

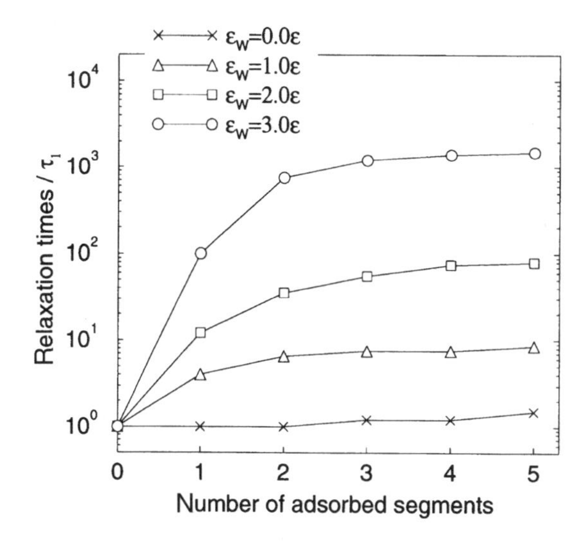

Fig. 15. Relaxation times of the first Rouse mode of confined pentamers (h=10) as a function of the number of contacts for all the simulated wall affinities, from Bitsanis I, Pan C (1993) J Chem Phys 99: 5520. The data indicate clearly a dramatic increase of the relaxation time inside the solid-oligomer interface with increasing ε_W. The origin of these "glassy" dynamics is attributed to the slowdown of segmental motions inside the adsorbed layer.

The relaxation times of the end-to-end vector time correlation function when calculated for the whole film (through the end-to-end vector time correlation function) exhibits a multimodal relaxation. The fast relaxation corresponds to the free chains located in the middle of the film and the slower modes to the relaxation of the adsorbed chains. This is concluded both by the relaxation times as calculated for those classes of chains separately, as well as by the ratio of the correlation function amplitudes of the fast versus the slow modes, which is very close to the ratio of the number of free versus adsorbed chains.

For the free chains the relaxation time is the same and very close to the bulk value. This happens even for the chains located inside the second fluid layer of wide gaps [38a] and the chains in much narrower confinements, where the space available for the free chains is comparable in size to the statistical segment.

The relaxation times of the adsorbed chains are higher compared to the bulk and vary significantly with wall affinity. The extent of the increase ranges from three orders of magnitude for the strongly physisorbing surfaces (ε_w=3) to less than ten times for weakly physisorbing surfaces (ε_w=1) and to almost a negligible degree (~ 1.3) for the neutral walls (Fig. 15).

The relaxation time of the end-to-end vector correlation for the adsorbed chains depends on the number of contacts. Chains with one or two contacts have most of their segments free and thus due to their bulk like dynamics the end-to-end vector can rapidly relax. On the other hand, for the chains with most of their segments adsorbed this process becomes very slow as the segment dynamics are very sluggish inside the solid oligomer interface (Fig. 15). For strong wall attractions (ε_w=2 or 3) the chains with more than three contacts relax with almost the same time constant. This insensitivity shows that the slowdown of the dynamics is caused by the densification inside the first layer rather than the magnitude of the surface-fluid interactions [38a,d].

For more realistic molecular structures, that involve bond angles and tetrahedral potentials such as simulations of confined alkane oligomers [43], similar behavior has been observed. There is a co-existence of fast and slow relaxing species, which were identified as the free and adsorbed molecules respectively, and for strongly physisorbing surfaces the first Rouse mode, as well as trans-to-gauche transition dynamics, slowed down by three orders of magnitude in comparison with the equivalent bulk. Moreover, the confined alkanes create liquid crystalline domains with smectic ordering and exhibit quite high diffusivities parallel to the confining surface, despite the huge slow down of their relaxation times [43a].

3.3.4
Cooperative Motion – TSC and DSC Measurements

The presence of cooperative motion of chain segments present in intercalated polymer chains can be examined using various analytical techniques such as Differential Scanning Calorimetry (DSC), thermally stimulated current (TSC) and dielectric spectroscopy. DSC measurements on an intercalated PEO, (M_w= 100,000)/montmorillonite hybrid (20 wt. % polymer), indicated the **absence of**

any thermal transitions corresponding to the glass or the melting transition of PEO ($T_g \sim -55$ °C and $T_m \sim 65$ °C). DSC studies of polystyrene intercalated in an organically modified layered silicate (Fig. 16) also indicated that the intercalated hybrid does not show a thermal transition corresponding to the glass transition of polystyrene over a temperature range of 50 – 180 °C [45]. Similar absence of a glass transition for polystyrene confined to zeolite cavities have also been observed by Frisch and coworkers [46].

Keddie and coworkers [47] have investigated the glass transition temperatures as a function of film thickness for thin films of polystyrene and poly(methyl methacrylate) on different substrates. For polystyrene on Si(111), they observed a decrease in the glass transition temperature (T_g) for films thinner than 400 Å independent of molecular weight. This decrease was attributed to the presence of a liquid-like layer at the free surface of the film. Similar results were also reported by Keddie et al. for PMMA on gold surfaces, wherein the surface polymer interaction is expected to be again weak. However, in the presence of a strongly interacting surface, the glass transition was found to dramatically increase with decreasing film thickness and was attributed to the presence of a layer near the surface wherein the mobility is greatly reduced [48] in agreement with computer simulation studies [49].

DSC measurements possess a relatively low sensitivity to weak glass transitions. This is particularly exaggerated in the case of the intercalated compounds

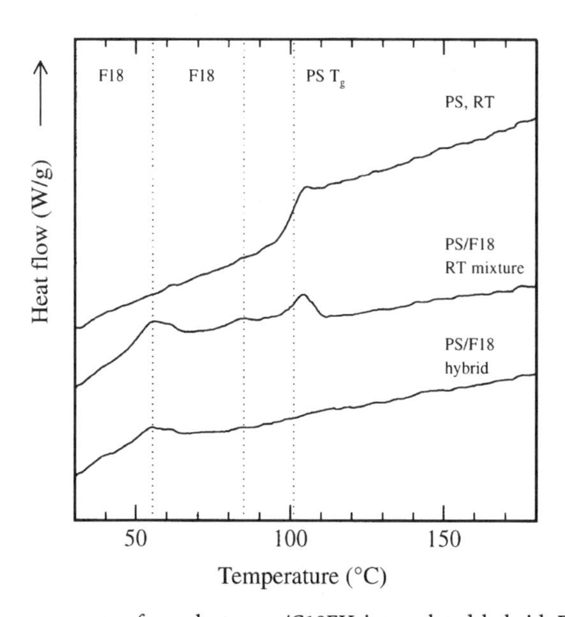

Fig. 16. DSC measurements of a polystyrene/C18FH intercalated hybrid. For comparison DSC traces for the neat polymer and for a physical mixture of the polymer and the organically modified silicate are also shown. There is an absence of any thermal transition for the intercalated polymer around the bulk T_g [45].

where there is a low volume fraction of the polymer. TSC techniques, on the other hand, have enhanced sensitivity to cooperative relaxations such as glass transitions [50]. To further evaluate the glass transition region of the intercalated PEO TSC was applied in two modes – 'global' and 'thermal sampling'. Due to the small dipole associated with polystyrene, similar measurements could not be performed with the intercalated polystyrene system. The 'global' TSC spectra have some similarity to a.c. dielectric loss spectra. As with the DSC measurements, no clear peak was observable in the global TSC spectra of the intercalated PEO in the temperature range of the bulk glass transition for the intercalated PEO. Thus, cooperative relaxations of the intercalated polymer are, at best, weak.

TSC measurements with 'thermal sampling' are very sensitive to cooperative relaxations even from a minor fraction of the overall relaxing species. Glass transition regions can be resolved using this method even if the relaxations are extremely broad or weak. The apparent activation energies, E_a, of relaxations contributing to the depolarization current for 100% PEO, the 20% PEO melt intercalate, and the 0% PEO pressed montmorillonite control pellet are shown in Fig. 17. The analysis consists of assigning cooperative glass transition like motions to the regions of departure of the values of E_a from the $\Delta S^{\ddagger}=0$ prediction, where $\Delta S^{\ddagger}$ is the activated state entropy. For non-cooperative 'secondary' transitions it has been observed that $\Delta S^{\ddagger}=0$. A single sharp peak, typical of amorphous and semicrystalline polymers, is seen at the T_g for the 100% PEO sample. However, for the intercalated PEO nanocomposite a distinct peak in E_a is not observed. Instead, a broad transition starting at about the nominal PEO T_g and ranging up to about 60 °C is seen.

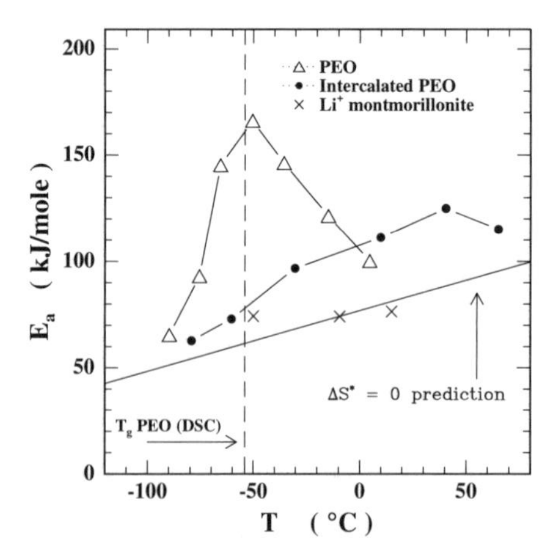

Fig. 17. Temperature dependence of the apparent activation energies E_a as determined by TSC-TS. Bulk PEO, the neat montmorillonite silicate and an intercalated 20wt. % PEO/montmorillonite hybrid are shown. The departure of E_a from the $\Delta S^{\ddagger}=0$ prediction line indicates the glass transition region. Figure adopted from [14].

Since no melting transition was observed in DSC, the fraction of 'amorphous' PEO chains in the intercalate is similar to the total PEO content of the hybrids (in this case ~20%). Therefore, the relaxation strength should be at a level where we can easily detect the 'cooperative' motions. However, the cooperative relaxations detected in the glass transition region (Fig. 17) are weaker than any measured before using the TSC-TS technique. Of the several systems studied with the TSC-TS technique, never before has such a broad transition region with a low degree of cooperatively been observed [50,51]. Based on the data in Fig. 17, it was concluded that the motions of the intercalated PEO chains are inherently non-cooperative relative to the cooperative T_g motions in less unconstrained environments such as the amorphous portion of the bulk polymer or the amorphous chains confined between crystal lamellae in pure semi-crystalline PEO. For semi crystalline PEO, the chains are confined in 'amorphous' gaps a few nanometers wide whereas the intercalated PEO occupies a gap of less than 1 nm!

Taken in context with previous investigations of polymer relaxations in confined environments the results from these polymer intercalates appear to indicate that cooperative motion precipitously decreases as polymers are confined to extremely narrow slits less than a few nanometers.

4 The Rheology of Polymer/Silicate Nanocomposites

Rheology of various polymer layered-silicate nanocomposites – intercalated, exfoliated and end-tethered exfoliated (prepared by in-situ polymerization from reactive groups tethered to the silicate surface), have been performed in a conventional melt-state rheometer in both oscillatory and steady shear modes. These experimental studies have provided insight into the relaxation of polymer chains when confined by the layers of inorganic silicates, as well as the role of shear in orienting the layered nanocomposites.

4.1 Linear Viscoelasticity

The steady-shear rheological behavior of a series of intercalated poly(dimethyl$_{0.95}$-diphenyl$_{0.05}$ siloxane)-layered silicate (dimethyl ditallow montmorillonite) nanocomposites (with varying silicate loadings) are shown in Fig. 18 [52]. The viscosity of the nanocomposites is enhanced considerably at low shear rates, and increases monotonically with increasing silicate loading (at a fixed shear rate). Furthermore, the intercalated nanocomposites display a shear-thinning behavior at low shear-rates, where the pure polymer displays a shear rate independent viscosity. At high-shear rates where the polymer displays shear-thinning behavior, the nanocomposites also display shear-thinning with the values of the viscosity (at least at the lower loadings of silicate) being comparable to that of the polymer itself. The same trends are also observed in linear dynamic oscillatory shear measurements, where the storage and loss moduli (G' and G''

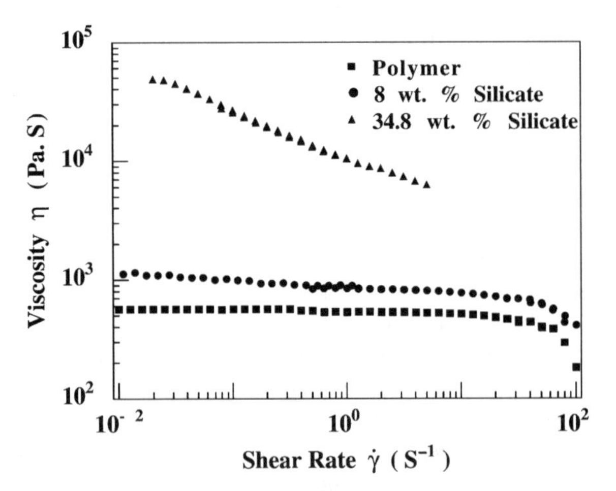

Fig. 18. The steady-shear rheological behavior for a series of intercalated nanocomposites of poly(dimethyl$_{0.95}$-diphenyl$_{0.05}$siloxane) with layered silicate (dimethyl ditallow montmorillonite) at 25 °C. The silicate loading is varied and are noted in the legend. From Ref. [5].

respectively) exhibit a plateau at low-frequencies. Furthermore, consistent with the steady shear measurements, a monotonic increase in both G′ and G″ at all frequencies is observed with increasing silicate content. It thus appears that the intimate contact between the polymer and the inorganic sheets leads to an alteration of the relaxation dynamics of the polymer, leading to the low-frequency plateau in the moduli and the low-shear rate non-Newtonian viscosity behavior.

In contrast, for a series of polydimethylsiloxane based **delaminated** hybrids (dimethyl ditallow montmorillonite), the steady-shear viscosity shows an increase with respect to that of the pure-polymer at low shear rates but still obeys Newtonian type behavior, even at the highest silicate loadings examined (Fig. 19). The increase in viscosity at the silicate loadings examined was roughly linear at all shear rates. Similar effects are also observed in linear dynamic oscillatory shear measurements, where the storage and loss moduli for the delaminated hybrids display similar frequency dependence as the pure polymer, with a monotonic increase in the magnitude of the moduli with increasing silicate loading [52]. It thus appears that in these delaminated hybrids, the relaxation of the polymer chains is not altered (at least within the sensitivity of the measurements) by the presence of the silicate layers. This rheological response is similar to the relaxation behavior of typical filled-polymer systems, wherein the relaxation dynamics of the polymer chains are not significantly altered by the presence of non-interacting filler particles [53].

In addition, two end-tethered delaminated hybrid systems prepared by in-situ polymerization – (a) Poly(ε-caprolactone)-montmorillonite (PCLC) and (b) nylon-6-montmorillonite (NCH) – wherein the polymer chains are end-tethered to the silicate surface via cationic surfactants [54] (Fig. 20), were also studied.

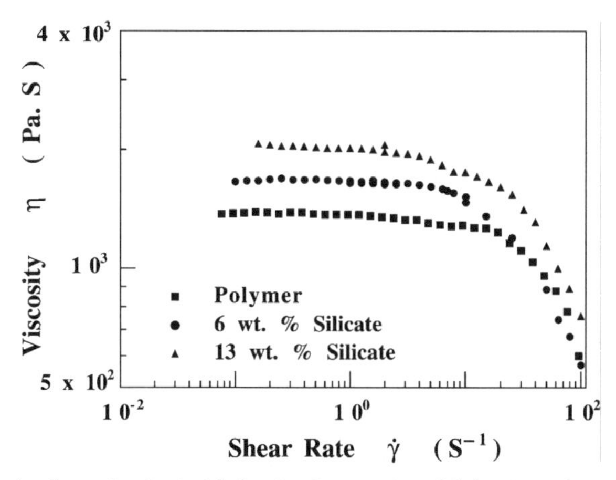

Fig. 19. The steady-shear rheological behavior for a series of delaminated nanocomposites of polydimethylsiloxane with dimethyl ditallow montmorillonite at 25 °C. The steady-shear viscosity shows an increase with respect to that of the pure-polymer at low shear rates but still obeys Newtonian type behavior, even at the highest silicate loadings examined. From Ref. [5].

The linear viscoselastic properties of the pristine nanocomposites, the effect of large amplitude oscillatory shear on the orientation of the nanocomposites and the viscoelastic properties of these oriented nanocomposites were examined.

The changing molecular weight of the polymer chains, a result of the synthetic scheme used in the preparation of the nanocomposites, poses some problems in the interpretation of the variation of the moduli with increasing silicate content. The average molecular weight decreased sharply at low silicate loadings and remained roughly constant for nanocomposites with more than 2 wt.% silicate. The linear viscoelastic response as measured by dynamic oscillatory measurements and quantified by G' and G'' for the five poly(ε-caprolactone) based nanocomposites examined is shown in Figs 21 and 22 respectively. Data were acquired at several temperatures using the lowest possible strain amplitudes (typically in the range of 0.1–5.0%) and shifted using the time-temperature superposition principle to form the master-curves presented in Figs 21 and 22. For all the nanocomposites examined, the data was somewhat restricted due to the alignment of the silicate layers by the application of large amplitude oscillatory shear (particularly required at low frequencies, to obtain force signals larger than the low limit of the transducer). This was significantly restrictive for the high silicate loading composites (PCLC5 and PCLC10) where at high temperatures and low frequencies, alignment of the layers would start to occur (as measured by a change in the rheological response consistent with alignment of the layers) at strain amplitudes as low as 10%. Only data verified to be in the linear regime are shown in Figs 21 and 22.

The storage moduli (Fig. 21) for the nanocomposites show a monotonic increase at all frequencies with increasing silicate content, with the exception of PCLC2, where at the highest frequencies, it has a slightly lower value than PCLC1. The loss moduli (Fig. 22), on the other hand, show a somewhat non-mo-

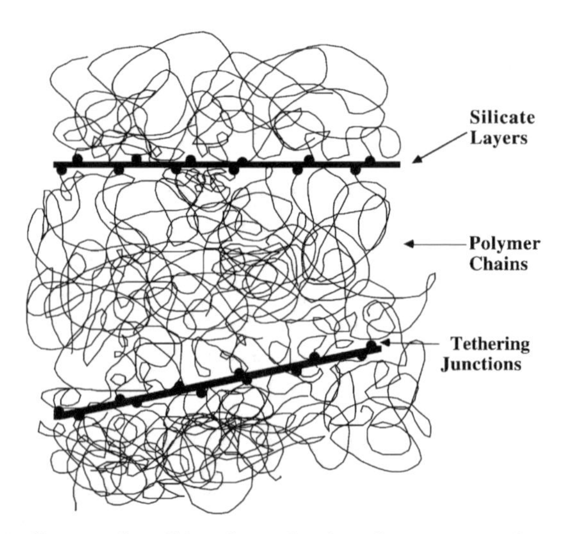

Fig. 20. Schematic diagram describing the end-tethered nanocomposites. The layered silicates are highly anisotropic with a thickness of 1nm and lateral dimensions (length and width) ranging from ~ 100 nm to a few microns. The polymer chains are tethered to the surface via ionic interactions between the silicate layer and the polymer-end. Adapted from Ref. [54].

notonic dependence, with the value for PCLC1 exceeding that for PCLC2, PCLC3 and PCLC5. However, the trend for the 2 – 10 weight % samples suggests that G″ increases with increasing silicate loading. It is worthwhile to note the trend of-decreasing molecular weight of the poly(ε-caprolactone)with increased silicate loading, with the largest decrease occurring from 1 to 2 weight % silicate loading, while the molecular weight of the polymer chains in the 2, 3, 5 and 10% nanocomposites is nearly the same.

For the molecular weights of the PCL samples examined here, it is expected that at the temperatures and frequencies at which the rheological measurements were carried out, the polymer chains should be fully relaxed and exhibit characteristic homopolymer like terminal behavior i.e., $G' \sim \omega^2$ and $G'' \sim \omega$ (based on the relaxation behavior of pure PCL, the data presented in Figs 21 and 22 are those corresponding to frequencies below the cross-over frequency associated with the transition from the plateau to the terminal relaxation of the polymer). While the polydispersity of the polymer chain lengths would affect this behavior, the effect is expected to be small in the dynamic regimes probed. The high-frequency-regime frequency dependence of the storage modulus decreases monotonically with increased silicate loading from $\omega^{1.65}$ for PCLC1 to $\omega^{0.5}$ for PCLC5.

The frequency dependence of G″ also progresses monotonically with silicate loading from $\omega^{1.0}$ for PCLC1 to $\omega^{0.65}$ for PCLC5. The frequency shift factors for the PCLC samples appear to be independent of the silicate loading consistent with the results for the frequency shift factors previously obtained for PCL [54].

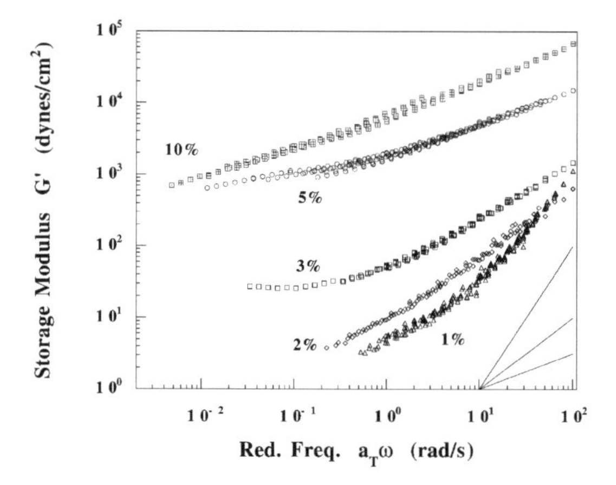

Fig. 21. Storage modulus (G') for PCL based silicate nanocomposites. Silicate loadings are indicated by percentual values in the figure. Master-curves were obtained by application of time-temperature superposition and shifted to T_0=55 °C. From Ref. [54].

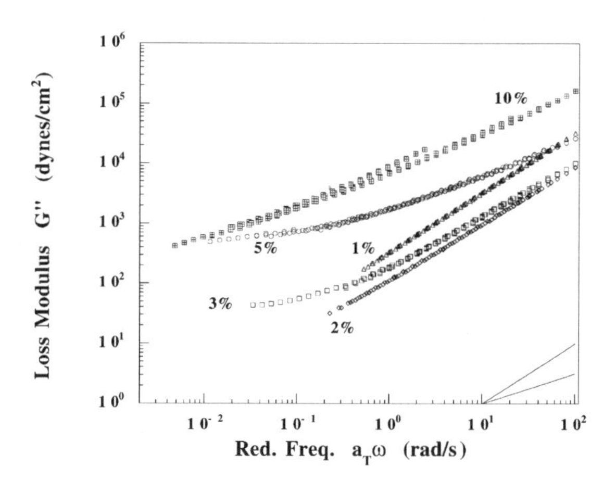

Fig. 22. Loss modulus (G″) for PCL bases silicate nanocomposites. The system is as described in Fig. 21. From Ref. [54].

Due to the low T_g of PCL (~ –60 °C) and the measurements being performed at temperatures greater than 55 °C (due to the crystallization of PCL), the temperature-dependent frequency shift factors are relatively small. The flow activation energy obtained by fitting the data at all silicate loadings is estimated to be 19 kJ/mol. Measurements on a higher molecular weight PCL homopolymer over the same temperature range yielded a value of 17 kJ/mol, in good agreement

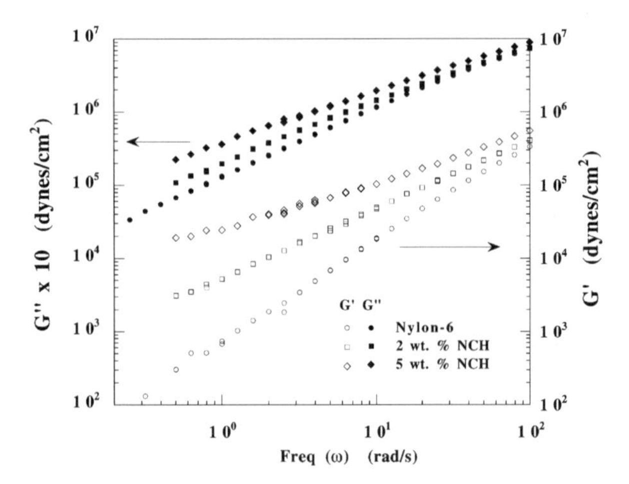

Fig. 23. Storage Modulus (G') and Loss modulus (G″) for nylon-6 silicate hybrids. Silicate loadings are indicated in the figure. All measurements were carried out at a single temperature of 235 °C. From Ref. [54].

with the results for the PCL nanocomposites. This implies that the temperature dependence of the relaxation being probed is that of the **polymer.** Since the silicate layers to which the polymers are tethered do not have a temperature-dependent relaxation, the only relaxation process probed as a function of temperature is that of the polymer segments. While the tethering must have some influence on the local dynamics of the monomers near the tethering junction, the temperature dependence of the rheological properties appears to indicate that the influence of the surface and the junction appear to be extremely local.

Similar rheological behavior was also observed with the nylon-6 silicate hybrids produced by Ube Chemical Company (Fig. 23). G′ and G″ recorded at 235 °C for a pure nylon-6, a 2 weight % nylon-6 nanocomposite and a 5 weight % nylon-6 nanocomposite [5] are shown. Both G′ and G″ exhibit a monotonic increase with increasing silicate loading at all frequencies. Unlike the PCL nanocomposites, the molecular weight of the polymer matrix is **nearly the same** for all three nylon samples. This result further suggests that the inconsistency observed in the PCL nanocomposites with the 1 weight % sample showing a higher G′ value than that for the 2 weight % sample may be caused by the abrupt decrease in molecular weight in the latter sample.

In close analogy to the PCL based nanocomposites, the terminal zone dependence of G′ and G″ for the 2 and 5 weight % samples, show non-terminal behavior with power-law dependencies for G′ and G″ much smaller than the expected 2 and 1 respectively. Furthermore, like the PCL based nanocomposites, there also appears to be a gradual decrease in the power-law dependence of G′ and G″ with increasing silicate loading.

The linear viscoelastic measurements of the end-tethered nanocomposites reveal several features unique to these materials. Time-temperature superposition can be applied to the low-amplitude oscillatory shear rheological response, with temperature shift factors being similar to that of the pure polymer. Holding the molecular weight relatively constant (the nylon-6 series and PCLC2 – PCLC10) reveals that at any given frequency the magnitude of the storage and loss moduli increase monotonically with increasing silicate loading. Furthermore, the frequency dependence of the high-frequency behavior of G′ and G″ shows a gradual change with increasing silicate loading – from homopolymer-like behavior at low silicate loading to increasingly non-terminal behavior at higher silicate contents and finally plateauing at G′ and G″~ $\omega^{0.5}$ for silicate loadings of greater than 5 weight %. Finally, at the very lowest frequencies accessed, these nanocomposites exhibit a low-frequency behavior that has both G′ and G″ almost independent of frequency.

The frequency dependence of G′ and G″ for these end-tethered nanocomposites can be contrasted to those observed in a series of delaminated hybrids with no end-tethering. Non-terminal flow behavior has been observed in filled-polymer systems exhibiting yield phenomena, but only in cases wherein the filler and polymer are actively interacting and in a dynamic regime controlled by much larger length scales (i.e., lower frequencies) than those observed in this study. Furthermore, the deviations from homopolymer-like behavior in these systems have been observed at relatively high filler-loadings. At similar high silicate loadings in either exfoliated or intercalated hybrids (not tethered to the silicate surface), significant non-linear rheological behavior is observed thereby complicating the data analysis [55]. The analogies with filled polymer systems may have particular relevance to the low-frequency behavior of the storage modulus in the PCLC nanocomposites and are discussed in greater detail below.

Non-terminal low-frequency rheological behavior has also been observed in ordered block copolymers and smectic liquid-crystalline small molecules [56,57]. Several hypotheses have been suggested to explain the observed rheological behavior in these systems [56–62]. Koppi et al. have suggested that undulations and defects in the layers might contribute to the low-frequency viscoelastic response in layered block copolymers [56]. Other ideas include that the domain structure of the ordered mesophases is responsible, due to the dynamic processes on both the microscopic and mesoscopic length scales. It is also well documented that topological defects also affect rheological properties dramatically, particularly in the low-frequency terminal regime [57]. Effects of domain collective dynamics were studied by Kawasaki and Onuki [58], who demonstrated that overdamped second-sound modes in an orientationally disordered lamellar phase could result in anomalous low-frequency rheological behavior. Rubinstein and Obukhov [59] also obtained the same result by considering diffusion-controlled annihilation of defects in a disordered lamellar system. The results of Larson et al. [57] with smectic small molecules and short ordered block copolymers has suggested that the non-terminal low-frequency response is due to the long-range domain structure and the presence of defects.

The silicate layers in the hybrids are highly anisotropic with lateral dimensions ranging from 100 to 1000 nm, and even when separated by large distances (i.e., when delaminated) cannot be placed completely randomly in the sea of polymer (as seen in Fig. 20). Furthermore, the majority of the polymer chains in the hybrids are tethered to the surface of the silicate layers. Thus, it can be expected that there are domains in these materials, even above the melting temperature of the constituent polymers, wherein some long-range order is preserved and the silicate layers are orientated in some preferred direction. Furthermore, this long-range order and domain structure is likely to become better defined at the higher silicate contents, where the geometrically imposed mean distance between the layers becomes less than the lateral dimensions of the silicate layers and thus forcing some preferential orientation between the layers. However, there is likely to be considerable polydispersity effects in terms of the orientation and the distance between the silicate layers. Many such randomly oriented grains make up the entire sample leading to the presence of a macroscopically disordered material. Thus in general, the material possesses a layered structure, with grains wherein the silicate layers are oriented in a preferred direction leading to the presence of grain boundaries and the concomitant presence of defects. However, for delaminated hybrids with no end-tethering, an increase in the moduli at all frequencies is observed and classical homopolymer like terminal behavior. Hence it appears that delamination alone is not sufficient to produce the non-terminal flow behavior.

The non-terminal low-frequency behavior observed in the PCL and nylon-6 based nanocomposites could also be attributed to the retardation of molecular relaxation processes produced by the tethering of one-end of these molecules to the silicate surface. Witten et al. [62] have suggested that tethering of the polymer molecules is expected to create an energetic barrier to the reptation motion, which leads to a dramatic increase in the relaxation time and hence a shift of the terminal relaxation to very low frequencies. Since the molecular weight of the samples examined here is small, it is expected that the dominant relaxation mode would be Rouse-like, which should not be drastically slowed by the tethering of one-end of the chains. Furthermore, at all silicate loadings, most of the chains are tethered to the silicate layers and any effect of the tethering should be discernible at all loadings [63]. However, it is clear that the terminal-zone behavior gradually changes with increased silicate loading, saturating at about 5 weight %.

The presence of a transition is manifested from the change of slope observed at high frequency to a more flattened behavior at low frequencies. The change is more pronounced in the case of G' than G''. The low frequency response is indicative of a 'pseudo solid-like' behavior and is clearly seen in the PCL samples with silicate loading greater than 3 weight %. Similar rheological response at low-frequencies has been observed in triblock copolymers in the ordered state and has been attributed to the quasi-tethering of the unlike polymer segments in their respective microdomains [64]. A solid-like response has also been observed in conventionally filled polymer systems in which there were strong interactions between the polymer and the filler and has been attributed to the presence of yield phenomena in these systems [65]. Thus, the presence of the silicate layers and the lack of complete relaxation of the chains contribute to the pseudo solid-like response at low frequencies (pseudo solid-like as G′ does not exceed G″ by orders of magnitude as would be expected from a true solid).

4.2
Alignment of Nanocomposites

Application of large strain-amplitude oscillatory shear leads to a shear-aligned sample. These measurements were carried out on the 3, 5 and 10% samples and the during-shear moduli show a decrease with continual shearing and finally reach a plateau value. With the exception of the first few cycles of shear for the 3 and 5 weight % samples, the modulus decreases monotonically and the stress signal remains sinusoidal. In the first few cycles of shear for the 3 and 5 weight % samples, the moduli show a maximum, before monotonically decreasing. The small-strain moduli after shear-alignment for PCLC10 carried out at T=70 °C, ω =1 rad/s, γ_0 = 120%, and time=3 h are shown in Fig. 24. First, both the storage and loss moduli for the aligned sample are considerably lower than those for the initially unaligned sample. Secondly, the frequency dependence of both G′ and G″ for the aligned samples are much stronger and start to resemble those of free homopolymers. The small-strain modulus results observed for the PCLC samples before and after shear alignment are in close analogy with those observed for block copolymers as well as small molecule smectic liquid-crystals [56,57]. The temperature-dependent frequency-shift factors for the aligned and initially unaligned sample are within experimental errors identical. That large-amplitude oscillatory shear can significantly alter the (small-strain) linear viscoelastic

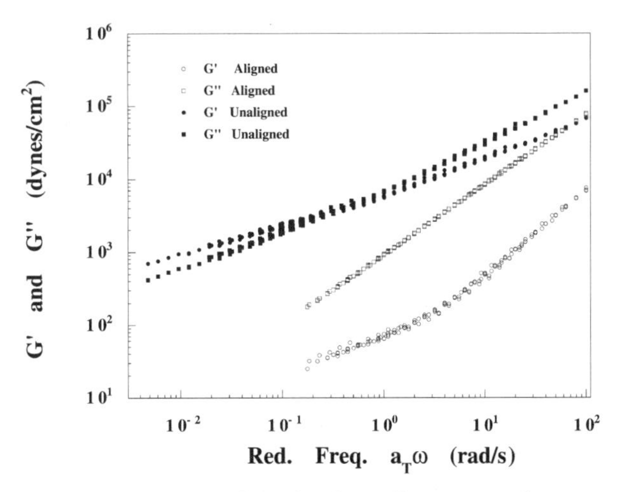

Fig. 24. G' and G″ obtained using small amplitude oscillatory strain for PCLC10 before and after large-amplitude shear. See text for conditions of large amplitude shear. Master curves obtained by application of time–temperature superposition and shifted to T_0=55 °C. From Ref. [54].

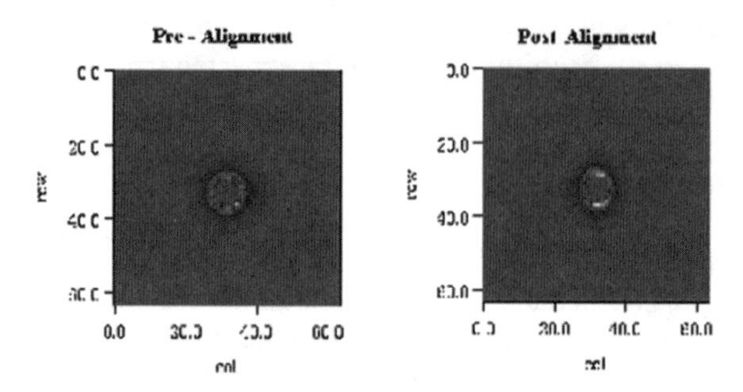

Fig. 25. Preliminary SANS measurements on post large amplitude oscillatory shear confirm the presence of global alignment of the silicate layers.

response, indicates that there is some mesoscopic arrangement of the silicate layers, which is organized by the application of large amplitude shear. Preliminary SANS measurements on post large amplitude oscillatory shear confirm the presence of global alignment of the silicate layers as seen in Fig. 25. Similar alignment effect has been also observed in injection molded nylon-6 hybrids as observed by the Toyota group. TEM and SAXS measurements on extruded nanocomposites revealed highly oriented samples at the edges, with relatively poorer order in the region of lowest shear, i.e. the central plane.

4.3 Strain – Hardening of Polymer Brushes

Strain sweeps at fixed frequencies (ω) and temperature were carried out on all aligned samples. Complex viscosity (η^*) and phase angle (tanδ) as a function of strain amplitude for PCLC10 at ω=1 rad/s and 3 rad/s (T=55 °C) during both increasing and decreasing strain amplitude cycles are shown in Fig. 26. At all strain amplitudes the torque was confirmed to be sinusoidal, thereby allowing for interpretation in terms of standard viscoelastic parameters. The complex viscosity at low strain amplitudes is **independent** of strain amplitude and is found to be dominated by the viscous response. However, progression to higher strain amplitudes leads to an increase in η^* with the elastic component becoming more prominent as seen by the decrease in the tanδ. At the highest strain amplitudes η^* appears to saturate, with a value much higher than that observed at low strain amplitudes. A slight hysteresis in the transition from high to low viscosity is also observed when the strain amplitude is decreased from a large value to a small value (as compared to the case where the strain amplitude is increased from low to high values).

Similar results were obtained on all three samples over a range of frequencies and temperatures. The upturn in the viscosity (and the downturn in tanδ) can be seen to occur at a **critical strain amplitude** over a wide range of temperatures and frequencies (for a given silicate loading). Three important features are observed in the rheological response of all three samples – (a) the process is reversible; (b) there

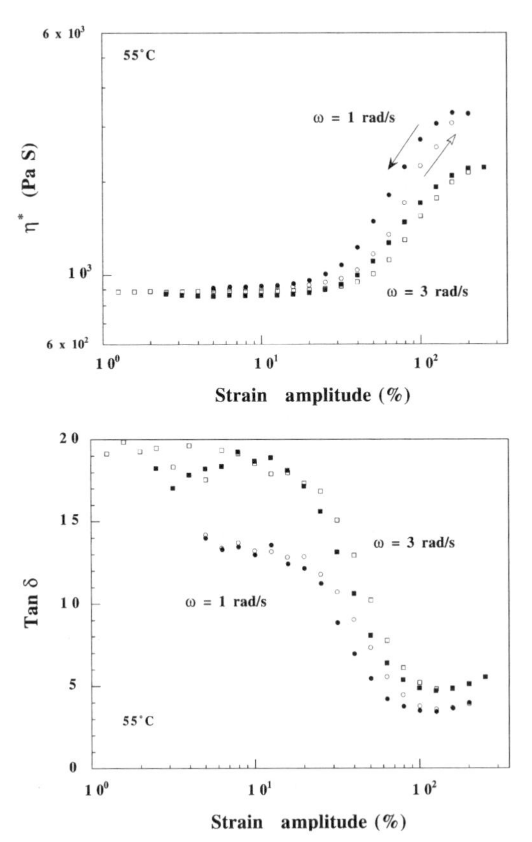

Fig. 26. Complex viscosity (η^*) and phase angle ($\tan\delta$) as a function of strain amplitude for PCLC10 at ω=1 rad/s and ω=3 rad/s (T = 55 °C) during both increasing and decreasing strain amplitude cycles. Open symbols were obtained with increasing strain amplitude and filled symbols with decreasing strain amplitude.

is a critical strain amplitude for the transition; and (c) the elastic component to the rheological response becomes more important with increasing strain amplitude.

Typically homopolymers in a shear flow exhibit decreasing viscosity with increasing shear rate. Also, tethered polymer solutions in a surface force apparatus, shear thin at a critical velocity presumably due to the slip of the tethering mica-layers [66]. These systems have also shown a dramatic increase in normal force beyond a critical Weissenberg number, and this has been attributed to hydrodynamic instabilities, flow induced brush thickening as well as shear induced diffusion [67,68]. In contrast, in these aligned end-tethered nanocomposites we observe an increase in viscosity with increasing shear strain amplitude with the transition occurring at modest strain amplitudes.

Based on the density of grafting, the systems examined are expected to be in the 'strong' brush regime, wherein the chains are strongly stretched away from the tethering surface, even in the absence of any external force. Upon application of shear-strain (beyond the critical strain amplitude), the tethered polymers are expected to completely unwind in response to the applied shear [69–72]. Based on the limitations of a 'melt-brush' system, wherein the tethered polymer has to fill all space, the chains stretch. This stretching occurs at a **critical displacement** that only depends on the geometrical restrictions of the system of interest i.e., the distance between tethering surfaces. This is borne out in the observation that the critical displacement decreases with increasing silicate content. It is expected that as the spacing between layers is decreased, the equilibrium chain structure is further distorted, and therefore the critical shear displacement required for the chain stretching is decreased. This is an unexpected result in light of solution based experiments, where the systems exhibit a **decrease in viscosity beyond a critical velocity** and have been theoretically attributed to the disentangling of chains from adjacent tethering surfaces and consequent slip of the layers. Further, mechanisms such as shear induced diffusion or brush thickening would require a frequency/temperature dependence to the observed phenomena and are also predicated on the presence of space filling solvent. The experimental results of the melt polymer brush system clearly show that the mechanism of flow in a melt brush are dramatically different than in solution-brushes and these effects manifest as opposite rheological responses with the controlling parameter for the transition being a critical displacement for the melt-brushes as opposed to a critical velocity for the solution-brushes [73].

Doyle et al [67] have performed Brown ian dynamics simulations of solvent filled polymer brushes under steady and oscillatory shear flows. Under steady flow they observe shear thinning behavior for both the normal force and the shear viscosity. Under oscillatory flow, they found extreme shear thickening when the ratio of the Weisenberg number to the frequency ~ 3. They believe that shear induced diffusion is the cause for this shear thickening. They also find the behavior to be more predominant at high frequencies, where the molecules of the chain effectively do not feel the tethering sites and can hence move affinely. Due to this affine motion the molecules collide with the neighbors and experience an effect similar to shear-induced diffusion as seen in theory of free suspensions. While this interpretation is useful in explaining some of the anomalous rheological responses of polymer brushes in a solvent, studied by the surface forces apparatus, it is clearly incapable of physically describing the dynamical processes observed in melt brushes [73].

5 Conclusions

Polymer silicate nanocomposites offer unique possibilities as model systems to study confined polymers or polymer brushes. The main advantages of these systems are: (a) the structure and dynamics of nanoconfined polymer chains can be conveniently probed by conventional analytical techniques (such as scattering, DSC, NMR, dielectric spectroscopy, melt rheology); (b) a wide range of different polymers can be inserted in the interlayer or end-grafted to the silicate surfaces, through proper surface modification; and (c) since the confining surfaces are atomically smooth and the interlayer distances have a narrow distribution of values, very well defined system geometries are involved.

Work to date has revealed the following: the mass transport of polystyrene (nearly non-polar) entering into the layered silicate appears to be unhindered by the confinement of the chains and exhibits similar diffusion coefficient and temperature dependence to those of the polymer self-diffusion. The non-bulklike dependence of the intercalation diffusion coefficient on polymer-surface energetics and on polymer molecular weight is the focus of ongoing investigations, and will bring further insight into the nature and mechanisms of polymeric motions in ultra-confined geometries and near surfaces, in the presence of a large chemical potential gradient.

The local and global dynamic behavior of the polymer chains in the nanocomposites are markedly different from the bulk. Locally, intercalated PEO chains exhibit greater flexibility along the chain backbone as compared to those in the bulk. A coexistence of fast and slow relaxation modes for a wide range of temperatures, in conjunction with atomistic simulations of confined coils, mirrors the rich local dynamics behavior of these confined systems [74]. Intercalated PS and PEO chains display marked suppression – or absence – of cooperative dynamics such as exhibited at the glass transition. In addition, intercalated PEO chains do not display a melting transition, but behave similarly to amorphous chains in the melt even at temperatures significantly below the melting temperature of bulk PEO.

On a global scale, the linear viscoelastic behavior of the polymer chains in the nanocomposites, as detected by conventional rheometry, is dramatically altered when the chains are tethered to the surface of the silicate or are in close proximity to the silicate layers as in intercalated nanocomposites. Some of these systems show close analogies to other intrinsically anisotropic materials such as block copolymers and smectic liquid crystalline polymers and provide model systems to understand the dynamics of polymer brushes. Finally, the polymer melt-brushes exhibit intriguing non-linear viscoelastic behavior, which shows strain-hardening with a characteric critical strain amplitude that is only a function of the interlayer distance. These results provide complementary information to that obtained for solution brushes using the SFA, and are attributed to chain stretching associated with the space-filling requirements of a melt brush.

6
References

1 (a) Giannelis EP, Mehrotra V, Tse OK, Vaia RA, Sung T (1992)in Synthesis and Processing of Ceramics: Scientific Issuesl, Rhine WE, Shaw MT, Gottshall RJ, Chen Y, eds MRS Proceedings: Pittsburgh, PA, 249: 547
(b)Wang MS, Pinnavaia TJ (1994) Chem Mater 6: 2216
(c) Aranda P, Ruiz-Hitzky E (1992) Chem Mater 4: 1395

(d) Messersmith PB, Stupp SI (1992) J Mater Res 7: 2599
(e) Nazar LF et al. (1991) in Solid State Ionics II, MRS proceedings:Pittsburgh
(f) Lemmon JP, Lerner MM (1994) Chem Mater 6: 207
(g) Wu J, Lerner MM (1993) Chem Mater 5: 835
(h) Ding Y, Jones DJ, Maireles P, Roziere J (1995) Chem Mater 7: 562
(i) Mehrotra V, Giannelis EP (1990) in Polymer Based Molecular Composites, Shaefer DW, Mark JE, eds MRS Proceedings: Pittsburgh, PA, 171
Mehrotra V, Giannelis EP (1991) Solid State Commun 77: 155
Mehrotra V, Giannelis EP (1992) Solid State Ionics 51: 115

2 (a) Pinnavaia TJ (1983) Science 220: 365
(b) Theng BKG (1974) The Chemistry of Clay-Organic Reactions, John Wiley and Sons: NY
(c) Theng BKG (1979) Formation and Properties of Clay Polymer Complexes, Elsevier: NY 1979
(d)Solomon DH, Hawthorne DG (1991) Chemistry of Pigments and Fillers Krieger: Malabar, FL
(e) Whittingham MS, Jacobson AJ eds (1982) Intercalation Chemistry Academic Press: NY

3 (a) Cao G, Mallouk TEJ (1991) Solid State Chem 94: 59
(b) Pillion JE, Thompson ME (1991) Chem Mater :3 777
(c) Kanatzidis MG, Wu CG, Marcy HO, DeGroot DC, Kannewurf CR (1990) Chem Mater 2: 222
(d) Liu YJ, DeGroot DC, Schindler JL, Kannewurf CR, Kanatzidis MG (1991) Chem Mater 3: 992
(e) Kanatzidis MG, Wu CG, DeGroot DC, Schindler JL, Benz M, LeGoff E, Kannewurf CR (1993) in NATO ASI: Chemical Physics of Intercalation II, Fisher J eds Plenum Press
(f) Kato C, Kuroda K, Takahara H (1981) Clays and Clay Minerals 29: 294
(g) Ogawa M, Kuroda K, Kato C (1989) Clay Sci 7: 243
(h) Fukushima Y, Okada A, Kawasumi M, Kurauchi T, Kamigaito O (1988) Clay Min 23: 27
(i)Liu YJ, Schindler JL, DeGroot DC, Kannewurf CR, Hirpo W, Kanatzidis MG (1996) Chem Mater 8: 525

4 Lan T, Kaviratna PD, Pinnavaia TJ (1995) Chem Mater 7: 2144
5 Krishnamoorti RK, Vaia RA, Giannelis EP (1996) Chem Mater 8: 29
6 Giannelis EP (1996) Advanced Materials 8: 29
7 Vaia RA, Vasudevan S Krawiec W, Scanlon LG, Giannelis EP (1995) Adv Mater 7: 154
8 Vaia RA, Ishii H, Giannelis EP, (1993) Chem Mater 5: 1694;
Vaia RA (1995) Doctoral Thesis, Cornell University: Ithaca
9 Vaia RA, Teukolsky RK, Giannelis EP (1994) Chem Mater 6: 1017
10 Burnside SD, Giannelis EP (1995) Chem Mater 7: 1597
11 Vaia RA, Giannelis EP (1998) Macromolecules 30: 8000
12 Vaia RA, Jandt KD, Kramer EJ, Giannelis EP (1995) Macromolec 28: 8080; Vaia RA, Jandt KD, Kramer EJ, Giannelis EP (1996) Chem Mat 8: 2628
13 Wong S, Vasudevan S, Vaia RA, Giannelis EP, Zax D (1995) J Am Chem Soc 117: 7568; Wong S, Vasudevan S, Vaia RA, Giannelis EP, Zax D (1996) Solid State Ionics 86: 547
14 Vaia RA, Sauer BB, Tse OK, Giannelis EP (1997) J Polym Sci Part B: Polym Phys 35: 59
15 Krawiec W, Scanlon LG, Fellner JP, Vaia RA, Vasudevan S, Giannelis EP (1995) J Power Sources, 54: 310
16 Brindley SW (1980) Crystal Structure of Clay Minerals and their X-ray Diffraction, Brown G eds, Mineralogical Society: London; Güven N (1988) in Hydrous Phyllosilicates, Bailey SW eds, Reviews in Mineralogy Vol. 19, Mineralogical Society of America: Washington DC 497: 560
17 Messersmith PB, Giannelis EP (1994) Chem Mater 6: 1719
18 Messersmith PB, Giannelis EP (1995) J Polym Sci Part A: Polym Chem 33: 1047

19 Usuki A, Kawasumi M, Kojima Y, Okada A, Kurauchi T, Kamigaito O (1993) J Mat Res 8: 1174; Usuki A, KojimaY, Kawasumi M, Okada A, Fukushima Y, Kurauchi T, Kamigaito O (1993) J Mat Res 8: 1179
20 Yano K, Usuki A, Karauchi T, Kamigaito O (1993) J Polym Sci, Part A: Polym Chem 31: 2493
21 Kojima Y, Usuki A, Kawasumi M, Okada A, Kurauchi T, Kamigaito O (1993) J Polym Sci Part A: Polym Chem 31: 983; Kojima Y, Usuki A, Kawasumi M, Okada A, Kurauchi T, Kamigaito O, Kaji K (1994) J Polym Sci Part B: Polym Phys 32: 625; ibid (1995) 33: 1039
22 de Gennes PG (1979) Scaling Concepts in Polymer Physics, Cornell University, Press: Ithaca, NY
23 Jinnai H, Smalley MV, Hasimoto T, Koizumi S (1996) Langmuir 12: 1199
24 Carrado KA, Thiyagarajan P, Elder DL (1996) Clays and Clay Minerals 44: 506
25 Muzny CD, Butler BD, Hanley HJM, Tsvetkov F, Peiffer DG (1996) Matt Lett 28: 379
26 Vaia RA (1995) Doctoral thesis, Cornell University, Ithaca, NY;
Vaia RA, Giannelis EP, Macromolecules 30: 7990
27 Scheutjens JMHM, and Fleer GJ., (1979) J. Phys. Chem 83 : 1619; (1980) J Phys Chem, 84: 178
28 Benshaul A, Szleifer I., Gelbart WM (1984) Proc. Nat. Acad. Sci., 81: 4601; Carignano MA, Szleifer I (1993) J. Chem. Phys., 98 : 5001
29 (a) Skipper NT, Sposito G, Chang FRC (1995) Clays and Clay Minerals 43: 285; ibid (1995) 43: 294
(b) Chang FRC, Skipper NT, Sposito G (1995) Langmuir 11: 2734
(c) Karaborni S, Smit B, Heidug W, Urai J, van Oort E (1996) Science 271: 1102
(d) Delville A, Sokolowski S (1993) J Phys Chemie 97: 6261
(e) Manias E, Chan YK, Giannelis EP, Panagiotopoulos AZ, in preparation
30 (a) Barrer RM, Craven RJB (1992) J Chem Soc, Faraday Trans 88: 645
(b) Breen C, Deane T, Flynn JJ, Reynolds D (1987) Clays and Clay Miner 17: 198
(c) Kärger J, Ruthven D (1992) Diffusion in Zeolites, J. Wiley & Sons, NY
31 Green PF, Kramer EJ (1986) J Mater Res 1: 202; Antonietti M, Coutandin J, Grütter R, Sillescu RH (1984) Macromolecules 17: 798
32 Manias E, Genzer J, Chen H, Kramer EJ, Giannelis EP, submitted for publication
33 Manias E, Subbotin A, Hadziioannou G, ten Brinke G (1995) Molecular Physics 85: 101
34 Frank B, Gast AP, Russel TP, Brown HR, Hawker C (1985) Macromolecules 29: 6531
35 Green PF, Palmstrom CJ, Mayer JW, Kramer EJ (1985) Macromolecules 18: 501
36 Reiter GA, Demirel L, Granick S (1994) Science 263: 1741; Overney RM (1995) Trends in Polymer Science 3: 359; Pelletier E, Montfort JP, Lapique F (1994) J Rheol 38: 1151
37 Brik ME, Titman JJ, Bayle JP, Judeinstein P (1996) J Polym Sci Part B: Polym Phys 34: 2533
38 (a)Bitsanis IA, Pan C (1993) J Chem Phys 99: 5520; (b) Mansfield KF, Theodorou DN (1989) Macromolecules 22: 3143; (c) Manias E, Subbotin A, Hadziioannou G, ten Brinke G (1995) Molecular Physics 85: 1017; (d) Manias E (1995) Ph D thesis, University of Groningen, Groningen
39 Manias E, Hadziioannou G, ten Brinke G (1996) Langmuir 12: 4587
40 Toxvaerd S (1981) J Chem Phys 73: 1998; Magda JJ, Tirell M, Davis HT (1985) J Chem Phys 83: 1888; Bitsanis I, Magda JJ, Tirell M, Davis HAT (1987) J Chem Phys 87: 1733; Schoen M, Diestler DJ, Cushman JH (1987) J Chem Phys 87: 5464; Schoen M, Cushman J, Diestler D, Rhykerd C (1988) J Chem Phys 88: 1394; Schoen M, Rhykerd CJ, Diestler DJ (1989) Science 245: 1223; Bitsanis I, Somers S, Davis HT, Tirell M (1990) J Chem Phys 93: 3427; Thompson PA, Robbins MO (1990) Phys Rev A 41: 6830; Thompson PA, Robbins MO (1990) Science 250: 792; Diestler D, Schoen M, Hertzner A, Cushman J (1991) J Chem Phys 95: 5432; Lupkowski M, van Swol F (1991) J Chem Phys 95: 1995; Somers S, Davis TH (1992) J Chem Phys 96: 5389; Schoen M, Diestler DJ, Cushman JH (1994) J Chem Phys 100: 7707

41 Mansfield KF, Theodorou DN (1989) Macromolecules 22: 3143; Thompson PA, Grest GS, Robbins MO (1992) Phys Rev Lett 68: 3448; Manias E, Hadziioannou G, Bitsanis I, ten Brinke G (1993) Europhysics Lett 24: 99; ibid (1996) 33: 371 Bitsanis I, Pan C (1993) J Chem Phys 99: 5520; Daivis P, Evans DJ (1994) J Chem Phys 100: 541; Koopman DC, Gupta S, Ballamudi RK, Westerman-Clark GB, Bitsanis I (1994) Chem Eng Sci 49: 2907; Manias E, Hadziioannou G, ten Brinke G (1994) J Chem Phys 101: 1721; Cieplak M, Smith ED, Robbins MO (1994) Science 265: 1209; Thompson PA, Grest GS, Robbins MO (1995) J Israel Chemie 35: 93
42 Bitsanis IA, ten Brinke GJ (1993) J Chem Phys 99: 3100
43 (a)Koopman DC, Gupta S, Ballamudi RK, Westerman-Clark GB, Bitsanis I (1994) Chem Eng Sci 49: 2907
(b) Ballamudi RK, Bitsanis IJ (1996) Chem Phys 105: 7774
44 Doi M, Edwards SF (1986) The theory of polymer dynamics §4.1.2 Oxford
45 In the case of intercalated polystyrene hybrids, the polystyrene was deintercalated and examined using DSC and GPC. The GPC traces before and after were identical, suggesting no degradation or cross-linking. Furthermore, the T_g (a sensitive function of molecular weight in this range) for both pure and deintercalated PS was identical with superimposable DSC traces. On the other hand, PEO undergoes thermal degradation when heated above its melting temperature. Using FTIR we find that some degradation has taken place in the intercalated PEO samples. Since the techniques used to probe polymer dynamics (NMR, DSC and TSC) are very localized (smaller than 2–3 nm), the conclusions should be largely unaffected by the slight thermal degradation of PEO.
46 Frisch HL, Xue YP (1995) J Polym Sci Part A: Polym Chem 33: 1979
47 Keddie JL, Jones RAL, Cory RA (1994) Europhys Lett 27: 59; Keddie JL, Jones RAL, Cory RA (1994) Faraday Disc 98: 219
48 Wallace WE, van Zanten JH, Wu W, (1995) Phys Rev E 52: R3329; Forrest JA, Dalnoki-Veress K, Stevens RJ, Dutcher JR (1996) Phys Rev Lett 77: 2002; ibid 77: 4108
49 Mansfield KF, Theodorou DN, (1991) Macromolecules 24: 6283
50 Sauer BB, Hsiao BS (1993) J Poly Sci: Polym Phys 31: 917; Sauer BB, DiPaolo NV, Avakian P, Kampert WG, Starkweather HW (1993) J Poly Sci, Polym Phys 31: 1851; Sauer BB, Beckerbauer R, Wang L (1993) J Poly Sci: Polym Phys 31: 1861; Sauer BB, Avakian P (1992) Polymer 33: 5128
51 Cheng SZ, Wu ZQ, Wunderlich B (1987) Macromolecules 20: 2802; Cheng SZ, Cao MY, Wunderlich B (1986) Macromolecules 19: 1868
52 Krishnamoorti R, Giannelis EP; Unpublished data
53 Enikolopyan NS, Fridman ML, Stalnova IU, Popov VL (1990) Adv Polym Sci 96: 1
54 Krishnamoorti R, Giannelis EP (1997) Macromolecules , 30: 4097
55 Krishnamoorti R, Giannelis EP, Unpublished data
56 Rosedale JH, Bates FS (1990) Macromolecules 23: 2329; Koppi KA, Tirrell M, Bates FS, Almdal K, Colby RH (1993) J Phys II (Paris) 2: 1941
57 Larson RG, Winey KI, Patel SS, Watanabe H, Bruinsma R (1993) Rheologica Acta 32: 245; Patel SS, Larson RG, Winey KI, Watanabe H (1995) Macromolecules, 28: 4313
58 Kawasaki K, Onuki A (1990) Phys Rev 42: 3664
59 Rubinstein M, Obukhov SP (1993) Macromolecules 26: 1740
60 Halperin A, Tirrell M, Lodge TP (1992) Adv Polym Sci 100: 31
61 Ohta T, Enomoto Y, Harden JL, Doi M (1993) Macromolecules 26: 4928; Doi M, Harden JL, Ohta T (1993) Macromolecules 26: 4935
62 Witten TA, Leibler L, Pincus P (1990) Macromolecules 23: 824
63 Recently we have conducted rheological experiments wherein the end-tethered PCL nanocomposites were blended with pure PCL homopolymer. Rheological behavior, particularly the terminal zone slopes, obtained for 5% and 10% (obtained by blending equal weight fractions of PCL homopolymer with a 10 weight % PCL and 20 weight % PCL respectively) were found to be similar to those obtained from the as-prepared nanocomposites.

64 Watanabe H, Kuwahara S, Kotaka T (1984) J Rheol 28: 393; Adams JL, Graessley WW, Register RA (1994) Macromolecules 27: 6026
65 Agarwal S, Salovey R (1995) Polym Eng and Sci 35: 1241
66 Reiter G, Demirel AL, Granick S (1994) Science 263: 1741; Klein J, Perahia D, Warburg S (1991) Nature 352: 143; Granick S (1996) MRS Bulletin 21: 33; Cai LL, Peanasky J, Granick S (1996) Trends in Polym Sci 4: 47; Granick S, Demirel L, Cai L, Peanasky J (1995) Israel J of Chem 35: 75
67 Doyle PS, Shaqfeh ESG, Gast AP (1997) Phys Rev Lett 78: 1182
68 Fytas G, Anastasiadis SH, Seghrouchni R, Vlassopoulos D, Li JB, Factor BJ, Theobald W, Toprakcioglu C (1996) Science 274: 2041
69 Rabin Y, Alexander S (1990) Europhys Lett 13: 49; de Gennes PG (1979) Scaling Concepts in Polymer Physics, Cornell University Press: Ithaca, NY; Joanny JF (1992) Langmuir 8: 989; Subramanian G, Williams DRM, Pincus P (1996) Macromolecules 29: 4045; Milner ST (1991) Science 251: 905; Semenov AN (1995) Langmuir 11: 3560; Brochard-Wyart F (1993) Europhys Lett 23: 105; Brochard-Wyart F, Hervet H, Pincus P (1994) Europhys Lett 26: 511
70 Israelachvili JN, Tabor D (1972) Proc R Soc London A331: 19; Klein J (1983) J Chem Soc, Faraday Trans 1 79: 99; Overney RM (1995) Trends in Polym Sci 3: 359; Israelachvili JN (1992) Surf Sci Rep 14: 109
71 Klein J (1996) Ann Rev Mat Sci 26: 581
72 Szleifer I, Carignano MA (1996) Adv Chem Phys 94: 165
73 Krishnamoorti R, Giannelis EP (1998) Langmuir, in press
74 Hacket E, Manias E, Giannelis EP (1998) J Chem Phys, 108: 7410

Received: January 1998

Normal and Shear Forces Between Polymer Brushes

Gary S. Grest[1)]

Corporate Research Science Laboratories, Exxon Research & Engineering Company, Annandale, New Jersey 08801, USA

Surface-polymer interactions are important in many technological applications, including colloidal stablization and adherence. Recently there has been considerable progress in understanding these interactions and the resulting forces between polymer-bearing surfaces. End-grafted polymers, commonly referred to as polymer brushes, are one example of a polymer-surface complex which has many interesting properties. In this article, recent progress in understanding the normal and shear forces between polymer brushes is reviewed with emphasis on the contributions from molecular simulations. These simulations show that under steady-state shear flow, some of the individual chains of a polymer brush stretch in the direction of flow while most are buried inside of the brush and are not affected by the shear flow. The height of the brush is only weakly dependent on the shear rate in contrast with several theoretical models. When two surfaces bearing end-grafted chains are brought into contact the normal force increases rapidly with decreasing plate separation, while the shear force is in most cases significantly smaller, particularly for large compressions. However, for weak compression, the range and the magnitude of the shear force depends on both the solvent quality and shear rate. These results, first observed experimentally using the surface force apparatus and recently confirmed in simulation, suggest a way to dramatically reduce the frictional force between two surfaces. For small relative velocity of the two surfaces, the surfaces slide pass each other with almost no change in the average radius of gyration of the chains or the amount of interpenetration of chains from the two surfaces. However, for large shear rates, there is significant stretching and some disentanglement of the chains.

1 **Introduction** 150

2 **Static Properties of a Polymer Brush** 152

3 **Flow Past a Polymer Brush** 155

4 **Forces Between Compressed Brushes** 168

4.1 Normal Force 168
4.2 Shear Force 172

5 **Summary and Concluding Remark** 176

6 **Appendix – Simulation Models and Methodology** 178

7 **References** 180

1 Present address: Sandia National Laboratory, Albuquerque, NM 87112-1411, USA

Advances in Polymer Science, Vol.138

1 Introduction

Grafting or adsorbing polymers onto surfaces is often used to modify surface forces and has important technological applications in many areas, including colloidal stabilization [1,2] and adherence [3]. Recent experiments suggest that adsorbed or grafted polymers may be useful as lubricants [4]. In highly confined geometries, such as between rubbing surfaces, polymer melts show a strong rubber-like elasticity that is not characteristic of the bulk when the film thickness is less than about five times their unperturbed radius of gyration [5]. This confinement induced freezing of liquid films has been observed in a wide range of systems from simple hydrocarbon liquids to long entangled chains using the surfaces forces apparatus (SFA). A number of recent computer simulations [6–9] has elucidated the behavior of molecularly thin films under shear and have shown that the pervasive phenomena of stick-slip motion is generally associated with phase transitions between distinct sliding and static states. Polymeric fluids show a strong tendency to slip at the wall-fluid interface compared to simple fluids [7,10], with the magnitude of the slip increasing as the chain length increases at least up to the entanglement length.

Although there has been a number of computer simulation studies of small molecules in confined geometry under shear and how they modify friction (see [7] for a recent review), there have only been a few studies of polymeric systems and most of these on end-grafted systems. Systems of adsorbed polymers are characterized by very long structural relaxation times and hysteresis effects. For this reason adsorbed chains are more difficult to simulate than end-grafted ones. Khare et al. [10] recently simulated a weakly adsorbed melt of entangled polymers under steady state shear. Their results are in qualitative agreement with experiments using the SFA [5]. Here I will focus on end-grafted chains and compare the results of recent computer simulations with both experiment and theory. The structure formed when one end of a polymer chain is tethered to a two-dimensional surface is referred to as a 'brush' [11,12]. Polymer brushes are made by attaching a functional group to one end of a chain that can then bind to the surface. The chain can either have a chemical (i.e. covalent) attachment, in which case the binding energy can be quite high (several hundred k_BT) [13,14] or a physi-adsorbed bond, in which the binding energy is of order 10 k_BT [15,16]. Diblock copolymers, in which one of the blocks (usually the shorter one) adsorbs strongly to the surface, while the other does not, also form brushes [17–21], as well as a diblock copolymer at the air-liquid interface of a selective solvent [22–24]. A qualitatively similar system is formed by a symmetric diblock copolymer in the strong segregation limit of the lamellar phase, where the junctions between the two segments of the copolymer lie on a plane [25]. Although the junction points are not fixed to the surface, and are free to move, their movement is strongly damped by the repulsive interaction with other chains.

When the surface density ρ_a of grafted chains increases above a critical density, the chains overlap and stretch away from the grafting surface to avoid over-

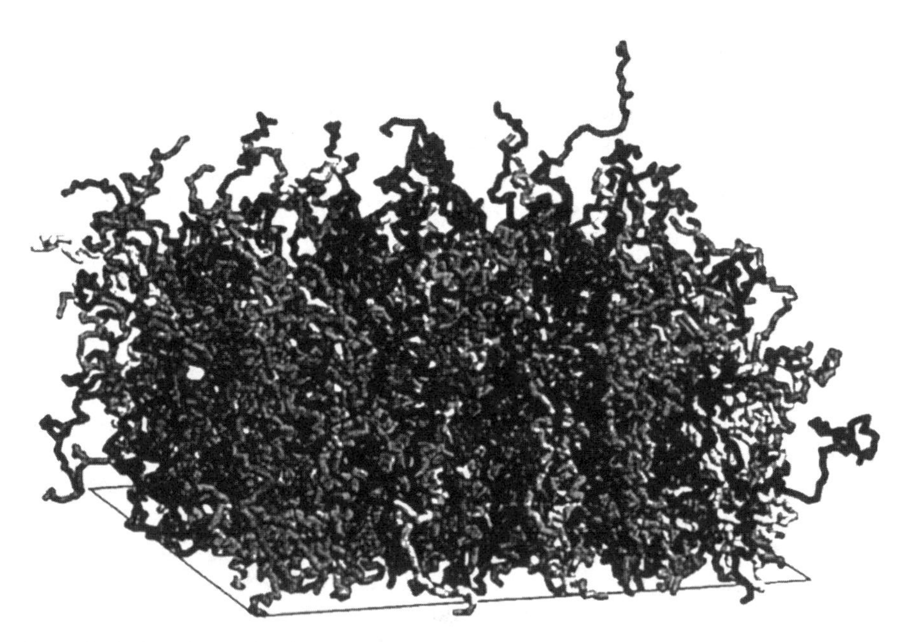

Fig. 1. Typical configuration for a polymer brush of chain length $N=100$ immersed in a good solvent (athermal, continuum) for $\rho_a=0.05\ \sigma^{-2}$. The total number of chains is 200. This figure were rendered by Raster3D [118] using sticks to better visualize the polymer chains.

crowding, as shown in Fig. 1 for a brush in a good solvent. The stretching effect of the tethering is much stronger for a brush than it is for polymers grafted to say a point as in a star polymer since for the latter the blob size for the chain increases with increasing distance from the grafting point. For a wet brush in the presence of a solvent, the stretching is due to the attraction between the polymer and the solvent. The amount of stretching is determined both by the quality of the solvent and by the binding energy of the end group to the grafting surface. For coverages above an overlap coverage, the end-to-end distance of end-grafted polymer chains scales linearly with the chain length N.

Over the last 10 years there have been a large number of experimental, theoretical and numerical simulations on the properties of polymer brushes. The static properties of polymer brushes are now very well understood and have been reviewed extensively elsewhere [26–29]. In this article I will concentrate on more recent results for polymer brushes in a shear flow. Accordingly, the next section on the static properties will be brief. In Section III, the hydrodynamic penetration depth for the solvent into the brush will be discussed for shear flow past the brush and for two surfaces approaching each other. In Section IV, the normal and shear forces between two surfaces bearing end-grafted chains will be discussed. Two processes, interpenetration and compression, are found to occur concurrently. The origin of the reduced friction observed in recent SFA experiments is discussed in comparison to recent molecular simulations. In the last section, a brief summary and outlook for future work will be presented. Details of the model and methodology used in the numerical simulations are presented in the Appendix. All of the results discussed here are for mono-dispersed brushes, although there have been a few measurements [30,31] as well as some self-consistent field (SCF) results [32] for the normal force between brushes with bimodal molecular weight distributions.

2
Static Properties of a Polymer Brush

The scaling properties of end-grafted chains were first studied by Alexander [11] and de Gennes [12]. They showed that grafting polymer chains to a surface leads to a deformation of the chains from their equilibrium configurations as shown in Fig.1. This deformation is a result of the competition between the entropic elastic energy of the chain and the monomer-monomer interaction. Alexander [11] assumed that the chains are uniformly stretched and that the density of the monomers is constant up to a height h from the grafting surface. The free energy can then be written using a Flory approximation as the sum of elastic stretching energy of Gaussian chains and binary monomer-monomer interactions [12]. Minimizing this free energy with respect to h, one finds that for high enough coverage that the chains overlap, h scales as

$$h \sim aN\rho_a^x \tag{1}$$

where a is a coefficient of the order of the monomer size, N is the chain length and ρ_a is the coverage (chains per unit area). The exponent x depends on the solvent quality. In a good solvent $x = 1/3$, while in a Θ and poor solvent, $x = 1/2$ and 1 respectively. For low coverage, in the so-called mushroom regime, the chains do not overlap sufficiently to cause swelling. The overlap coverage ρ_a^* separating the mushroom and stretched regimes depends both on N and solvent quality [12]. One can alternatively view the brush as a stack of blobs [12]. This approach gives $h \sim N\rho_a^{(1-\nu)/2\nu}$ which identical to Eq. 1 if one uses the Flory value ν=3/5 for the good solvent. This result is rather striking, as the thickness of the grafted layer increases *linearly* with N, significantly different than the $N^{0.59}$ for a self-avoiding walk in a good solvent or $N^{1/2}$ for a random walk at the Θ temperature.

Numerical and analytic SCF theories [27] showed that the density of the monomers is not uniform. Milner et al. [33,34] and Zhulina et al. [35] independently proposed an analytical theory based upon an observation by Semenov [36]. When the stretching is strong, a chain of length N whose free end is at a specified point, fluctuates very little around the 'most favorable' configuration that minimizes the classical action. The relative amplitude of this fluctuation diminishes with increasing N. As $N \rightarrow \infty$, this configuration dominates. For binary interactions, analytical solutions for the properties of the brush can be obtained. The brush height was found to have the same scaling with N and ρ_a as predicted by

Alexander [11], although the numerical coefficient is slightly larger. This agreement in the overall size scaling confirms the SCF assumption that the stretching diminishes the fluctuations around the most favorable configuration. The monomer density $\rho(z)$ in this theory varies with the distance z from the grafting distance in a good solvent as

$$\rho(z) = \frac{3N\rho_a a}{2h}\left[1 - \left(\frac{z}{h}\right)^2\right] \tag{2}$$

This parabolic density profile is valid in the limit of very long N and moderate ρ_a, where binary interactions dominate.

Numerous experimental studies and computer simulations have been carried out during the last several years to check the results of these two theoretical approaches (for a recent review see [28]). The brush height can be obtained from force measurements between two brushes, since the brushes first interact when the distance between their respective grafting surfaces is 2h [18,37]. The inner structure of the brushes has been probed by small angle neutron scattering [13,14,38,39] and neutron reflectivity [21,23,24,40,41]. All these studies, as well as a number of simulations [28], give results that are consistent with the analytical SCF predictions.

Figure 1 shows a typical configuration of an off-lattice brush consisting of 200 chains in a good solvent with N=100. The details of the simulation method and model are presented in the Appendix. The scaling $h \sim aN\rho_a^x$ has been confirmed in several simulations [42–47].These showed that provided ρ_a is above the critical overlap density $\rho_a^* \sim N^{-6/5}$ (good solvent) and $\rho_a^* \sim N^{-1}$ (Θ solvent), the brush size, as measured by the first moment of the density distribution $\langle z \rangle$ or the z-contribution to the mean squared radius of gyration, $\langle R_{Gz}^2 \rangle$, approaches the predicted scaling form. Results for a continuum bead-spring model in a good and Θ solvent are shown in Fig. 2. When $\rho_a < \rho_a^*$, h is almost independent of ρ_a, as expected in the mushroom regime. For large ρ_a, Grest [47] has shown that there is another N-dependent threshold, ρ_{a1}, above which the brush size scales as $N\rho_a^x$ with x approximately 1/2 in a good solvent. This behavior, seen in curve (a) of Fig. 2, is in agreement with Raphaël and coworkers [48,49] who showed that at high surface densities where three body interactions dominate over two body terms, there is a second scaling regime with x=1/2. Experimentally this regime is difficult to study, as such high grafting densities are not easily accessible.

Detailed tests of the analytical SCF theory of Milner et al. [33,34] and Zhulina et al. [35] have been carried out using both molecular dynamics and Monte Carlo methods [28]. As an example of the results, in Fig. 3, results for the density of a brush in a good solvent is shown for three value of the surface coverage ρ_a. The data collapse onto a single curve, which can be fitted reasonably well by a parabola. Near the tail of the brush, the data collapse fails dues to finite size effects. The magnitude of these finite size corrections decrease as N and ρ_a increase [50-52]. For higher coverage, the parabolic fit gets worse as higher order interactions

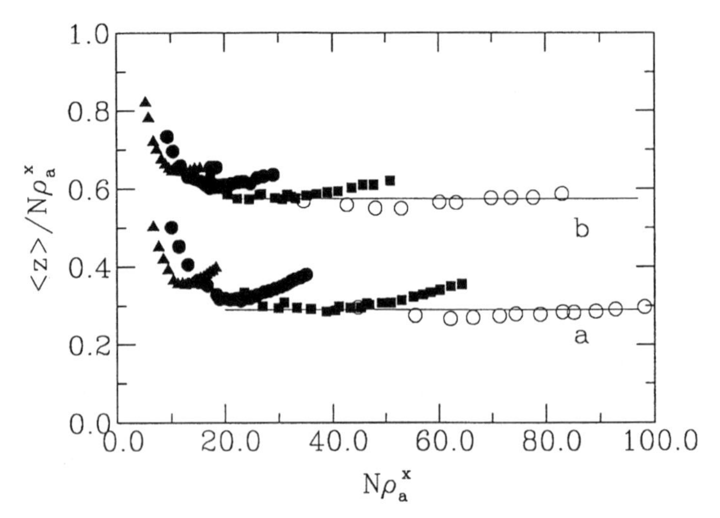

Fig. 2. Average mer distance from the host surface <z> versus $N\rho_a^x$ for a brush in (a) a good solvent with x=1/3 (T=4.0ϵ/k_B) and (b) a Θ solvent with x=1/2 (T=3.0 ϵ/k_B) from a molecular dynamics simulation in which the Lennard-Jones interaction between mers is truncated at r_c=2.5σ. The results for the Θ solvent are shifted by 0.2 for clarity. The results are for chain length N=25 (▲), 50 (●), 100 (■), and 200 (○). From ref. [47].

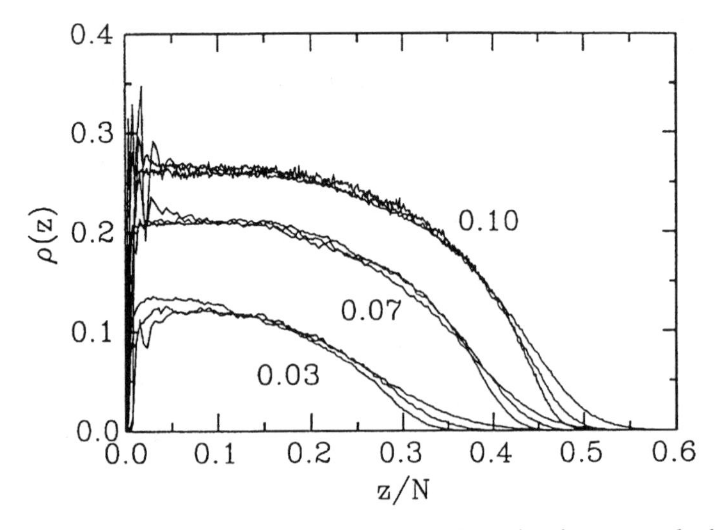

Fig. 3. Brush mer number density $\rho_1(z)$ versus z/N for polymer brushes in a good solvent for chain length N=50, 100 and 200 for the three values of $\rho_a\sigma^2$ shown. The smaller N have the longer tails due to finite size effects. From ref. [46]

play an important role. In this regime, the profiles become flatter than a parabola in agreement with Shim and Cates [53] and Lai and Halperin [54].

Another important difference between the scaling approach and the SCF theory is the location of the free ends of the chains. Unlike the implicit assumption of the scaling analysis that all the free ends are at the outer boundary of the brush, the density of the free ends is non-zero everywhere within the brush. Simulations have confirmed that the free ends have a finite probability of being anywhere in the brush [42–44].

3
Flow Past a Polymer Brush

The hydrodynamic penetration of a simple shear flow into a polymer brush was first studied by Milner [55]. Milner showed that the hydrodynamic penetration depth l_p is much larger for a parabolic density profile compared to what it would be if the profile was a step function. For a step function profile, one would expect that the hydrodynamic penetration length is essentially the mesh size ξ of a semidilute solution of the same concentration, i.e. $l_p=\xi=ar\to\rho^{-\nu/(3\nu-1)}$ where ρ is the average density. However, if the brush has a parabolic profile, the hydrodynamic penetration length varies depending on the distance from the grafting surface and diverges at the outer extremity. Thus one would expect that the flow will penetrate much further into the brush than for a step function profile.

To obtain the penetration depth quantitatively, Milner assumed that the brush profile was undistorted by the flow and invoked the Brinkman equation [56] for flow in a porous media,

$$\eta_s \nabla^2 \mathrm{v} = \frac{\eta_s}{\xi^2(\rho)} \mathrm{v} + \nabla P \tag{3}$$

to describe the flow field of the solvent inside the brush. Here η_s is the solvent viscosity and ξ depends on the local density ρ. The left hand side of the equation is the viscous force due to dissipation within the flowing liquid, which is balanced by the friction term arising from flow past the polymer segments and pressure gradient. Darcy's form, $\zeta=\eta_s/\xi^2$, has been used for the friction since the flow resembles that through a porous media. Note that the screening length ξ depends on the local density. For a simple shear, the pressure is constant and the transverse component of the velocity satisfies

$$\frac{\partial^2 \upsilon}{\partial z^2} = \frac{\upsilon}{\xi^2(\rho)} \tag{4}$$

Assuming the parabolic profile and that l_p is short enough that the density varies linearly around $z=h$, Milner [55] showed that this equation can be transformed into a modified Bessel equation and solved analytically for $\nu=1/2$. In Fig. 4, his results for the velocity profile for a brush with a parabolic and step profile are

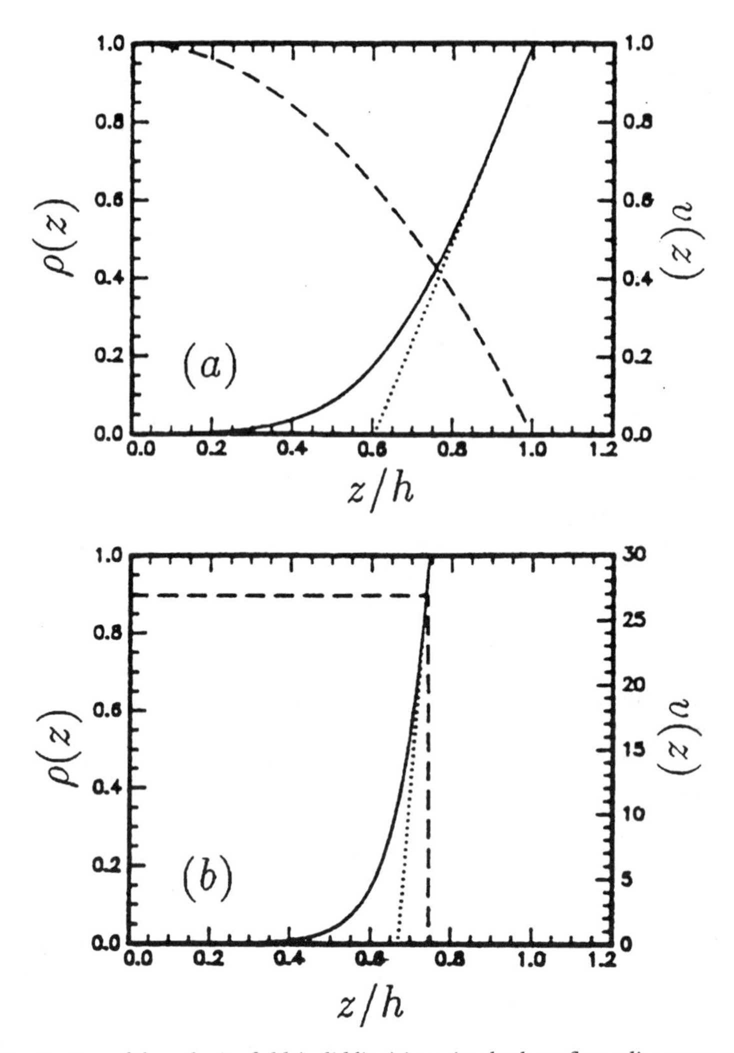

Fig. 4. Penetration of the velocity field (solid line) in a simple shear flow adjacent to a surface bearing end-grafted chains. The mer density $\rho(z)$ is the dashed curve. The velocity profile $v(z)$ outside the brush extrapolates to zero at the penetration depth (dotted curve). (**a**) Illustrates the parabolic brush and (**b**) the corresponding result for an Alexander-de Gennes brush. Here $h/\xi_0=30$, where ξ_0 is the correlation length at the grafting surface. From Milner [55].

compared. For a parabolic brush, Milner found that $l_p=1.04(h\xi_0)^{1/2}$, proportional to the geometric mean of the brush height and the smallest mesh size ξ_0, if swelling effects are ignored. However if swelling effects are included the result changes slightly, $l_p \sim \xi_0^{3/5}h^{2/5}$. In either case, the penetration depth is larger for a parabolic profile than for a step function profile, since in the latter case $l_p \sim \xi_0$.

The hydrodynamic penetration depth l_p has been measured from the lubrication force between polymer brushes using the SFA [4,57–59]. In these experiments, the equilibrium thickness h of each layer is first determined from quasi-static measurements [18,37]. Then the lubrication force can be determined by bringing two surfaces together at a finite velocity $\dot{D}$ or by applying a small amplitude oscillatory normal force to one of the surfaces. The instantaneous lubrication force F_H arises from the pressure gradient in the annular region between a curved surface (radius of curvature R_c) and a flat surface as they are brought closer together. From F_H, an effective mobility G can be defined as

$$G=6\pi^2 R_c^2 \dot{D} / F_H \tag{5}$$

For two polymer coated surfaces, a simple extension of Reynolds formula for the lubrication force gives [4,60]

$$F_H = -6\pi^2 R_c^2 \frac{\eta_s}{\left(D-2L_H\right)} \dot{D} \tag{6}$$

where $L_H=(h-l_p)$ is the hydrodynamic thickness. This gives

$$G=\left(D-2L_H\right) / \eta_s \tag{7}$$

Thus a plot of G versus D for large D should be a straight line with a slope inversely proportional to the solvent viscosity, while the extrapolation of this line to G=0 gives L_H. Klein et al. [57] found that for polystyrene in a good solvent, the results satisfied Eq. 7 very well for large D but gave a penetration depth l_p which is almost equal to the polymer correlation length $\xi_0 \simeq s$, where s is the mean spacing between grafting sites. This result is in contradiction with the theoretical prediction of Milner [55], which for the case studied predicted a value which would have been approximately four times larger. However the high shear rates associated with the solvent leaving the gap could have distorted the parabolic profile, leading possibly to an underestimate of l_p [4,57]. For polybutadiene in a good solvent (tetradecane), Dhinojwala and Granick [58] found that l_p was much larger $l_p/s \simeq 3.5$ and in very good quantitative agreement with the predictions of Milner. For polystyrene in Θ solvent (trans-decalin), Cho et. al [59] found a smaller penetration length as expected. Their result is in qualitative agreement with the prediction of Milner using his result for no swelling (ν=1/2). As the opposing surfaces are moved into contact, the hydrodynamic thickness L_H decreases monotonically as D decreases [57–59].

Another approach to study the penetration depth is to use computer simulations of simple shear flow. Computer simulations can check the validity of certain assumptions implicit in the theories, such as the assumption that the shear flow does not distort the density profile. However to do this solvent molecules must be included explicitly in the simulation. Almost all previous simulations of polymer brushes have modeled the solvent as a continuum to save CPU time.

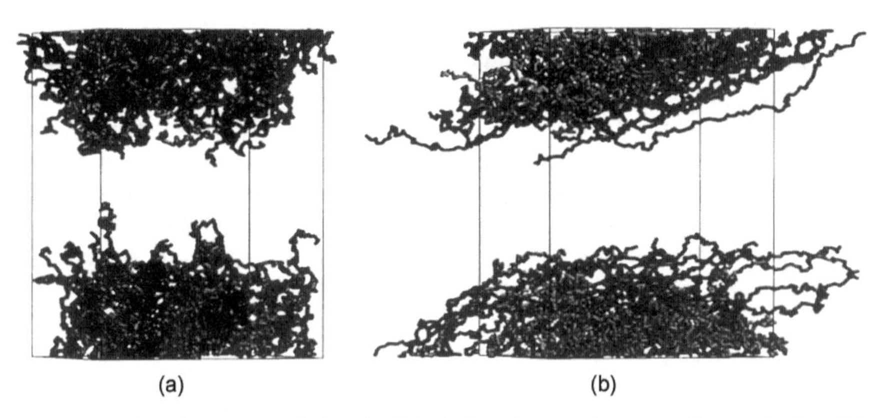

Fig. 5. Typical configurations of a brush of chain length N=100 immersed in a melt of mobile dimers for ρ_a=0.03σ^{-2}. The solvent is not shown for clarity. The shear velocity is v_w=0 (**a**) and 0.2σ/τ (**b**). The dimensions of the cell are $\mathcal{A}$=40.8$^2\sigma^2$ and D=64.9σ. Result from molecular dynamics simulations of ref. [63].

Recently there has been a few simulations of brushes with the solvent introduced explicitly. Grest [61] studied the equilibrium properties of brushes of long chains, N=100, immersed in a melt of mobile polymers ranging in chain length from a single monomer to N_f=40 (see Appendix for details). In a steady state shear, Peters and Tildesley [62] studied a brush of short chains (N=20) in a solvent of single mers while Grest [63,64] studied longer chains (N=100) in a solvent of dimers. In Fig.5, a typical configuration of a polymer brush is shown for ρ_a=0.03σ^{-2} for v_w=0 and 0.2σ/τ, where v_w is the shear velocity of the top surface. The later turns out to be a very large shear as discussed below. For $v_w \lesssim \sigma/\tau$, there is no detectable change in the brush profile or end-to-end distance of the grafted chains. For larger velocities however, the brush chains become strongly stretched along the direction of flow as v_w increases. As can be seen from Fig. 5b, this also has the effect of slightly decreasing the brush height. The small decrease in the brush height for large v_w can also be seen from the density profiles for v_w=0.02 and 0.2σ/τ shown in Fig. 6. Thus although some of the chains become very stretched for large v_w, the mer density for the brush depends only weakly on the shear rate. That only some of the chains are stretched is seen clearly in Fig. 7, where results for the end density distributions $\rho_e(z)$ and $\rho_e(r)$ are shown. Here z is the height above the grafting surface and r is the distance from the grafting point. A large fraction of the chain ends are buried deep inside the brush and do not feel the effect of the shear even for large v_w. The results shown in Fig. 7 quantify the visual observation of Fig. 5b.

The solvent velocity profiles are shown in Fig. 6. Away from the brush, the velocity profiles are linear as expected. From this region, the shear rate $\dot{\gamma}$ can easily be determined from the slope of $v(z)$. In the vicinity of the brush, the solvent velocity decays as expected from the discussion above. Although the chains are

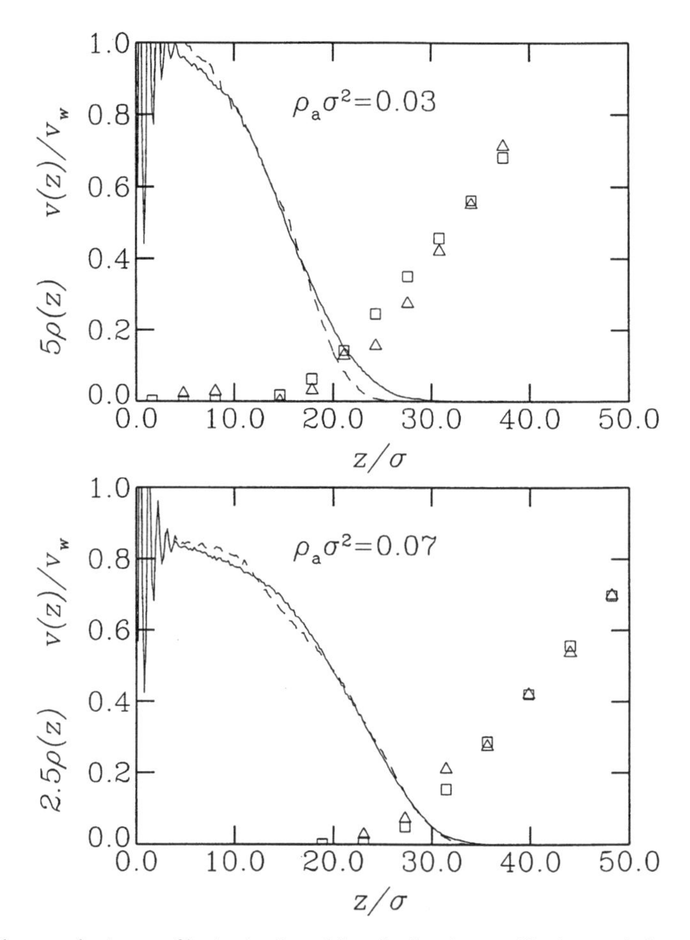

Fig. 6. Solvent velocity profile (points) and brush density profile (curves) for a polymer brush under shear flow for two values of ρ_a. The chain length of the brush is N=100, while that of the solvent is N_f=2. Results are shown for $v_w=0.02\sigma/\tau$ (solid line and triangles) and $0.2\sigma/\tau$ (dashed line and squares). $\rho(z)$ has been multipled by a factor of 5 and 2.5 respectively, to put the data on the same scale. From ref. [63].

not long enough to give a definite test of the Milner prediction for l_p, the results do show that l_p is larger than the correlation length ξ_0, where the latter can be determined from the monomer density at the surface. Using the fact that the ratio of the predictions for the models for l_p scales as $(h/\xi_0)^{2/5}$ in a good solvent [55], one finds that for the present case the predictions from the two models only differ by about a factor of 2. Thus longer chain lengths, which are prohibitive at this time due to the large amount of CPU time required, are needed to unambiguously distinguish the predictions for the parabolic and step function profiles.

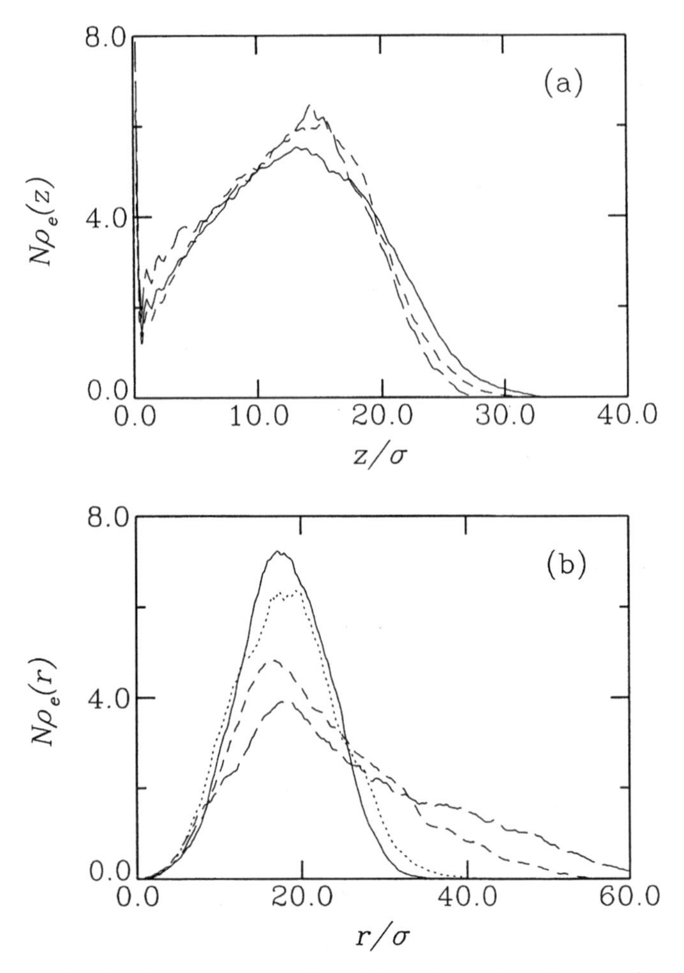

Fig. 7. Free end distribution for brushes of length N=100 in a shear flow as a function of the height z above the grafting surface and a distance τ from the grafting point. The shear rate $\dot{\gamma}$=0 (solid line), 7.2×10^{-4} (dotted line), 3.6×10^{-3} (short dashed line), and $7.2\times10^{-3}\tau^{-1}$ (long dashed line). Results are for $\rho_a\sigma^2$=0.03. From ref. [63].

This idea that the solvent flow field can be approximated by the Brinkman equation has been used in several recent simulations of a polymer brush in simple shear flow. In these simulations, the solvent is not included explicitly but it's effect is modeled using the Brinkman equation. Lai and Binder [65] and Lai and Lai [66], using a bond fluctuation lattice model, and Miao et al. [67], using a continuum model, studied the properties of a dense polymer brush in a flow field by modifying the standard Metropolis Monte Carlo transition probability to take into account the effective force acting upon the brush chains by the moving solvent. Lai and Binder assumed a parabolic profile for the brush density and determined the height h self-consistency. Miao et al. [67] self-consistently determined the density profile and did not assume a parabolic profile in order to more accurately account for the tail region near the outer boundary of the brush. Doyle et al. [68] carried out a similar procedure using a Brownian dynamics simulation for a bead-spring model.

The results of these simulations are in very good agreement with the molecular dynamics simulations with the solvent included explicitly. Lai and Binder [65], Lai and Lai [66] and Miao et al. [67] all found that the brush height is only weakly dependent on the shear rate in agreement with the molecular dynamics simulations. Doyle et al. [68] found a larger decrease in the brush height under shear, but their chains were extremely short (N=20), which probably enhanced the effect of the shear. Fig. 8 shows the results for the stretching of the grafted chains from the simulations of Miao et al. [67]. Fig. 8a shows the average position of monomer n along the chain. Although $<z(n)>$ remains essentially unchanged as the shear rate increases, $<x(n)>$ increases approximately linearly with $\dot{\gamma}$. The same stretching is also clearly seen in the increase in the mean squared radius of gyration $< R_G^2 >$ shown in Fig. 8b. Note that all of the increase in $< R_G^2 >$ comes from the component in the direction of shear flow, x. The other two components are relatively insensitive to the shear rate, though $< R_{Gz}^2 >$ decreases slightly. Similar results were found by Lai and Binder [65] in their lattice simulations and in the molecular dynamics simulations by Grest [63].

Theoretically the situation is less clear. The results of these simulations for a brush in a steady state shear flow differ considerably from some recent theoretical predictions by Barrat [69], Kumaran [70], Harden and Cates [71] and Aubouy et al. [72]. All of these theories suggest that the height should increases under shear. The only theoretical prediction which found that the height does not increase under shear is that by Rabin and Alexander [73]. Most of these predictions [69–71,73] are based on the Alexander-de Gennes model [11,12] in which the brush density profile is assumed to be a step function and all the free ends are at the outer surface of the brush. In these calculations, the chains in the brush are represented as a string of excluded-volume blobs and the free energy per chain is written as a sum of an elastic term involving the conformation of deformed Gaussian string of blobs and an osmotic term involving interactions between blobs. In the original calculation of Rabin and Alexander [73] and the subsequent calculation of Barrat [69], the chains were assumed to stretch uniformly under shear. The difference in the two calculations arises from how the osmotic contribution to the force was handled. Harden and Cates [71] later generalized this model to allow for non-uniform stretching of the blobs in the presence of strong shear flow, as shown in Fig. 9. In this calculation [71], the internal structure of the layer and the velocity profile is determined self-consistently subject to the constraint that all the ends are at the outer layer. The only theoretical model which includes the fact that some of the chains are stretched while others are not effected by the shear is by Aubouy et al. [72]. They allow for two types of chains, those which are exposed to the flow and those which are protected from

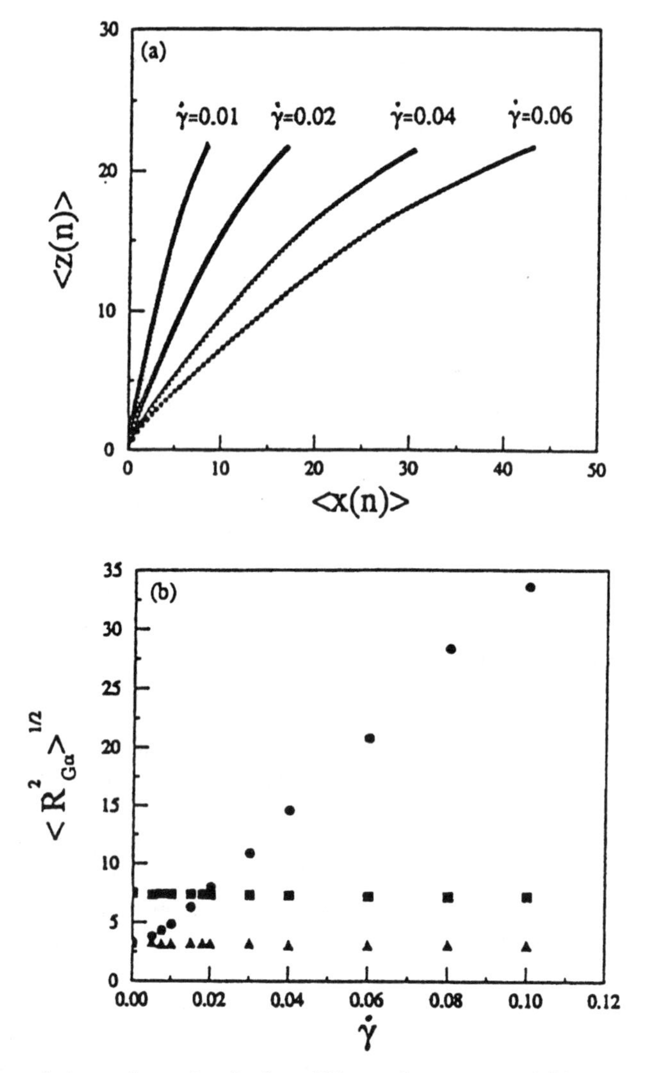

Fig. 8. Average chain conformation for four different shear rates and (b) components of the mean squared radius of gyration, (R^2_{Gx}) (●), (R^2_{Gy}) (▲), and (R^2_{Gz}) (■) as a function of the shear rate $\dot{\gamma}$. These results are from the Monte Carlo simulations of Miao et al. [67].

the flow inside the brushes. For each class of chains, the density profile is still assumed to be a step function and the free ends are assumed to be at the outer layer of their respective profiles. In this model, the fraction of chains exposed to the flow decreases with increasing shear rate. However this leads to a higher susceptibility of the layer to the shear rate and an onset of significant swelling at a lower shear rate than when all the chains are treated equivalently. At high shear rates,

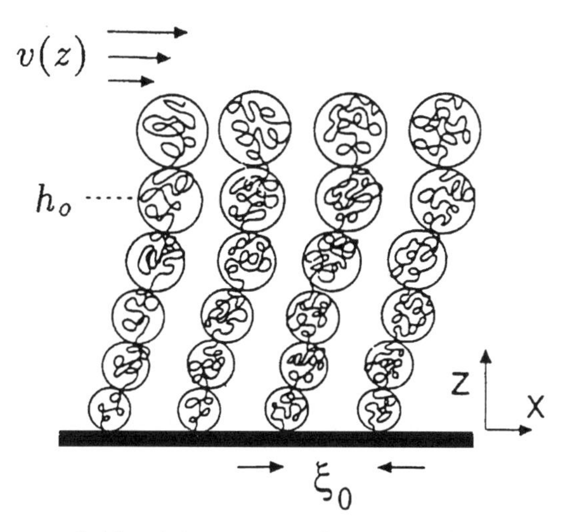

Fig. 9. Sketch of a extended brush in a strong solvent shear flow, with $\xi(z)<\xi_o$ and $h>h_o$, where ξ_o is the equilibrium blob size and h_o is the equilibrium brush height in the absence of flow. From ref. [71].

the relative swelling is actually larger than predicted by Harden and Cates [71]. Barrat [69] and Harden and Cates [71] both predict an asymptotic swelling of approximately 25% for $\dot{\gamma}\tau_z \gg 1$, where $\tau_z = 4\pi^3\eta_s\xi_0^3 / k_B T$ is the characteristic relaxation time of a sphere of hydrodynamic radius ξ_0 in a solvent of viscosity η_s. Kumaran [70] suggested a slightly larger (33%) increase in the brush height from hydrodynamic interactions. However as clearly seen from Fig. 5, the structure of a brush, in steady shear flow is very different than that shown in Fig. 9. For the case shown in Fig. 5b, the shear rate is $\dot{\gamma} = 4.1\times10^{-3}\tau^{-1}$. The solvent viscosity for chains of length N_f=2 can be estimated from the non-equilibrium molecular dynamics simulations by Kröger et al. [74] who studied a homopolymer melt for the same model for $N_f \geq 10$. From these results, $\eta_s \sim 1.4\epsilon\tau/\sigma^3$ for N_f=2. Assuming $\xi_0 \simeq s \sim \rho_a^{1/2}$, then for $\rho_a\sigma^2$=0.03, $\tau_z \simeq 3.23\times10^4\tau$. Thus the simulation results are well within the regime where swelling would be predicted by these theoretical calculations as one would expect from the configuration shown in Fig. 5. Thus one can only conclude from these studies, that while this simple model gives the correct scaling behavior for the brush height h, it is inadequate to describe the shear rate dependence of h. To date there has been no theoretical prediction for the properties of a brush under shear using the self-consistent field approach, which correctly predicted not only the scaling dependence of h on chain length N and coverage ρ_a, but also that $\rho(z)$ is parabolic and the ends are distributed throughout the brush.

Most of the discussion above has been for brushes in a good solvent. Williams [75] has applied the blob model for polymer brushes to grafted chains in a poor solvent. In this case, the blobs no longer repel but attract, leading to a reduction in the brush height in a steady shear flow. Williams found that if the grafting density is low or the solvent quality is fairly poor, the brush height decreases only above a critical shear rate. However for large ρ_a or in a marginally poor solvent, the brush is predicted to shrink as soon as the shear is applied. To date there have been no experiments or simulations to test these predictions, which could be very interesting since they would tell us more about the importance of three body interactions.

Experimentally, the only direct measure of the brush height in a shear flow has been the neutron reflectivity measurements by Nguyen et al. [76]. In these experiments in which no change in the brush height h was observed, the shear was produced by flowing a solvent past the polymer brush, with a maximum shear rate of 7500 s^{-1}. However for the grafting densities used in this study, the maximum shear was not large enough to reach the regime $\dot{\gamma}\tau_z \gg 1$. An indirect measurement of the brush height by Webber et al. [77] also found that h was independent of shear rate for $\dot{\gamma}$ in the range 10^3–$10^4 s^{-1}$. By comparing the permeability through a channel with and without an absorbed diblock copolymer on the surface, they could estimate the hydrodynamic penetration length l_p, which did not change over the shear rates measured. Both of these results are in agreement with the present simulations in that they find that the height of the polymer brush is not affected by shear for the range of shear rates.

Although a clear picture (at least from simulation and experiment, if not theory) in steady state shear is beginning to emerge, in oscillatory shear the results are not as clear. This is partially due to the limited number of studies in this case. Using the SFA, Klein et al. [16] found that if the surfaces were either weakly compressed or uncompressed but separated by a small gap, there was an extra normal force at high velocity as shown in Fig. 10. These experiments suggested an increase in the brush height of an uncompressed layer when a fluid flows past. As seen in Fig. 10, the increase occurred only above a critical shear rate. These experiments were carried out at oscillatory frequencies of a few hundred Hertz with a maximum displacement of up to a few μm. The fact that the onset of the extra normal force is shear rate dependent suggests that it is probably arises from an interplay between the chain relaxation rate and the applied shear rate. Assuming that the additional force arises from chain stretching, then one plausible explanation for these results is that chain stretching only occurs when the applied shear rate is much larger than the chain relaxation.

Doyle et al. [68] studied in oscillatory shear the case of one surface bearing end-grafted chains compressed by a second, bare surface. In this simulation the solvent is not included explicitly but its velocity field is solved for self-consistently. Doyle et al. did observe an additional normal force for very high frequencies. However the additional force occurs for maximum displacements of the moving surface which are comparable to or less than the mean spacing s between grafting sites. In this case, the beads move affinely and are not affected by the fact that the chains are tethered to the surface. The increase in the normal force arises from collisions of beads with their neighbors which drives a normal motion of the brush analogous to shear-induced diffusion in free suspensions

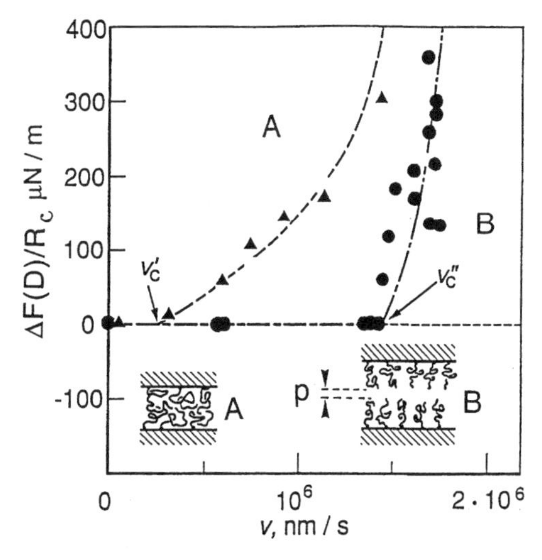

Fig. 10. Variation of the excess normal force ΔF induced by sliding two curved surfaces bearing end-grafted polystyrene (M_W=1.41×10^5) in a good solvent (toluene) as a function of the shear velocity v. In equilibrium the brushes first come into contact at 2h=125 nm. Curve A and cartoon A correspond to a compressed brush (D=95 nm); curve B, cartoon B are for non-overlapping brushes (D=155 nm). From ref. [16].

[78]. In the experiments by Klein etal [16], the maximum displacement of the surfaces (up to 3 μm) is large compared to the mean distance between grafting sites (8.5 nm for molecular weight M_W=1.41×10^5 polystyrene). Thus the effect observed by Doyle et al. [68] in their simulations is probably not applicable to the experiments of Klein et al. [16].

To clarify the situation for oscillatory shear, I performed some molecular dynamics simulations [79] for brush in an explicit solvent for surfaces which are not in contact, as in Fig. 5. To allow for hydrodynamic interactions, the viscosity damping term Γ in the equations of motion Eq. 12 (see Appendix) was zero except for the first ten monomers of each brush chain (counting from the grafting point). No damping term was added to the solvent molecules. The velocity of the top plate was set to v_w=U cos(ωt). Results are shown in Fig. 11 for the average end-to-end distance $R(t)$ for two cases with the same value for the maximum velocity U=0.26σ/τ but with different values for the maximum displacement X= 100σ and 300σ. The lower curve in each figure is the displacement of the top wall scaled by a value of 50 and 150 respectively. For comparison, the square root of the mean end-to-end distance $\langle R^2\rangle^{1/2}$=18.8σ for v_w=0 and 28.7σ for v_w=0.2σ/τ, the case shown in Fig. 5b under steady state shear. From the estimates of the relaxation times discussed above, it is clear that these two systems the average shear rate $\dot{\gamma}$ is in the regime $\dot{\gamma}\tau_z \gg 1$. Note that the peak in $R(t)$ occurs between the point of maximum wall velocity v_w=U and the turning point, v_w=0. The corresponding results for the brush height $z(t)$ are shown in Fig. 12 for the same two

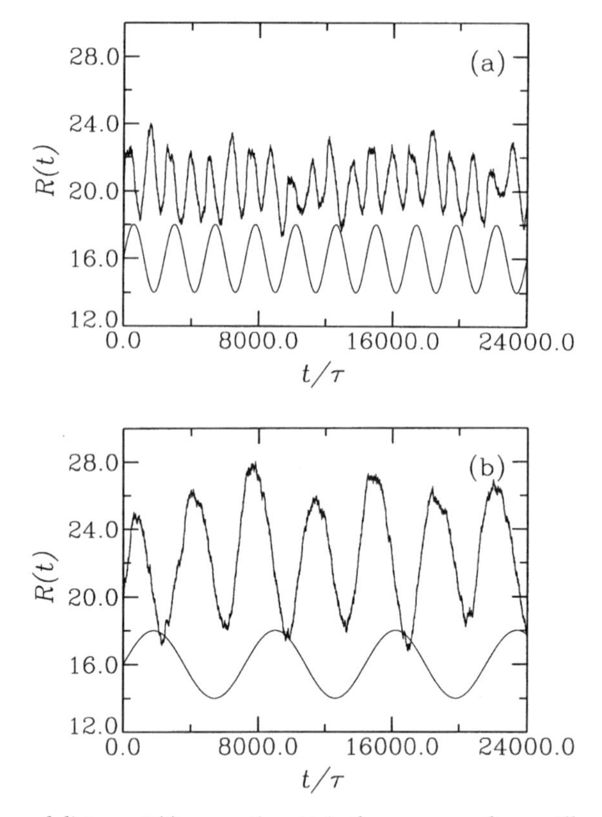

Fig. 11. End-to-end distance $R(t)$ versus time t/τ in the presence of an oscillatory shear (upper curve). The top wall has a velocity $v_w=U\cos(\omega t)$, where $U=0.26\sigma/\tau$. The maximum displacement X is (**a**) 100σ and (**b**) 300σ. The lower curve in each case is the displacement of the top wall divided by a factor of 50 and 150 respectively. Results are from a molecular dynamics simulation for chains of length N=10,0 in a solvent of dimers [79].

cases. For the high frequency case, $\langle z \rangle=8.85\sigma$ which is unchanged from its value for no shear, $v_w=0$. The time averaged density profile for this case is nearly identical to that for $v_w=0$. There is no evidence for an increase in the brush height. Results for the lower frequency (larger maximum displacement) case, show a decrease in the average brush height, $\langle z \rangle=8.3\sigma$. Three other cases studied, $(U,X)=(0.26\sigma/\tau, 20\sigma)$, $(0.52\sigma/\tau, 100\sigma)$, and $(0.52\sigma/\tau, 300\sigma)$, also showed no increase in the brush height.

While the results presented in Fig. 11 and 12 are clearly in the regime $\dot{\gamma}\tau_z \gg 1$, the question is how close do they correspond to the experimental situation studied by Klein et al. [16]. From earlier work on mapping the equilibrium simulations to the experiments of Taunton et al. [37], we [46] found that $\sigma \simeq 1.5$ nm for the case of polystyrene of $M_W=1.41\times10^5$ in toluene. (The analysis was done for a continuum solvent simulation but the results will not change sig-

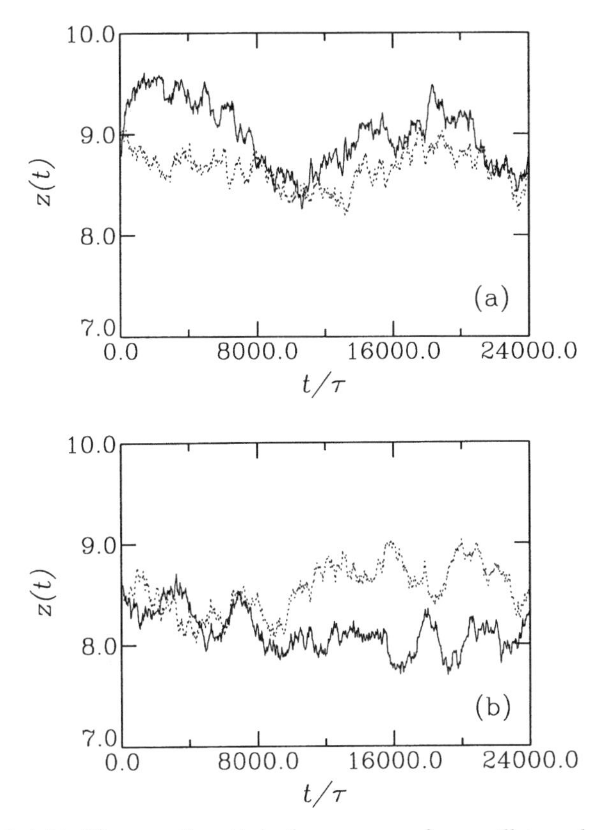

Fig. 12. Brush height $z(t)$ versus time t/τ in the presence of an oscillatory shear for the two cases presented in Fig. 11. The solid curve is for the upper (moving) surface and the dotted line is for the lower surface. From ref. [79].

nificantly for this case.) From comparisons for the shear forces for highly compressed brushes, discussed in the next section, I estimate that $\tau \simeq 1.5 \times 10^{-6}$ s for this case. Thus the maximum displacements of X=100 and 300σ correspond to approximately 150 and 450 nm respectively with a frequency $\nu = \omega / 2\pi \simeq 270$ and 90 Hz respectively. These are well within the experimental range studied by Klein et al. [16] in which the maximum displacement covered a range of values with the largest being about 1500 nm (quoted in their paper as 3000 nm peak to peak) and frequencies up to 540 Hz. While simulations at yet higher velocities and larger displacements are definitely of interest to completely cover the experimental range of parameters, these are not possible at present due to limitations on computer resources. The results presented here each took approximately 1000 hours of CPU time on one processor of our Cray J916 (about the speed of the older Cray XMP processor).

Finally, there also remains the possibility that the extra normal force observed by Klein et al. [16] may arise for other reasons than an increase in the brush height. Since the experiments are in an open system, solvent entering and leaving the gap between the two curved surfaces may contribute to the observed extra normal force. This could explain the fact that there is a critical shear rate for the extra normal force to appear. Sufficed to say, the origin of the excess normal force under an oscillatory shear remains an open issue. Further experiments and simulations are needed to clarify the origin of this effect.

4 Forces Between Compressed Brushes

4.1 Normal Force

When two surfaces with end-grafted chains are brought into contact, they normally repel each other. This repulsion, which is a result of the osmotic interaction between the polymer segments, is the basis for colloidal stabilization [1,80,81]. Experimentally the force between polymer brushes has been studied using the SFA for both the normal [18,37] and shear forces [82–87]. While this is a very powerful and sensitive technique for measuring forces between surfaces separated by tens to hundreds of angstroms, it does not give any spatial information on the conformations of the polymer chains. Because of the small distances involved, it is difficult to obtain such information experimentally. Simulations in the other hand can give detailed information about these interactions, including the force-distance profile, changes in the monomer density as a result of compression and the amount of interpenetration of the two brushes. This last quantity is usually ignored in theoretical treatments, as it is expected to vanish for infinitely long chains [88,89]. In experimentally relevant systems [46] however, the amount of interpenetration will be finite due to the finite length of the polymer chains. Interpenetration affects the interaction between grafted surfaces, as well as the shear rheology of compressed brushes [4,82–87].

As two brushes are brought into contact, two processes are found to occur concurrently: interpenetration and compression. When the distance D between the brushes equals twice the maximum extent h_{ext} of each, the parabolic profiles of both brushes begin to overlap and the density increases everywhere between the two grafting surfaces. At small separations the overall density becomes almost uniform in the gap between the grafting surfaces as shown in Fig. 13}. The amount of interpenetration at a given separation decreases as N increases. One can quantify the amount of interpenetration in several ways. The quantity $I(D)$, where

$$I(D)=\int_{D/2}^{D}\rho_l(z)dz\Big/\int_0^D\rho_l(z)dz \tag{8}$$

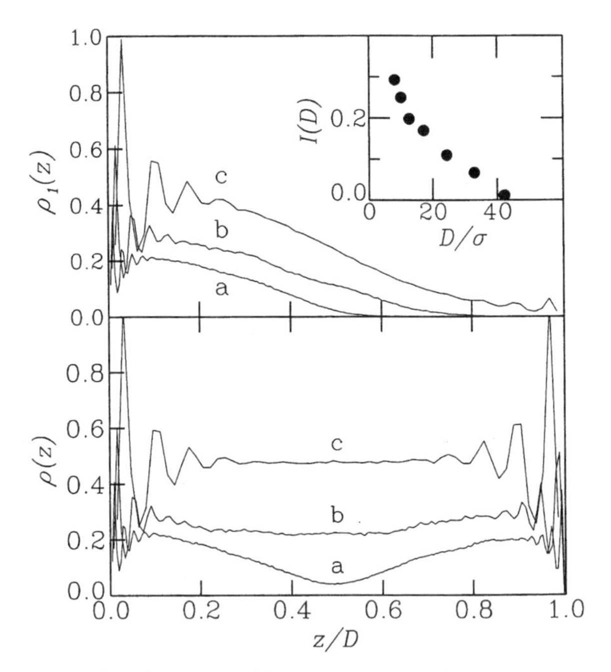

Fig. 13. Brush mer number density profile versus distance from the grafting surface z for polymers of length N=100 and solvent of length N_f=2 at $\rho_a\sigma^2$=0.03 for (a) D=42.1σ, (b) D=24.3σ, and (c) D=12.6σ. Mers from opposite surfaces first begin to overlap at $D \simeq 60\sigma$. The total brush mer number density $\rho(z)$ is shown in the lower panel and the mer density $\rho_1(z)$ for one brush is in the upper panel. The inset shows the amount of interpenetration $I(D)$ versus D. From ref. [63].

was suggested in ref. [90]. Here $\rho_1(z)$ is the contribution of one of the brushes to the overall density and z is the distance measured from the grafting surface of that brush. The inset of Fig. 13 shows a typical result for $I(D)$ from my simulations for chains of length N=100 in a solvent of dimers. Results of the SCF theory of Milner et al. [33] together with a scaling argument by Witten et al. [89] can be used to show that $I(D)$ in a good solvent has the following scaling form with N and ρ_a,

$$I(D)N^{-2/3}h_{\text{par}}^{4/3} \sim x^{-4/3}\left(1-x^3\right) \tag{9}$$

where $x=D/2h_{\text{par}}$. Here h_{par} is the brush height derived from the parabolic density profile predicted by the SCF theory. In addition to giving the interpenetration as a function of the separation for a given brush, this expression also shows that $I(D)$ at a given separation decreases as $N^{-2/3}\rho_a^{-4/9}$. Since the decay with N is rather slow, the amount of interpenetration is expected to be significant for experimental systems. This scaling form was confirmed by Murat and Grest [90].

Chakrabarti et al. [91] used a different measure of the amount of interpenetration but also found that their results agreed with the self-consistent field theory.

The force between grafted polymers is repulsive at all separations. There are also no hysteresis effects, indicating fast equilibration in both experiments [18,37,82,83] and simulations [90–95]. For continuum simulations, forces between the brushes can be calculated in a straightforward manner from the virial [96], while on a lattice the force can be obtained by periodically testing whether shifting the plate separation from D to $D\pm1$ causes an overlap of two mers. For high densities, which occur for strong compression, several intermediate simulations are needed to determine the probability that D can be shifted by one lattice spacing [92,93]. A similar approach has been used to determine the force between brushes modeled as a freely jointed tangent hard-sphere (or pearl-necklace) chain [94]. The force per unit area $f(D)$ between the two plates is then found by subtracting the pressure at $D=2h_{ext}$. The force between two surfaces bearing end-grafted chains reaches its equilibrium value very rapidly, in contrast to other equilibrium properties such as the free end distribution, which is much slower to relax [90]. Because the SFA experiments use two crossed cylinders and not two flat surfaces as in the simulations, the two cannot be compared directly. However in the Derjaguin approximation [97], the interaction energy per unit area,

$$E(D)=\int_{2h_{ext}}^{D}\Pi_{\perp}(D')dD' \tag{10}$$

between two flat surfaces is proportional to the normal force F divided by the radius of curvature R_c, F/R_c, for two crossed cylinders. Here $\Pi_{\perp}$ is the normal osmotic pressure. Scaling arguments [88,98] and SCF theory [33] predict that this energy should scale as $E(D)\sim h\rho_a^y\tilde{E}(D/2h)$, with $y = 3/2$ and $4/3$, respectively. The difference originates from the dependence of the osmotic pressure on the monomer density in the two theories. These predictions are obtained assuming that there is no interpenetration, so that bringing the two brushes into contact results only in their compression. The repulsive force is then calculated from the increase in the free energy due to the compression. This turns out to be a very good approximation as the force between two surfaces each bearing end grafted chains is approximately the same as the force between two surfaces, one bare and the other bearing end-grafted chains, after one scales the distance D by the number of surfaces n_s bearing polymer [86,99–101]. Martin and Wang [102] found using SCF theory that the force profiles are relatively insensitive in a good solvent to the amount of interpenetration of the chains. Fig. 14 shows molecular dynamics simulation results (solid line) for $E(D)$ scaled as suggested by the scaling theory for $N=100$ compared to the experimental results of Taunton et al. [37] for terminally attached polystyrene chains of various M_W [90]. Since the energy scale in this simulation is arbitrary, the line is vertically shifted to lie on the experimental points for comparison. The overall agreement between the simulation and the experimental results is excellent. This fit between the molecular dynamics simulations and experiments at intermediate values of ρ_a can be utilized [46] to map the simulation to real physical systems. It is important to note that

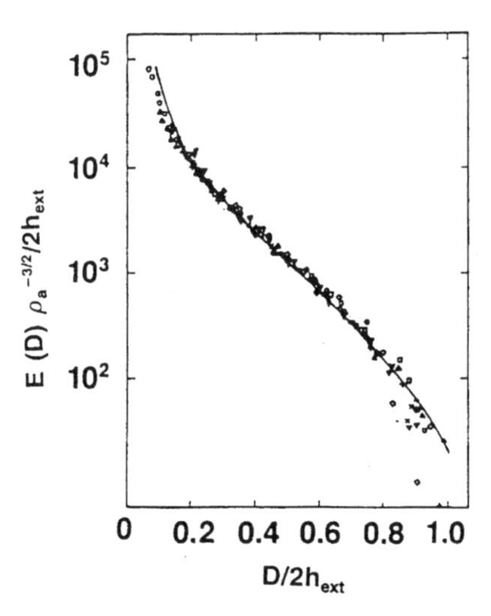

Fig. 14. Scaled energy per unit area versus $D/2h_{ext}$ from the molecular dynamics simulations of Murat and Grest [90] for chains of length N=100 for $\rho_a\sigma^2$=0.03 (full line) compared to SFA results of Taunton et al. [37] (data points) for polystyrene in a good solvent (toluene) for a variety of molecular weights. The simulation results have been shifted vertically by an arbitrary amount to overlap with the experimental data.

the fit is achieved using only one adjustable parameter, the vertical shift due to the arbitrary energy scale of our simulations. The length scale used in the scaling of D is taken from simulations of a single brush.

Although the data collapse shown in Fig. 14 uses the functional form predicted by the scaling analysis, a comparable collapse of the data is obtained using the prediction of the SCF theory. This is also the case with experimental data [37]. While the experimental results can be fit satisfactory with either the SCF or the scaling theories if both the coverage ρ_a and scale of the force are taken as adjustable parameters, the difference in the fits shows up strongly if one uses known parameters. Milner [37] has shown that the parabolic profile, with a small polydispersity (M_W/M_N), can account for the measured force profiles with no adjustable parameters, whereas the step function profile in the Alexander-de Gennes model has too short a range for the onset of the force.

This very good agreement between the simulations and the theoretical predictions are expected to break down at high compression as the average mer density ϕ approaches unity. In this limit the osmotic pressure diverges logarithmically with $(1-\phi)$, rather than increase as a power law as assumed by both the scaling and SCF theories. This was seen in the simulation of Murat and Grest [90] for high coverage even for moderate values of $D/2h$. Such a trend, however, has not been seen experimentally since all of the SFA experiments are carried out at relatively low ρ_a. This is particularly true for the experiments of Taunton et al. [37] in which case the binding energy of the end attachment is only about 6–8 k_BT. Accordingly ρ_a decreases as N increases, making large value of ρ_a unachievable.

While the interaction between polymer brushes is almost always repulsive, in a mixed solvent it is possible to have attractive interactions [52,102,104–106]. In a pure solvent the interaction is always repulsive since removal of the solvent to the reservoir is always accompanied by an increase in the unfavorable steric interactions between the brushes. However in a mixed solvent, the relative concentrations of the solvent species within the brush can vary, allowing different amounts of the two species to leave the brush. For the case of a mixed solvent containing identical chains of two different chain lengths, the longer chains will be more strongly repelled from the brush than the shorter ones. The longer chains will therefore leave the gap between the two surfaces before the smaller ones, resulting in some circumstances in an attractive minimum over a limited range in the compression force. This attractive minimum is found only for strongly stretched chains, when the grafted and free polymer profiles are symmetric around their intersection point [52].

4.2 Shear Force

The shear force between brushes has been measured for compressed brushes for both good [82,86,87] and Θ solvents [83–85,87]. Results of Klein et al. [82] for the normal and shear forces for polystyrene in a good solvent (toluene) are shown in Fig. 15 for several molecular weights and coverages. Note that the brushes have to be highly compressed up to a factor of 5 to 6 before the shear

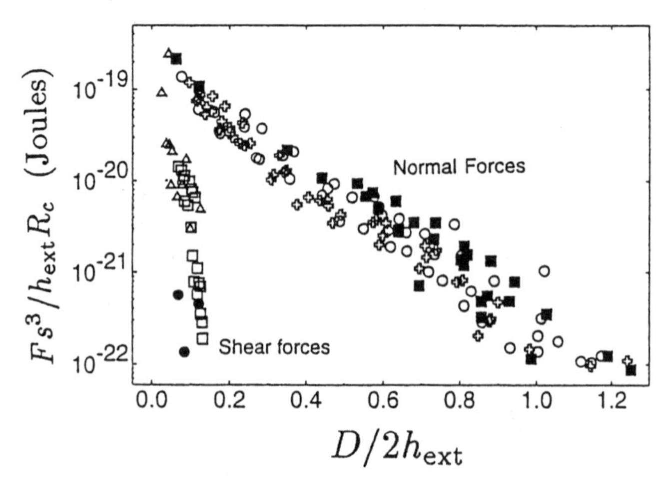

Fig. 15. Normal and shear forces F between compressed polymer brushes sliding pass each other in a good solvent for a variety of systems. From Klein et al. [82].

force is above the resolution of the measurements for all shear velocities studied (v_w=15–450 nm/s). Thus the brushes appear to provide an extremely efficient lubricant for reasonable loads, which are 1–2MPa at the highest compressions. This result agrees with recent work by Pelletier et al. [86] and Kilbey et al. [87] for adsorbed polystyrene/polyvinylpyridine who both found that they were unable to detect a shear stress for compressions up to about $D \simeq h_{max}$. Pelletier et al. used very small deformations, while Kilbey et al. used much larger deformation up to about 2μm, similar to that of Klein et al. [82]. For a Θ solvent [83–85,87], the shear and normal forces set in at comparable separations D particularly for large shear rates. For weak compressions, the elastic component of the shear force can exceed the repulsive normal force for large shear rates. For brushes in the mushroom regime, however, the shear force is predominantly dissipative except at large compression.

Recently I have carried out two sets of simulations for brushes in a good solvent to gain more insight into this interesting phenomena. In the first set of simulations, the solvent was treated as a continuum [95], while in the second set of runs the solvent was included explicitly [63] as discussed in the previous section. The results for the two cases were very similar. Here I concentrate on the explicit solvent results. The brush density profile for this case is shown in Fig. 13. The only difference with the continuum solvent case [95] is the extra structure near the surface due to packing of the mers of the bead-spring chain which are nearly hard spheres. The shear stress f is obtained from the xz component of the microscopic pressure-stress tensor, $f=-P_{xz}$. Because the fluctuations in the shear force are much larger than the average force, long runs are required to determine f. In Fig. 16, the dependence of the shear force f on the sliding velocity v_w is shown for three values of D. For small v_w and large D, f is observed to increase approximately linearly as would be expected for a Newtonian fluid. In this regime there is little change in the mean squared radius of gyration $< R_G^2 >$ or the end-to-end distance $<R^2>$. There also is no change in $\rho_1(z)$compared to the results shown in Fig. 13. Beyond the Newtonian regime, f increases sublinearly for large v_w, in analogy with earlier observations [7,107], for small molecules confined between two bare surfaces. One may call this the non-Newtonian regime. As seen from $< R_G^2 >$ the chains stretch in this regime and there is some disentanglement of the polymers from the two surfaces, particularly for small D and large v_w, as shown in Fig. 17 [63]. In this case for D=8.1σ, $I(D)$ decreases by a factor of two for v_w=0.02σ/τ compared to its value for v_w=0. Although $I(D)$ decreases significantly, the normal osmotic pressure decreases by only about 1%. This reduction in the amount of interpenetration is consistent with the predictions of Joanny [108] for lubrication of molten polymer brushes. In the limit of very large chain length, where the amount of interpenetration is much smaller (see Eq. 9), Joanny found that above a critical sliding velocity the chains can no longer sustain viscous stress. The crossover of the shear force to a non-Newtonian regime for large is partially a consequence of this disentanglement.

Comparison of the normal osmotic pressure $\Pi_\perp$ and the shear force in this regime is shown in Fig. 18 for chains of length N=100 for three values of v_w. These

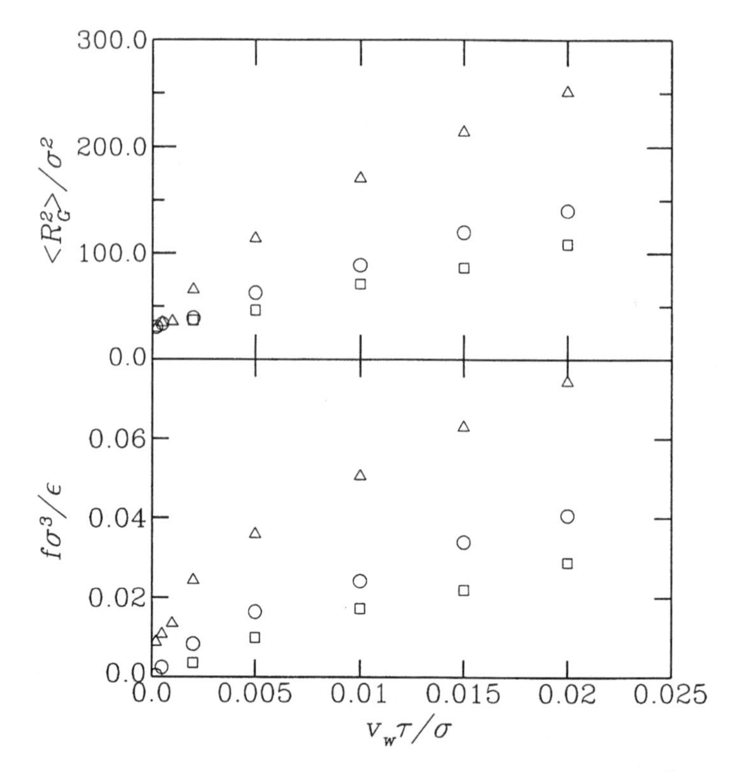

Fig. 16. Shear stress f per unit area and mean squared radius of gyration $< R_G^2 >$ as a function of the relative velocity of the two surfaces for N=100 and N_f=2. The data is for $\rho_a\sigma^2$= 0.03 and D=24.2σ (□), D=17.1σ (○), and D=8.1σ (△). From ref. [63].

results are in qualitatively good agreement with the experiments shown in Fig. 15. The simulation results do not increase quite as rapidly with decreasing plate separation as in the experiment except for very small v_w. Note that since the shear force f is a continuously varying function of velocity even for high shear rates, a true coefficient of friction cannot be defined. As discussed above, the normal forces in the SFA experiment which are on curved cylinders cannot be compared directly to the simulation on flat plates. While a relationship between the two has been derived within the Derjaguin approximation as discussed above, the relationship between the shear forces in the two cases is less clear. It is important to note that the range of D over which f is large enough to detect is much larger in the simulation than in the experiment. This is almost assuredly due to the fact that the wall velocities accessible in the simulations are much larger than in the experiment. Comparing the experimental and simulation results, the latter is fit very well with the lowest velocity, v_w=2.0×10$^{-4}\sigma/\tau$ used in the simulation. Thus it is very likely that all of the present experimental SFA results are only in the low velocity Newtonian-like regime. However since there is

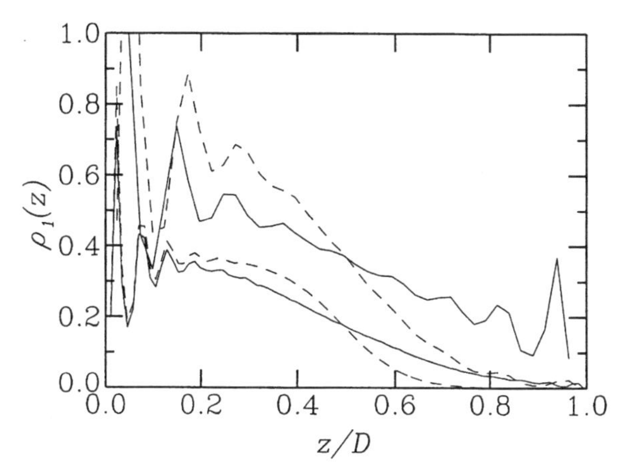

Fig. 17. Monomer density profile $\rho_1(z)$ versus distance from the plate z for D=8.1σ (upper pair of curves) and for D=17.1σ (lower set of curves) for v_w=0 (solid line) and v_w=0.02σ/τ (dashed line) for N=100 and N_f=2 at $\rho_a\sigma^2$=0.03. From ref. [63].

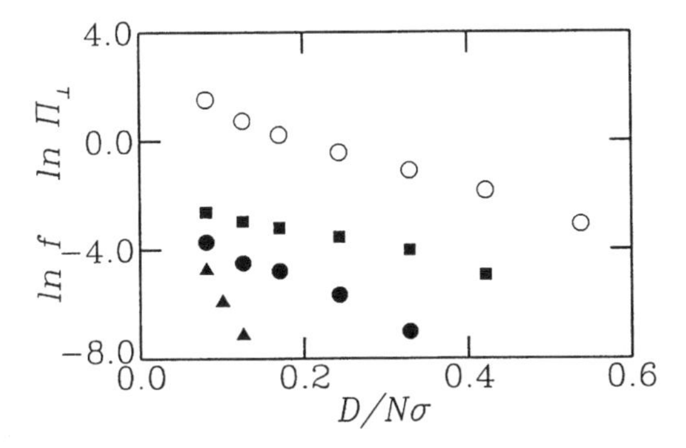

Fig. 18. Semilog plot of the normal osmotic pressure $\Pi_\perp$ (○) and shear force f versus the plate separation D between polymer brushes of 100-mers in a solvent of free dimer molecules for ρ_a=0.03σ^{-2}. The shear force is shown for v_w=2.0×10^{-4} (▲), 2.0×10^{-3} (●), and 2.0×10$^{-2}\sigma/\tau$ (■). From ref. [63].

presently no experimental data available for the velocity dependence of f, it is not possible to determine with certainty which part of the shear force-velocity curve the experiments access.

These results for the shear force can be used to map the bead-spring model simulations to experiment. Equating the velocity v_w=2.0×10$^{-4}\sigma/\tau$ with a typical value of 200 nm/s used in the experiments [82] (actual range 15–450 nm/s) for M_W=1.41×10^5 molecular weight polystyrene in toluene, gives $\sigma/\tau\simeq 10^6$ nm/s. Using $\sigma\simeq$1.5 nm determined from our previous mapping [46] of the equilibrium

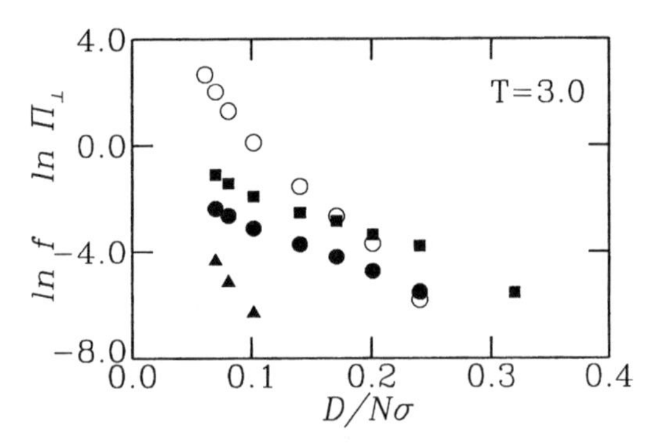

Fig. 19. Semilog plot of the normal osmotic pressure $\Pi_\perp$ (○) and shear force f versus the plate separation D between polymer brushes of 100-mers at ρ_a=0.03σ^{-2} for $T=T_\theta$. The shear force is shown for v_w=0.001 (▲), 0.01 (●), and 0.1σ/τ (■). From ref. [79].

brush to these same experiments, gives $\tau \simeq 1.5\times10^{-6}$s. This value was used in the previous section to compare the simulations and experiments [16] for non-overlapping brushes. Note that this mapping is only approximate and definitely is not universal, as it depends on the specific polymer, its chain length, and the solvent quality.

For a Θ solvent [83–85,87], the shear and normal forces set in at comparable separations D in contrast to what occurs in a good solvent. In fact for weak compressions, the elastic component of the shear force can even exceed the repulsive normal force for large shear rates. In Fig. 19, results from a molecular dynamics simulation for the normal force is compared to the shear force for three different wall velocities [79]. For simplicity, the simulations were carried in a continuum solvent. The effect of the solvent was included by extending the range of the interaction to r_c=2.5σ as discussed in the Appendix. In agreement with experiment, for large enough shear rate the shear force can even exceed the normal force. To compare these results to that of a good solvent with the same interaction, the simulations were repeated at a higher temperature, 4.0ϵ/k_B($T_\Theta\simeq$ 3.0ε/k_B). Surprisingly, the shear forces in the two case were nearly identical. The main difference in the two cases was in the normal force, which as expected extended to larger separations D in the good solvent compared to the Θ solvent. This study suggests that simulations are an excellent way to study the effect of solvent quality on the shear properties of brushes. Including the solvent monomers explicitly, which was not done in these first numerical studies of the effect of solvent quality, is also likely to be of interest.

5 Summary and Concluding Remarks

This review summarizes our present understanding of the forces, both normal and shear, between polymer brushes with emphasis on the contributions from computer modeling. The equilibrium properties of an uncompressed brush predicted by self-consistent field theory have been confirmed by a number of experiments and simulations. Under steady state flow, some of the chains stretch in the direction of flow while many others are buried within the brush and are only weakly affected by the flow. Molecular dynamics simulations in which the solvent molecules are included explicitly as well as several Monte Carlo simulations in which the effect of the solvent is treated self-consistently show that the brush height decreases slightly with increasing shear rate in accordance with the prediction of Rabin and Alexander [73] and in contrast to several theoretical models [69–72] which predict a strong increase. These simulations strongly suggest that the simple Alexander-de Gennes model of a brush which assumes that the density profile is uniform and all the ends are at the outer surface is inadequate for explaining the properties of polymer brush under shear. Further work, both theoretical and experimental, are needed to clarify the case of oscillatory shear where the interplay of the relaxation of the brush with the shear rate can lead to very interesting effects, including a coupling of the normal and shear forces.

When two surfaces bearing an end-grafted polymer are compressed, the brushes interpenetrate only weakly and easily slide pass each other in a good solvent with immeasurably small shear force at low compression. This is seen both in the experiments by Klein et al. [82] and in two recent molecular dynamics simulations [63,95]. Most of the shear occurs near the midplane between the two surfaces. For higher compressions two regimes could be identified from the simulations, depending on the shear rate. For low shear rates, the brush density profiles and chain conformations are relatively unchanged even for large compressions. In this Newtonian-like regime, the shear force increases approximately linearly with velocity. However for large shear rates, the chains become strongly stretched and the shear force increases sublinearly with shear velocity. The experimental results for a brush in a good solvent [82] appear to be in the Newtonian-like regime, but further experiments which explore the velocity dependence of the shear forces is needed to determine this for certain. As the solvent quality is decreased, the nature of the shear forces changes as it becomes more elastic. For a Θ solvent [79,83–85] the shear elastic force can even be larger than the normal force for large shear rates.

The effect of solvent quality on the lateral forces has only just begun to be investigated [79] using numerical simulations and is an interesting problem for future research. These simulations, along with the study of adsorbed polymers under shear which experimentally do not show a comparable reduction in the sliding friction [4], will be very important in improving our understanding of these systems. Experimentally, there are a number of interesting questions, including the velocity dependence of the shear forces observed in the simulations, which should be investigated. A direct measurement of the chain stretching, presumably by scattering, is an important challenge for experimentalist which will be very helpful in developing a better understanding of how two surfaces bearing end-grafted chains slide pass each other.

Acknowledgement. I thank my collaborator, M. Murat, for his contributions to the simulation work and for a critical reading of the manuscript.

6
Appendix – Simulation Models and Methodology

Polymer brushes have been studied extensively using both lattice and continuum models. Two types of models have been widely used to study polymer on a lattice [28,109]. The first is the standard lattice algorithm in which the polymer is represented by a succession of N steps which are subject to the constraint that no lattice site is visited more than once. Chakrabarti and co-workers [43,91,110] have used this model in many of their simulations of polymer brushes. The other is the bond fluctuation model [111] in which each monomer consists in three dimensions of eight lattice sites. In addition to the excluded volume interaction, the bond length l is allowed to vary subject to a maximum extension to avoid bond crossings. This method is significantly faster than the standard lattice algorithms and has been used by Lai and Binder [44,45,65], Dickman and Hong [92] and Dickman and Anderson [93] to study static properties of polymer brushes including the force between two compressed brushes. A variant on this model in which each monomer consists of only one site in a simple cubic lattice and the bond lengths are allowed to fluctuate among certain allowed values has been used by Shaffer [112] to study polymer brushes. For shear properties lattice models are not applicable and one needs to use off-lattice models.

One model which has been extensively used to model polymers in the continuum is the bead-spring model. In this model a polymer chain consists of N beads (mers) connected by a spring. The easiest way to include excluded volume interactions is to represent the beads as spheres centered at each connection point on the chain. The spheres can either be hard or soft. For soft spheres, a Lennard-Jones interaction is often used, where the interaction between monomers is

$$U^0(r)=\begin{cases} 4\varepsilon\left[\left(\frac{\sigma}{r}\right)^{12}-\left(\frac{\sigma}{r}\right)^{6}-\left(\frac{\sigma}{r_c}\right)^{12}+\left(\frac{\sigma}{r_c}\right)^{6}\right] & \text{if } r\le r_c; \\ 0 & r>r_c. \end{cases} \tag{11}$$

Here ϵ characterizes the strength of the interaction, σ is the unit of length, and r_c is the range of the interaction. If the potential is truncated at $r_c=2^{1/6}\sigma$ then the potential is purely repulsive. This is a very efficient model for studying good solvent conditions and is often referred to as athermal since temperature has only a very small effect on the results. Alternatively, one can extend the range of interaction to model the effect of solvent quality. In some simulations the cut-off was set of infinity but it is computationally much more efficient to truncate the interaction at a finite range, such as $r_c=2.5\sigma$. In this case the Θ-temperature $T_\theta \simeq 3.0\epsilon/k_B$ [46]. By changing T, the relative importance of the attraction between mers and thereby the effective quality of the solvent can be changed without explicitly introducing solvent particles. A similar approach as been used to study polymer brushes for different solvent qualities on a lattice using Monte Carlo methods [45,113].

In addition to the excluded volume interaction, one also needs to add an attractive interaction binding the beads along the chain and for polymer brushes a wall-monomer interaction. The specific form of these interactions is not critical, except that the interaction between bonded beads should be such that the maximum extent of the bond is small enough such that bond crossing is inhibited. For the simulation results presented here, the finite extendible nonlinear elastic (FENE) model first introduced by Bird et al. [114] with suitably modified parameters [46,115] to avoid bond crossing was used.

Now that we have settled on a model, one needs to choose the appropriate algorithm. Three methods have been used to study polymers in the continuum: Monte Carlo, molecular dynamics, and Brownian dynamics. Because the distance between beads is not fixed in the bead-spring model, one can use a very simple set of moves in a Monte Carlo simulation, namely choose a monomer at random and attempt to displace it a random amount in a random direction. The move is then accepted or rejected based on a Boltzmann weight. Although this method works very well for static and dynamic properties in equilibrium, it is not appropriate for studying polymers in a shear flow. This is because the method is purely stochastic and the velocity of a mer is undefined. In a molecular dynamics simulation one can follow the dynamics of each mer since one simply solves Newton's equations of motion for mer i,

$$m\frac{d^2\mathbf{r}_i}{dt^2} = -\nabla U_i - m\Gamma\frac{d\mathbf{r}_i}{dt} + \mathbf{W}_i(t) \tag{12}$$

where m is the mass of each monomer. To take into account the effect of the solvent, a viscous damping and random white noise term have been added to the equation of motion, where Γ is the friction coefficient that couples the mers to a heat bath. The random part of the mer-heat bath coupling is given by a white-noise term $\mathbf{W}_i(t)$ which satisfies the fluctuation dissipation theorem. In the limit that the inertia term dominates over the viscous damping and noise term, I refer to this method as molecular dynamics or more properly stochastic molecular dynamics. For equilibrium simulations without shear, a convenient value of the viscous damping is $\Gamma=0.5\tau^{-1}$, where $\tau=\sigma(m/\epsilon)^{1/2}$. In this case the monomeric friction from collisions is significantly larger than that coming from the viscous damping term in Eq. 12 [115]. In the overdamped limit, in which the inertia term on the right hand side of Eq.12 can be ignored, one has Brownian dynamics. However for dense polymer systems, this method is not very efficient [116] since very small time step are needed to keep the algorithm stable.

The equations of motion of the mers can be integrated using a variety of techniques. In the work presented here, I used a velocity-Verlet algorithm [96]. The time step Δt depends on the range of interaction, the temperature and the viscous damping. In the purely repulsive case, with no shear and $\Gamma=0.5\tau^{-1}$, one can use a time step as large as Δt=0.012τ. For larger values of r_c or if one sets the viscous damping Γ=0, Δt must be reduced to 0.006–0.008τ to keep the algorithm stable.

The standard geometry for studying polymer brushes, either on a lattice or in the continuum consists of M linear chains of length N+1 attached randomly to a flat surface, usually taken as the z=0 plane as shown in Fig. 1. The first mer of each chain is usually firmly attached to the grafting surface and does not move, although there have been a few simulations in which the grafting point is free to move in the grafting plane [47]. Periodic boundary conditions are used in the x and y directions and the length of the cell in these two directions were equal, $L_x = L_y$. The surface coverage $\rho_a = M/A$, where $A = L_x L_y$. However to include solvent molecules explicitly, the simulation cell must be closed. This is done by introducing a second surface in the xy plane a distance L_z from the first as shown in Fig. 5. The volume of the cell is $V = AL_z$. Enough free solvent, in this case short chains of length N_f of identical mers as the polymer brush, was then added to bring the overall density of the system to a specified value. For the results presented in this paper, I used $\rho = 0.85\sigma^{-3}$, the same density used by Kremer and Grest [115] in their studies of the dynamics of polymer melts for the same model.

Shear is imposed by sliding the top surface at a fixed velocity v_w in the x-direction [7,63]. The simulations are performed in an ensemble in which the total number of polymer brush and solvent molecules, the surface coverage ρ_a, and the separation of the plates D are held constant. To avoid any possible biasing of the flow in the shear direction (x), only the y component of the velocity was coupled to the reservoir for $v_w \neq 0$ in contrast to the equilibrium case in which the all three components were coupled to the heat bath. Thus Langevin noise and frictional terms were added to the equations of motion in all three directions for $v_w = 0$, though only in the y-direction for nonzero shear velocities. However since introducing a viscous damping screens hydrodynamic interactions [117], some runs were made in which only the first 10 mers of each grafted chain and none of the solvent chains were coupled to the thermal reservoir. No significant differences in the two cases were observed [63].

7 References

1. Napper DH (1983) Polymeric Stabilization of Colloidal Dispersions. Academic, London
2. Russell WB, Saville DA, Schowalter WR (1989) Colloidal Dispersions. Cambridge University Press, Cambridge
3. Brown HR (1993) Macromolecules. 26: 166; (1996) MRS Bulletin 21(1): 24
4. Klein J(1996) Ann Rev Mat Sci 26: 581
5. Hu HW, Granick S (1992) Science 258: 1339
6. Thompson PA, Robbins MO (1990) Science 250: 792
7. Thompson PA, Grest GS, Robbins MO (1992) Phys Rev Lett 68: 3448; Thompson PA, Robbins MO, Grest GS (1995) Israel J Chem 35: 93
8. Manias E, Bitsanis I, Hadziioannou G, ten Brinke G (1996) Europhys Lett 33: 371
9. Manias E, Hadziioannou G, ten Brinke G (1996) Langmuir 12: 4587

10. Khare R, de Pablo JJ, Yethiraj A (1996) Macromolecules 29: 7910
11. Alexander S (1977) J Phys Paris 38: 983
12. De Gennes PG (1980) Macromolecules 13: 1069
13. Auroy P, Auvray L, Léger L (1991) Macromolecules 24: 5158
14. Auroy P, Auvray L (1992) Macromolecules 25: 4134
15. Taunton HJ, Toprakcioglu C, Fetters LJ, Klein J (1990)Macromolecules 23: 571
16. Klein J, Perahia D, Warburg S (1991) Nature 352: 143
17. Ansarifar MA, LuckhamPF (1988) Polymer 29: 329
18. Patel SS, Tirrell M (1989) Ann Rev Phys Chem 40: 597
19. Gast AP, Munch MR (1989) Polymer Comm 30: 324
20. Parsonage E, Tirrell M, Watanabe H, Nuzzo RG 1987 (1991)Macromolecules 24
21. Field JB, Toprakcioglu C, Ball RC, Stanley HB, Dai L, Barford W, Penfold J, Smith G, Hamilton W (1992) Macromolecules 25: 434
22. Granick S, Herz J (1985) Macromolecules 18: 460
23. Kent MS, Lee LT, Farnoux B, Rondelez F (1992) Macromolecules 25: 6240
24. Kent MS, Lee LT, Factor BJ, Rondelez F, Smith GS (1995) J Chem Phys 103: 2320
25. Bates FS (1991) Science 251: 898
26. Halperin A, Tirrell M, Lodge TP (1991) Adv Polym Sci 100: 31
27. Milner ST (1991) Science 251: 905
28. Grest GS, Murat M (1995) In: Binder K (ed) Monte Carlo and Molecular Dynamics Simulations in Polymer Science. Oxford University Press, New York, p 476
29. Szleifer I, Carignano MA (1996) In: Prigogine I, Rice SA (ed) Advances in Chemical Physics. Wiley, New York, Vol. 94, p. 165
30. Watanabe H, Tirrell M (1993) Macromolecules 26: 6455
31. Dhoot S, Watanabe H, Tirrell M (1994) Colloid Surf A 86: 47
32. Dan N, Tirrell M (1993) Macromolecules 26: 6467
33. Milner ST, Witten TA, Cates ME (1988) Macromolecules 21: 2610
34. Milner ST, Wang ZG, Witten TA (1989) Macromolecules 22: 489
35. Zhulina EB, Borisov OV, Pryamitsyn VA (1990) J Colloid Interface Sci 137: 495
36. Semenov AN (1985) Sov Phys JETP 61: 733
37. Taunton HJ, Toprakcioglu C, Fetters LJ, Klein J (1988) Nature 332: 712; (1990) Macromolecules 23: 571
38. Cosgrove T (1990) J Chem Soc Faraday Trans 86: 1323
39. Auroy P, Auvray L (1993) J Phys II France 3: 227
40. Field JB, Toprakcioglu C, Dai L, Hadziioannou G, Smith G, Hamilton W (1992) J Phys II France 2: 2221
41. Perahia D, Weisler D, Satija SK, Fetters LJ, Sinha SK, Milner ST (1994) Phys Rev Lett 72: 100
42. Murat M, Grest GS (1989) Macromolecules 22: 4054
43. Chakrabarti A, Toral R (1990) Macromolecules 23: 2016
44. Lai PY, Binder K (1991) J Chem Phys 95: 9288
45. Lai PY, Binder K (1992) J Chem Phys 97: 586
46. Grest GS, Murat M (1993) Macromolecules 26: 3108
47. Grest GS (1994) Macromolecules 27: 418
48. Raphaël E, Pincus P, Fredrickson GH (1993) Macromolecules 26: 1996
49. Aubouy M, Fredrickson GH, Pincus P, Raphaël E (1995) Macromolecules 28: 2979
50. Milner St (1990) J Chem Soc Faraday Trans 86: 1349
51. Wijmans CM, Scheutjens JMHM, Zhulina EB (1992) Macromolecules 25: 2657
52. Wijmans CM, Zhulina EB, Fleer GJ (1994) Macromolecules 27: 3238
53. Shim DFK, Cates ME (1989) J Phys, Paris, 50. 3535
54. Lai PY, Halperin A (1991) Macromolecules 24: 4981
55. Milner ST (1991) Macromolecules 24: 3704
56. Brinkman HC (1947) Appl Sci Res A1: 27

57. Klein J, Kamiyama Y, Yoshizawa H, Israelachvili JN, Fredickson GH, Pincus P, Fetters LJ (1993) Macromolecules 26: 5552
58. Dhinojwala A, Granick S (1997) J Chem Soc Faraday Trans 92: 619
59. Cho YK, Dhinojwala A, Granick S (1997) J. Polym. Sci. Polym. Phys. Ed. 35: 2961
60. Fredrickson GH, Pincus PA (1991) Langmuir 7: 786
61. Grest GS, J Chem Phys 105: 5532
62. Peters GH, Tildesley DJ (1995) Phys Rev E 52: 1882; (1996) ibid 54: 5493
63. Grest GS (1997) In: Drake JM, Klafter J, Kopelman R (ed). Dynamics in Small Confined Systems III. Materials Research Society, Pittsburgh, Vol. 464, p 71
64. Grest GS (1997) Current Opinion Colloid Interface Science 2: 271
65. Lai PY, Binder K (1993) J Chem Phys 98: 2366
66. Lai PY, Lai CY (1996) Phys Rev E 54: 6958
67. Miao L, Guo H, Zuckermann MJ (1996) Macromolecules 29: 2289
68. Doyle PS, Shaqfeh ESG, Gast AP (1997) Phys Rev Lett. 78: 1182; 1998 Macromolecules 31: 5474
69. Barrat JL (1992) Macromolecules 25: 832
70. Kumaran V (1993) Macromolecules 26: 2464
71. Harden JL, Cates M (1996) Phys Rev E 53: 3782
72. Aubouy M, Harden JL, Cates ME (1996) J Phys II France 6: 969
73. Rabin Y, Alexander S (1990) Europhys Lett 13: 49
74. Kröger M, Loose W, Hess S (1993) J Rheol 37: 1057
75. Williams DRM (1993) Macromolecules 26: 5806
76. Nguyen D, Clarke CJ, Eisenberg A, Raifailovich MH, Sokolov J, Smith GS (1997) J Appl Cryst 30: 680
77. Webber RM, Anderson JL, Jhon MD (1990) Macromolecules 23: 1026
78. Leighton D, Acrivos A (1987) J Fluid Mech 177: 109; (1987) 181: 415
79. Grest GS (unpublished)
80. Witten TA, Pincus P (1986) Macromolecules 19: 2509
81. Stuart MAC, Cosgrove T, Vincent B (1986) Adv Colloid Interface Sci 24: 143
82. Klein J, Kumacheva E, Mahalu D, Perahia D, Fetters LJ (1994) Nature 370: 634; Klein J, Kumacheva E, Perahia D, Mahalu D, Warburg S (1994) Faraday Discuss 98: 173
83. Granick S, Demirel AL, Cai LL, Peanasky J (1995) Israel J Chem 35: 75; Cai LL, Peanasky J, Granick S (1996) Trends Polym Sci 4: 47
84. Dhinojwala A, Granick S (1996) Macromolecules 29: 1079
85. Cai LL (1997) Nanorheology of Polymer Brushes, Ph D Thesis, University of Illinois
86. Pelletier E, Belder GF, Hadziioannouu G, Subbotin A (1997) J Phys II France 7: 271
87. Kilbey II SM, Bates FS, Tirrell M (unpublished)
88. De Gennes PG (1985) C R Acad Sci Paris 300: 839
89. Witten TA, Leibler L, Pincus P (1990) Macromolecules 23: 824
90. Murat M, Grest GS (1989) Phys Rev Lett 63: 1074
91. Chakrabarti A, Nelson P, Toral R (1994) J Chem Phys 100: 748
92. Dickman R, Hong DC (1991) J Chem Phys 95: 4650
93. Dickman R, Anderson PE (1993) J Chem Phys 99: 3112
94. Toral R, Chakrabarti A, Dickman R (1994) Phys Rev E 50: 343
95. Grest GS (1996) Phys Rev Lett 76: 4979
96. Allen MP, Tildesley DJ (1987) Computer Simulation of Liquids. Clarendon, Oxford
97. Israelachvili J (1994) Intermolecular & Surface Forces. Academic, London
98. Patel S, Tirrell M, Hadziioannou G (1988) Colloid Surf. 31: 157
99. Murat M, Grest GS (1996) Macromolecules 29: 8282
100. Dhinojwala A, Cai L, Granick S (1996) Langmuir 12: 4537
101. Strictly speaking, the force between two surfaces each bearing end-grafted chains is not always the same as the force between two surfaces, one bare and the other bearing end-grafted chains, after one scales the distance D by n_s. This is true only if the force between monomers on the chain is comparable to that between a monomer and the

wall or if the compression is high in which case the details of the monomer-wall interaction are not relevant. If the monomer-monomer and monomer-wall interactions are very different, then simply scaling the separation D by the number n_s of surfaces bearing polymer, does not necessarily give the same force profile for weak compressions. As a test, I simulated a brush in a Θ solvent in contact with a wall in which the monomer-wall interaction is dominated by the short range repulsive component while the long range attractive component is small compared to k_BT. When plotted versus D/n_s, the normal force between a bare surface and a surface bearing end-grafted chains extended out further than for that between two surfaces each bearing end-grafted chains. For high compression, the interactions between monomers dominated and scaling by D/n_s gave approximately the same force profile. By increasing the strength of the long range attractive component of the monomer-wall interaction, it is possible to find a range of D where there is an attractive interaction between the bare surface and the surface bearing the end-grafted chains. Earlier simulations by Murat and Grest [99] for a good solvent brush found that the force in the two cases agreed very well when plotted versus D/n_s even for weak compressions.
102. Martin JI, Wang ZG (1995) J Phys Chem 99: 2833
103. Milner ST (1988) Europhys Lett 7: 695
104. van Lent B, Israels R, Scheutjens JMHM, Fleer GJ (1990) J. Colloid Interface Sci 137: 380
105. Shull KR (1991) J Chem Phys 94: 5723
106. Hasegawa R, Aoki Y, Doi M (1996) Macromolecules 29; 6656
107. Granick S (1992) Science 253: 1374
108. Joanny JF (1992) Langmiur 8: 989
109. Binder K (1994) Adv Polym Sci 112: 181
110. Chakrabarti A, Nelson P, Toral R (1992) Phys Rev A 46: 4930
111. Carmesin I, Kremer K (1989) Macromolecules 21: 2819
112. Shaffer JS (1994) Phys Rev E 50: R683
113. Weinhold JD, Kumar SK (1994) J Chem Phys 101: 4312
114. Bird RB, Armstrong RC, Hassager O (1977) Dynamics of Polymeric Liquids. Wiley, New York, Vol. 1
115. Kremer K, Grest GS (1990) J Chem Phys 92: 5057
116. Neelov IM, Binder K (1995) Macromol Theory Simul 4: 1063
117. Dünweg B (1993) J Chem Phys 99: 6977
118. Bacon DJ, Anderson WF (1988) J Molec Graphics 6: 219; Merritt EA, Murphy MEP (1994) Acta Cryst D50: 869

Received: January 1998

Surface-Anchored Polymer Chains: Their Role in Adhesion and Friction

Liliane Léger[1,2], Elie Raphaël[1,3] and Hubert Hervet[1,4]

[1] Laboratoire de Physique de la Matière Condensée, URA 792 du CNRS, Collège de France, 11 Place Marcelin-Berthelot, 75231 Paris Cedex 05, France
[2] *e-mail: lleger@ext.jussieu.fr*
[3] *e-mail: elie@ext.jussieu.fr*
[4] *e-mail: hervet@ext.jussieu.fr*

Polymer surfaces and interfaces have many technological applications. In the present article we review some recent experiments conducted on model systems with the aim of understanding the role played by surface-anchored polymer layers in adhesion and friction. We also discuss some of the related theoretical models. The key parameter for both situations is the degree of interdigitation between the surface layer and the bulk polymer system (an elastomer in the case of adhesion, a molten polymer in the case of friction). We analyze how this degree of interdigitation governs the optimum enhancement in the adhesion energy between the solid wall and an elastomer, and how it is at the origin of the various wall slip regimes observed experimentally.

Keywords. Polymers, Interfaces, Grafted chains, Adsorbed chains, Polymer brushes, Adhesion, Friction, Chains pull out, Slip at the wall

1 Introduction 186

2 Formation and Characterization of the Surface Layers 187

2.1 Formation of the Surface Layers 187
2.2 Characterization of the Internal Structure of the Surface Layers . . 190

3 Interdigitation Between Surface-Anchored Chains and a Polymer Melt 196

3.1 Polymer Brushes in Polymeric Matrices 196
3.2 Irreversibly Adsorbed Layers in Polymeric Matrices 201

4 Role of Surface-Anchored Chains in Adhesion 202

4.1 Adhesion Between a Brush and an Elastomer 203
4.2 Adhesion Between a Pseudo-Brush and an Elastomer 207
4.3 Experimental Results 209

5 Role of Surface-Anchored Chains in Friction 212

5.1 The Three Friction Regimes 213
5.2 Modulation of the Interfacial Friction 218

6 Conclusions 221

7 References 222

1 Introduction

In recent years, a large amount of experimental and theoretical work has been devoted to surface-anchored polymer chains [1–3]. Particular attention has been paid to polymer chains tethered by one end to a solid surface. At high enough coverage, the polymer chains stretch away from the surface forming a polymer "brush" [4]. As shown by Alexander and de Gennes in their pioneering work on polymer brushes [5, 6], the behavior of such end-grafted chains is qualitatively different from that of free chains. For a polymer brush in a good solvent, for instance, the layer thickness L is expected to vary *linearly* with the index of polymerization of the chains, N, while the radius R of a free chain in a dilute solution varies as $R \propto N^{3/5}$. The Alexander-de Gennes model was followed more recently by self-consistent field (SCF) calculations [7–13] and by computer simulations [14–16]. Several review articles on the subject are now available [17–20].

It is commonly admitted that polymer brushes should be quite efficient in many technological applications including in particular the steric stabilization of colloidal suspensions [21–23]. Because of these potential practical applications, strong experimental efforts have been devoted to characterize the structure and the properties of polymer brushes [1, 24]. These efforts, however, have long been slowed down by the difficulty to reach sufficiently high grafting densities in good solvent. Indeed, as soon as the first grafted chains begin to overlap, they repel each other. This, in turn, creates a repulsive barrier for the other chains and the grafting reaction almost stops. The resulting layer then consists of adjacent "mushrooms" [4]. A way to get around this difficulty was proposed by Auroy and co-workers [25–27]. It consists in performing the grafting reaction in a relatively concentrated solution where a rather large number of chains are located in the vicinity of the surface and are susceptible of being grafted. The particular system used by Auroy et al. was α-ω-hydroxyl terminated polydimethylsiloxane (PDMS) chains end-grafted on previously modified porous silica beads (by esterification of most of the silanol sites of the surface by pentanol molecules). An important result of the work of Auroy et al., apart from the demonstration of the possibility of forming chemically end-grafted brushes with high grafting densities, is the fact that the grafting density can be adjusted via the volume fraction of free polymer in the reaction bath. A derived way of modifying the silica surface, adaptable to plane large areas, has been developed recently and will be detailed in Sect. 2.1.

Let us mention at this point that a number of studies have used block copolymers in order to test the Alexander-de Gennes description of polymer brushes. For example, surface force measurements have provided some global characteristics of the layers formed by block copolymers where one block adsorbs strongly on the substrate while the other block adsorbs negligibly [28, 29]. Diblock copolymer monolayers on the surface of a selective solvent have been studied by surface pressure measurements and by neutron reflectivity [30–32]. Block copolymers phases have also been investigated [33]. In these phases, the density of chains tethered to the interface is fixed by the equilibrium conditions of the self-assembly process and can not be chosen independently of the polymerization index, contrary to what can be done with chemically end grafted chains.

When a strongly attractive surface is exposed to a polymer melt, a certain amount of polymer becomes permanently bound to the surface. This occurs, for instance, when an untreated silica surface is exposed to a melt of PDMS [34]. If the polymer layer is then exposed to a good solvent, the layer swells. The resulting layer is made of loops, with a large polydispersity of loop sizes, reflecting the statistics of the chains in the melt state. The structure of such irreversibly adsorbed layers when swollen by a good solvent has been worked out by Guiselin [35]. Irreversibly adsorbed layers have been named "pseudo-brushes", because they are somewhat analogous to highly polydisperse brushes of loops. A comparison of brushes and pseuso-brushes in terms of adhesive and friction properties was seen as an interesting problem, and is one of the objectives of this paper.

The present paper is organized as follows. In Sect. 2 we present an experimental system for which both chemically end-grafted layers with high grafting densities and irreversibly adsorbed surface anchored layers can be formed (Sect. 2.1). We then discuss how the internal structure of these two kinds of surface layers can be analyzed and compared to the different available models (Sect. 2.2). In Sect. 3, we review what is expected for the interdigitation between such surface-anchored layers and a polymer melt, and compare these expectations with neutrons reflectivity results. In Sects. 4 and 5, we analyze the experimental data and the different models which allow to understand under which conditions such surface anchored layers can be used to promote adhesion or to reduce friction.

2 Formation and Characterization of the Surface Layers

2.1 Formation of the Surface Layers

There are several ways of forming surface layers of polymer chains, and various solid/polymer systems have been used. The silica/PDMS system is quite convenient since both end-grafted layers with high grafting densities (i.e., brushes) and irreversibly adsorbed layers (i.e., pseudo-brushes) can be formed with controlled molecular characteristics (polymerization index of the tethered chains and surface density), allowing a detailed investigation of the structure and properties of these two different classes of surface anchored polymer layers.

The irreversibly adsorbed layers are the easier to obtain. PDMS forms hydrogen bonds between the silanol sites of the silica surface and the oxygen atoms of the backbone of the chains, and thus spontaneously adsorbs on a clean silica surface. This adsorption is quasi permanent [36], and can be rendered permanent by using di-hydroxyl terminated PDMS chains. Incubation of the surface with a polymer solution (polymer volume fraction ϕ_0) at 120 °C during 12 h, leads to either condensation of the OH extremities of the chains on silanol sites of the silica, or to double hydrogen bond formation and the surface chains appear immobilized [37]. Then the layer can be washed by a good solvent in order to rinse away all the unattached chains, and dried. The monomers of the surface attached chains then collapse, and form a layer with a thickness directly related to the surface density of the anchored chains. The characteristics of these dry layers can be investigated through X-ray reflectivity techniques [38]. The electronic density inside the layer appears to be comparable to that of the polymer melt, and the dry thickness of the layer, h_0, is governed by both the polymerization index of the surface attached chains, N, and the volume fraction in the solution used for the incubation, ϕ_0. A collection of typical results for h_0, obtained through X-ray reflectivity or ellipsometry, on layers made with chains having polymerization indices in the range 10^2 to 10^4, is reported in Fig. 1, either for incubation from a melt (triangles, ϕ_0=1) or from a solution (squares and discs,

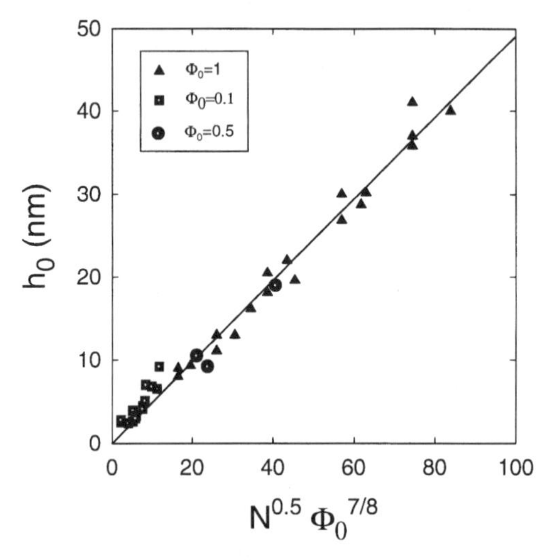

Fig. 1. Dry thickness h_0 of various PDMS layers irreversibly adsorbed on the silica surface of a silicon wafer, as a function of the scaling variable $N^{1/2}\phi_0^{\ 7/8}$. N is the polymerisation index of the surface anchored chains and ϕ_0 is the polymer volume fraction in the incubation bath. The triangles correspond to ϕ_0=1, i.e., to the maximum surface density of anchored chains for the particular molecular weight. The range of molecular weights used is 29.6 kg mol^{-1}$\leq M_w \leq$740 kg mol^{-1}, with M_w=mN, where m is the molar mass of the monomer, m=0.074 kg mol^{-1} for PDMS

$10^{-1} \leq \phi_0 < 1$). A clear scaling law is observed, with the scaling variable $N^{1/2}\ \phi_0^{7/8}$. The origin of this scaling law can be easily understood: when the reaction cell is filled, all the polymer chains located in a layer from the surface with a thickness comparable to the radius of the chains in the incubation solution can rapidly find the surface and bind, keeping the local concentration equal to the bulk polymer concentration. Further adsorption of additional chains implies an increase of the local monomer concentration close to the surface (the chains already attached being not able to go away) and corresponds to a repulsive osmotic pressure. Thus the adsorption process stops when all the chains initially located in the surface layer with a thickness comparable to the radius of the chains in the incubation bath have adsorbed. Note that this is quite different from what occurs in the case of *reversible* adsorption where the chains can permanently exchange between the surface layer and the bulk solution. In the case of reversible adsorption, the adsorbed layer can reach its thermodynamic equilibrium structure, which has been analyzed in terms of a self similar concentration profile [39–41]. For the irreversibly adsorbed layer, the structure is governed by the incubation bath and keeps track of the history of the layer formation. Such irreversibly adsorbed layers can however be put into an environment different from the incubation bath, keeping the surface density of anchored chains constant. Since the radius R of a chain in a semidilute solution scales like $R \cong aN^{1/2}\phi_0^{7/8}$ under good solvent conditions [42], one can easily count the number of monomers per unit surface area inside the surface layer and obtain for the adimensional surface chain density, σ, the following form: $\sigma \cong N^{-1/2}\phi_0^{7/8}$ (the number of chains per unit area, ν, is equal to σ/a^2, where a is the size of a monomer). Each chain containing N monomers of size a, the thickness of the dry surface layer is

$$h_0 \cong a N^{1/2} \phi_0^{\ 7/8} \tag{1}$$

The data reported in Fig. 1 follow this scaling law, and give for the typical size of the monomer, $a \approx 0.5$ nm, a quite reasonable value for PDMS.

In order to form high density end-grafted layers with the same silica/PDMS system, one has to chemically modify the silica surface in order to suppress the monomers adsorption, while still keeping the possibility of end-grafting. An elegant way to do so has been developed by Folkers et al. [43]. It consists in first forming on the silica surface a self assembled monolayer of a short PDMS oligomer made of four monomeric units, and terminated by a chlorine at one extremity and either a Si-H or a Si-vinyl at the other one. The corresponding synthesis and monolayer formation are described in details somewhere else [43]. On such a self assembled monolayer, PDMS hardly adsorbs: under incubation conditions similar to those used to form the surface adsorbed layers, only a layer with a typical thickness of 1 nm remains fixed on the surface (probably adsorbed chains in the holes of the monolayer), to be compared with the 20–50 nm deposited in the case of bare silica with the same polymer and same conditions. On the functionalized extremities of the molecules in the monolayer, mono vinyl or Si-H end terminated long PDMS chains can be grafted by a hydrosililation reaction,

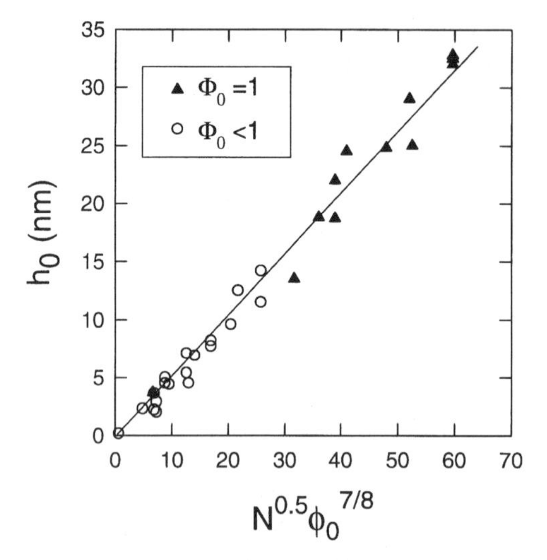

Fig. 2. Dry thickness h_0 f various PDMS layers end grafted on the modified silica surface of a silicon wafer, as a function of the scaling variable $N^{1/2}\phi_0^{7/8}$, with N the polymerisation index of the grafted chains and ϕ_0 he polymer volume fraction in the reaction bath. The range of molecular weights used for the grafted chains is 5 kg mol$^{-1} \leq M_w \leq 263$ kg mol^{-1}, $M_w = mN$, with $m = 0.074$ kg mol^{-1}

in the presence of platinum as catalyst [38, 43]. Again, and for reasons quite similar to what has been said for the adsorbed layers, the surface density in the grafted layer is governed by both the volume fraction in the reaction bath and the polymerization index of the PDMS chains used. The value of the dry thickness of the layer, h_0, determined after rinsing with a good solvent and drying, is a measurement of the surface density of grafted chains. Typical results for h_0, for a variety of PDMS chains and several volume fractions in the reaction bath are reported in Fig. 2. The same scaling law (Eq. 1) as for the irreversibly adsorbed chains is observed, with, at the accuracy of the experiments, the same prefactor, a≈0.5 nm. This means that the dry thickness of the layers, h_0, is a good measurement of the surface density of the surface anchored chains, but does not tell any information on the internal structure of the two categories of layers, irreversibly adsorbed and end-grafted, which, as we will see below, are quite different.

2.2 Characterization of the Internal Structure of the Surface Layers

In order to investigate the internal organization of the surface-anchored chains, we now consider the concentration profile of the layers when swollen by a good solvent. Such concentration profiles have been investigated in details theoretically, since the Alexander-de Gennes model, and we can hope to be able to dis-

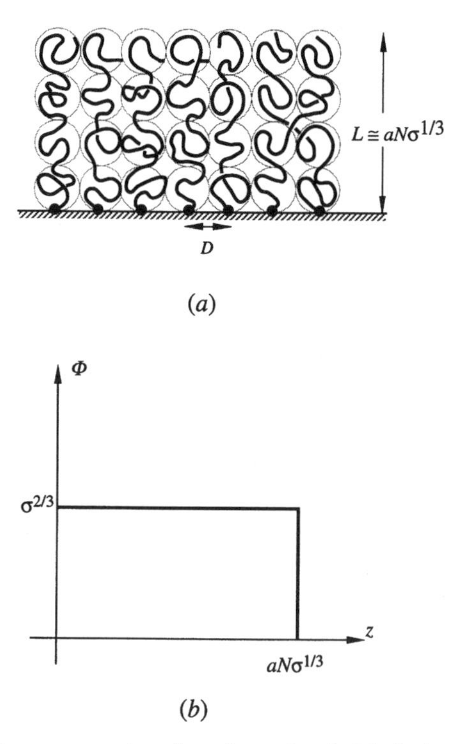

Fig. 3.a Schematic representation of a polymer brush. L is the layer thickness and D the average spacing between two grafting points. **b** The monomer density profile Φ vs distance from the grafting plane z according to the Alexander-de Gennes model. Figure adapted from [66]

criminate between different internal organizations and different models by comparing experiments and theoretical predictions.

Let us first briefly review the Alexander-de Gennes model [5, 6]. Specifically, we consider an ensemble of polymer chains, with degree of polymerization N, terminally grafted onto a flat surface and exposed to a solvent of low molecular weight. The number of terminally grafted chains per unit area is σa^{-2}, where a is the monomer size. The average distance between two grafting sites is given by $D = a\sigma^{-1/2}$. At low coverage, the different coils do not overlap (the so-called mushroom regime) and each chain occupies roughly a half-sphere with a radius comparable to the Flory radius of a free chain $R \cong aN^{3/5}$ (for simplicity reasons we assume the solvent to be athermal, i.e. $v \cong a^3$, where v is the excluded-volume parameter). When the grafting density, σ, becomes greater than the critical density $\sigma_0 \cong N^{-6/5}$ the different chains begin to overlap and form a brush of thickness L, as schematically presented in Fig. 3. The equilibrium structure of the brush results from the competition between the osmotic forces that tend to expand the layer and the restoring elastic forces. Accordingly, the free energy per chain can be written as $F = F_{os} + F_{el}$. In a Flory type approach, the elastic free energy is given by $F_{el}\cong kTL^2/(a^2N)$ while the osmotic free energy is given by $F_{os}\cong kTa^3N^2/(LD^2)$. By minimizing F we obtain

$$L \cong aN\sigma^{1/3} \tag{2}$$

and $F\cong kTN\sigma^{3/2}$. Thus the brush thickness increases *linearly* with N. It is possible to develop a more refined scaling analysis of the overlap regime $\sigma>\sigma_0$ [5, 6]. A grafted chain is subdivided into blobs of size D, each containing g_D monomers. Within one blob, the chain behaves like a free excluded volume chain; this leads to the relation $D\cong ag_D{}^{3/5}$. Different blobs repel each other and the brush is essentially a closely packed system of blobs. The monomer volume fraction, Φ, is given by $\Phi\cong g_D(a/D)^3\cong\sigma^{2/3}$. Since the volume per grafted chain is LD^2 the monomer volume fraction can be written as $\Phi\cong Na^3/LD^2$. Comparing the above two expressions for Φ we obtain the equilibrium brush height $L \cong aN\sigma^{1/3} \cong (\frac{N}{g_D})D$. This last result indicates that the chain can be viewed as a string of blobs almost fully stretched along the normal to the wall (see Fig. 3). The kT per blob ansatz leads to a free energy per chain $F\cong kTN\sigma^{5/6}$. Note that while the Flory analysis gives a correct estimate of the brush height, it overestimates the brush free energy.

The above Alexander-de Gennes model constrains the concentration profile to be steplike and all free ends to be at the same distance from the surface. This imposes strong restrictions on the allowed chains configurations. Some years ago, Hirz [7] and Cosgrove et al. [8] numerically solved the self-consistent field (SCF) equations describing the brush situation, but relaxing the above-mentioned restrictions. They found that in the case of a good solvent the concentration profile was essentially parabolic. This was rationalized analytically by Milner et al. [10, 11] and Zhulina et al. [12]. Much of the physics of the parabolic profile may be understood in terms of the following simple argument due to Pincus [44]. The mean-field free energy per unit area of the brush, H, may be written

$$\mathrm{H} = kT\int\frac{dz}{a^3}\left[\frac{1}{2}\phi^2(z) + \frac{1}{2Na^2}z^2\Psi(z) - \mu\phi(z)\right] \tag{3}$$

where $\phi(z)$ and $\Psi(z)$ are respectively the monomer concentration and the free end concentration at a distance z from the surface. The first two terms in the integral, respectively, represent the excluded-volume energy and the polymer stretching energy;μ is a Lagrange multiplier to fix N monomers per chain. The problem is to determine $\Psi(x)$. A crude approximation may be obtained by guessing that far enough from the surface $\Psi(x)\cong\phi(x)/N$. Now setting $\delta H/\delta\phi(c)=0$, we obtain a parabolic profile $\phi(x) \cong \left[\mu - \frac{1}{2}(x/Na)^2\right]$ where $\mu\cong(3^{2/3}/2)(a/D)^{4/3}$ and the brush thickness $L\cong Na(3a^2/D^2)^{1/3}$ is consistent with the Alexander-de Gennes result (Eq. 2).

We now turn to pseudo-brushes. Five years ago, an elegant scaling approach has been proposed by Guiselin [35] to describe the irreversible adsorption from a polymer melt onto a plane surface, assuming that the time necessary to saturate all the surface sites was small compared to that of chain diffusion in the bulk. The conformation of the chains is thus not modified during the adsorption process and the distribution of loops and tails is related to the statistics of the chain folding in the melt. Guiselin considered the following sequence of events: (i) the plane solid surface is put in contact with a polymer melt and all the monomers touching the surface adsorb instantaneously and irreversibly; (ii) the surface is then washed with a pure solvent and only the initially adsorbed chains are retained onto the surface. The layer thus formed can be visualized as a succession of loops and tails. In the Guiselin's approach, each loop made of $2n$ monomers is considered as two pseudotails of n monomers. The number (per unit area) of pseudotails made of more than n monomers is given by [35]

$$S(n) \cong a^{-2} n^{-1/2} \qquad (1<n<N) \tag{4}$$

where, as for the brush, N is the degree of polymerization of the adsorbed chains and a the size of the monomer. When swollen by a good solvent, the pseudo-tails tend to stretch normal to the surface in order to decrease the excluded volume interactions. Let us assume that the n-th monomers of all pseudotails larger than n are located at the same distance z from the surface. We can then consider n as a function of z. At a distance z from the surface, the average distance between two pseudotails is given by $D(z) \cong \left[S(n(z))\right]^{-1/2}$. Using Eq. (4) we get $D(z) \cong a n^{1/4}(z) \quad (1<n<N)$. By analogy with the Alexander-de Gennes' model for monodisperse polymer brushes (see above) we expect that, locally, at a distance z from the surface, the layer can be visualized as a closely packed system of units of size $D(z)$. In order to determine the function $n(z)$, we consider the number $g(z)$ of monomers within a volume $[D(z)]^3$. The function $n(z)$ satisfies the differential equation $dn/g(z)=dz/D(z)$. At a scale smaller than D, the chain behaves like a free chain and g is related to D by the relation $g \cong (D/a)^{5/3}$ and, consequently, $n(z) \cong (z/a)^{5/6}$. The volume fraction of monomers is then given by

$$\Phi(z) \cong \left(D(z)/a\right)^{-4/3} \cong \left(a/z\right)^{-2/5} \tag{5}$$

and the total extension of the layer scales as

$$L \cong a N^{5/6} \tag{6}$$

The internal structure of a Guiselin's pseudo-brush is schematically presented in Fig. 4. Similar arguments in the case of an irreversibly adsorbed layer pre-

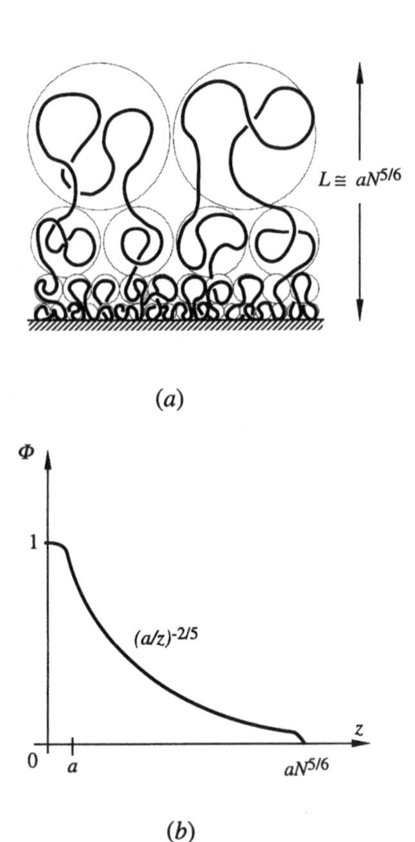

Fig. 4. a Schematic of a pseudo-brush. L is the layer thickness. **b** The monomer density profile Φ vs distance from the grafting plane z according to Guiselin. Figure adapted from [66]

pared by adsorption from a semidilute solution with a volume fraction ϕ_0, lead to an extension of the layer that scales as $L \cong aN^{5/6}\phi_0^{7/24}$ [35].

Systematic investigations of the concentration profile inside the two kinds of PDMS surface anchored layers have been performed through neutrons reflectivity techniques [45]. Octane was used as a good solvent and PDMS with a molecular weight of 92 kg mol^{-1}was either grafted or adsorbed on the surface of a silicon wafer, taking advantage of the contrast for neutrons between hydrogenated and deuterated species (either the solvent or the surface chains were deuterated). Fitting procedures of the reflectivity data have been developed in order to extract the concentration profile from the reflectivity curves, without assuming a priori the shape of the concentration profile, contrary to what is usually done [45]. The profile was divided into 10 or 20 slabs of equal thickness, and the index for neutrons of each slab was adjusted in order to obtain the best fit to the reflectivity curve. Despite the large number of adjustable parameters, this fitting pro-

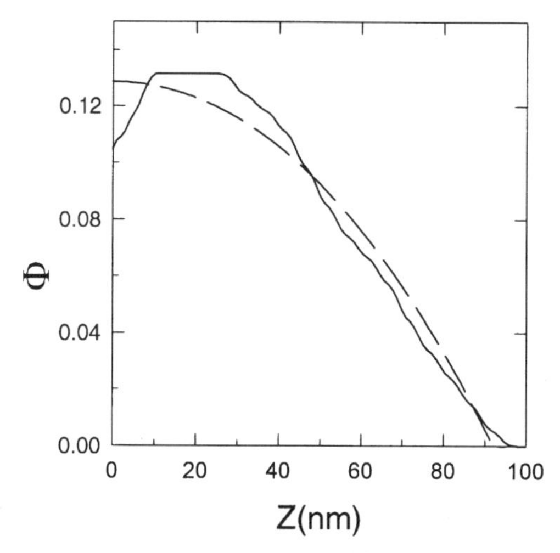

Fig. 5. Concentration profile inside a grafted PDMS layer swollen by a good solvent (octane). The molecular weight of the grafted chains is 92 kg mol^{-1} and the surface density in the layer is σ=0.011. The *full line* is the profile determined by neutrons reflectivity. The *dotted line* is the SCF result of Zhulina et al. [52] calculated for a surface density σ=0.011 and an excluded volume parameter v=0.8 a^3 (*a* is the size of the monomer, determined to be $a \approx 0.5$ nm by the slope of the scaling line in Fig. 2).

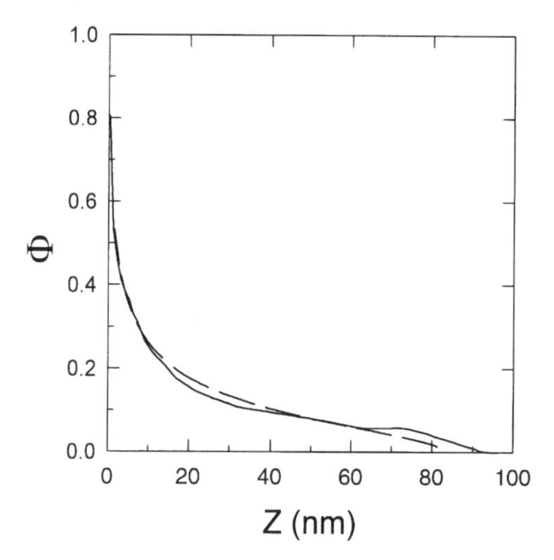

Fig. 6. Concentration profile inside an irreversibly adsorbed PDMS layer swollen by a good solvent (octane). The molecular weight of the adsorbed chains is 92 kg mol^{-1}. The surface density in the layer is σ=0.02. The *full line* is the profile determined by neutrons reflectivity. The *dotted line* corresponds to the theoretical profile calculated following the Guiselin's argument corrected to take into account the finite length of the surface chains [45]

cedure revealed to be quite efficient, provided one imposes a condition to favour smooth profiles and checks that the total surface excess of polymer, is equal to that measured independently by the dry thickness of the layer, h_0. In Figs. 5 and 6, typical concentration profiles for respectively a end-grafted layer (i.e. a brush) and an irreversibly adsorbed layer (i.e. a pseudo-brush) are reported.

In Fig. 5, the data points for the brush (full line) are compared to the calculated SCF parabolic profile of Milner et al. [10, 11] and Zhulina et al. [12]. In Fig. 6, the data points for the pseudo-brush (full line) are compared to the profile calculated by Guiselin [35]. Note that there are no adjustable parameters in the comparison between the theoretical predictions and the data points since the surface excess is imposed. The quite good agreement between the predicted and measured concentration profiles, and the very different shapes of the two curves for either the grafted or the irreversibly adsorbed layer are strong evidences of the fact that these layers have indeed quite different internal organization of the surface-anchored polymer chains. It is interesting to notice that, despite the good agreement between the experiments and the models for both kinds of layers, the two theoretical descriptions used are somewhat different: the parabolic profile is the result of an SCF analysis where the constrains on the allowed chains configurations imposed by the Alexander-de Gennes model are relaxed, while the Guiselin's description treats the layer at a degree of approximation comparable to the Alexander-de Gennes approach (the only difference being in the polydispersity of the pseudo-tails). While it clearly appears that the experimental concentration profile of the end-grafted layer cannot be described by the step profile of the scaling Alexander-de Gennes model (see Fig. 5), it seems that the Guiselin's model is sufficient to describe the experimental concentration profile of the irreversibly adsorbed layer (this might be due to the large polydispersity of loop sizes [11, 66]).

3 Interdigitation Between Surface-Anchored Chains and a Polymer Melt

In order to understand the role played by surface-anchored chains in adhesion and friction, it is essential to understand under which conditions a surface layer, when in contact with a melt, is penetrated by free chains. The question has been addressed theoretically mostly for polymer brushes, and more recently for Guiselin's pseudo-brushes. We want to review here some of these analysis, and compare the predictions of the models with the available experimental data.

3.1 Polymer Brushes in Polymeric Matrices

Most of the existing studies focus on brushes exposed to a low molecular weight solvent and less is known of the behavior of surface-tethered polymers in polymeric matrices [6, 46–56]. On the experimental side, nuclear reaction analysis [57, 58], neutron scattering [59–61], neutron reflectivity [62] and secondary-ion mass spectrometry [63] have provided some information concerning the structure of grafted chains immersed in a polymer melt or a solution of mobile polymer chains. The first computer simulation of a brush immersed in a polymer melt was presented recently by Grest [64].

The case where the grafted chains are in contact with a melt of shorter, chemically identical chains was studied theoretically by de Gennes some years ago [6]. His approach has been further developed by Leibler [46], Raphaël et al. [51], and Aubouy et al. [54]. Consider a brush made of chains, with degree of polymerization N, terminally grafted onto a flat surface and exposed to a polymeric solvent made of chemically identical chains, with degree of polymerization $P<N$ (the effect of a small non-zero χ between the P and the N chains has been discussed by Aubouy and Raphaël [53]) The number of terminally grafted chains per unit area is σa^{-2}, where a is the monomer size. The average distance between two grafting sites is given by $D=a\sigma^{-1/2}$. At sufficiently low σ, the grafted chains do not overlap (the so-called mushroom regime) and the layer thickness L is given by $L\cong aN^{1/2}$ for $P>N^{1/2}$ and by $L\cong aN^{3/5}P^{-1/5}$ for $P<N^{1/2}$ (see regions {1} and {2} in Fig. 7). As σ increases, the different chains begin to overlap for $D\cong L$. This defines an overlap concentration: $\sigma_{ov}\cong N^{-1}$ for $P>N^{1/2}$ and $\sigma_{ov}\cong N^{-6/5}P^{2/5}$ for $P<N^{1/2}$. In a Flory-type approach the free energy per chain is then given by [6]

$$\frac{F}{kT}\cong\frac{L^2}{a^2N}+\frac{a^3}{P}\frac{N^2}{LD^2} \tag{7}$$

The first term in Eq. (7) represents the elastic contribution. The second term corresponds to the effect of the screened two-body interactions. From Eq. (7) one can easily construct the (P,σ) diagram represented in Fig. 7. In region {3}, two-body interactions are relevant and the brush thickness is obtained by minimizing Eq. (7): $(L=aNP^{-1/3}\sigma^{1/3}$. In region {4}, repulsive interactions are not sufficient to swell the brush and the conformation of a grafted chain remains Gaussian: $L\cong aN^{1/2}$. At higher σ, we reach region {5} where the P chains are almost completely expelled from the brush: the brush is "dry" [6]. The brush thickness L is then related to σ by the requirement $\Phi=aN\sigma/L\cong1$ which leads to $L\cong aN\sigma$. The free energy per chain consists merely of an elastic term $F\cong kTL^2/a^2N\cong N\sigma^2$. The boundaries between regions {3} and {4} and between regions {3} and {5} are given by $\sigma\cong PN^{-3/2}$ and $\sigma\cong P^{-1/2}$, respectively.

According to the scaling theory [51, 54], region {3} has to be subdivided into two regions. In region {3a}, a grafted chain may be subdivided into spherical blobs of size D, each containing g_D monomers. Within one blob, the chain behaves like a free chain and can therefore be pictured as a self-avoiding walk of subunits called *melt blobs* [42, 51]. This leads to the relation $D\cong ag_D{}^{3/5}P^{-1/5}$. Different blobs repel each other and the brush is essentially a closely packed system of blobs. The volume fraction Φ of monomers belonging to grafted chains is given by $\Phi\cong g_D(a/D)^3\cong\sigma^{2/3}P^{1/3}$. Since on the other hand $\Phi=aN\sigma/L$, we obtain for the equilibrium brush thickness $L\cong aNP^{-1/3}\sigma^{1/3}$. This thickness can be rewritten as $L\cong(N/g_D)D$: the chain can be viewed as a string of blobs almost fully stretched

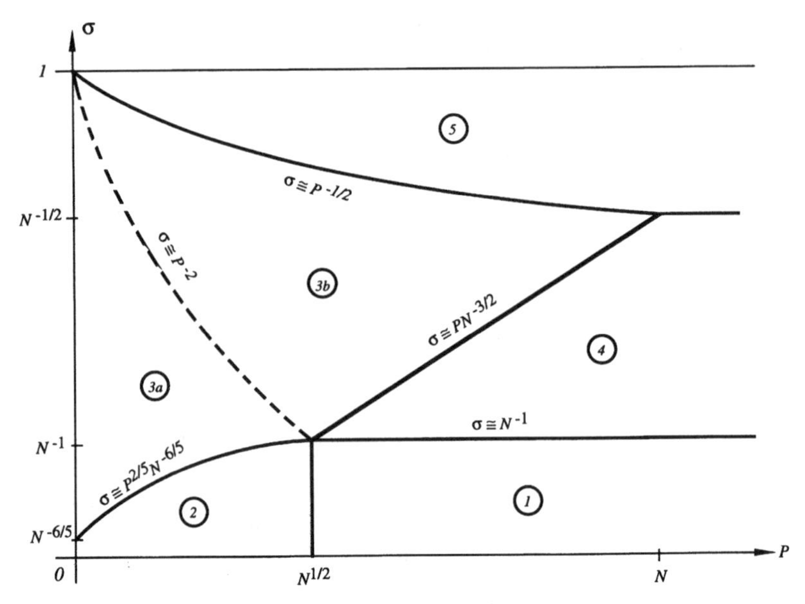

Fig. 7. Schematic (P, σ) diagram for a brush of polymerization index N exposed to a chemically identical melt of polymerization index P. At sufficiently low σ, the grafted chains do not overlap and the layer thickness L is given by $L \cong aN^{1/2}$ for $P>N^{1/2}$ (region {2}) and by $L \cong aN^{3/5}P^{-1/5}$ for $P<N^{1/2}$ (region {1}). As σ increases, the different chains begin to overlap for $\sigma \geq \sigma_{ov}$ where the overlap surface density is given by $\sigma_{ov} \cong N^{-1}$ for $P>N^{1/2}$ and by $\sigma_{ov} \cong N^{-6/5}P^{2/5}$ for $P<N^{1/2}$. In region {4}, two-body interactions are not sufficient to swell the brush and the conformation of a grafted chain remains Gaussian: $L \cong aN^{1/2}$. In region {3}, two-body interactions are relevant and the brush thickness is given by $L=aNP^{-1/3}\sigma^{1/3}$. At higher σ, region {5}, the P chains are almost completely expelled from the brush and the brush is "dry". The brush thickness L is then given by $L \cong aN\sigma$. The subdivision of region {3} into two regions (region {3a} and region {3b}) is discussed in the text. Figure adapted from [54]

along the normal to the wall. The kT per blob ansatz leads to a free energy per chain $F \cong kTNP^{-1/3}\sigma^{5/6}$. When the grafting density increases, the blob size D progressively decreases down to a point where it is of the order of the melt blob size $l_c \cong aP$. The crossover occurs for $D \cong aP$ that is for $\sigma \cong P^{-2}$(this grafting density corresponds to the broken line in Fig. 7). Therefore, the blob picture developed above is only valid in region {3a} of the (P,σ) diagram. In region {3b}, each chain can be viewed as a string of non-overlapping spherical subunits, of size Λ larger than D, almost fully stretched along the normal to the wall (see Fig. 8). At a scale smaller than Λ, the chain behaves like an ideal chain of $g_\Lambda \cong (\Lambda/a)^2$ monomers. It is important to notice that in a plane parallel to the wall, subunits of different chains do overlap and therefore the brush as a whole cannot be described as a closely packed system of subunits (see Fig. 8). The subunit size is given by $\Lambda \cong aP^{1/3}\sigma^{-1/3}$ [54]. Since the different subunits of a grafted chain repel each other, the brush thickness is given by $L \cong \left(N/g_\Lambda\right)\Lambda \cong aNP^{-1/3}\sigma^{1/3}$, in agreement with the Flory result. The free energy per chain is given by $F/kT \cong N/g_\Lambda \cong NP^{-2/3}\sigma^{2/3}$.

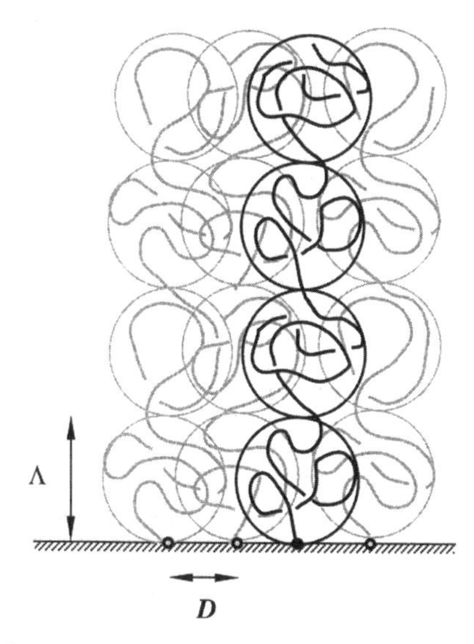

Fig. 8. Schematic of the brush in the region {3b} of the (P,σ) diagram of Fig. 7. Figure adapted from [66]

It is of some interest to build up a scaling picture for the high coverage regime $\sigma > aP^{-1/2}$ (region {5}). As in region {3b}, each chain can be viewed as a string of subunits of size Λ larger than D almost fully stretched along the normal to the wall. At a scale smaller than Λ, the chain behaves like an ideal chain (the corresponding number of monomers is given by $g_\Lambda \cong (L/a)^2$). Within a given subunit, there is on average a number $(\Lambda/D)^2$ of other chains. The subunit size Λ is given by $\Lambda \cong a(D/a)^2$. The kT per subunit ansatz leads to a free energy per chain $F \cong kTN/g_\Lambda \cong kTN(a/D)^4$ in agreement with the result found above.

Testing in details the (P,σ) diagram of Fig. 7 is a difficult task. Indeed, in order to check this diagram, one has to use: (a) a technique which gives contrast between the surface and the bulk chains, and (b) a reliable enough way of forming the brushes so that a large range of surface densities can be analyzed. Techniques allowing contrast all imply deuterated polymers, which are not always easy to obtain, and the real control of the surface density and of the brush structure (no adsorption) immediately renders the program difficult: each point in the diagram requires a synthesis and a control of the characteristics of the grafted layer, prior to any further investigation. The general tendency of the diagram of Fig. 7 is the following: increasing the surface density σ or increasing the polymerization index P of the bulk polymer (compared to the polymerization index of the grafted chains, N) lead both to an expulsion of the free chains from the brush, and thus to a decrease of the thickness of the brush profile. Such tendencies are indeed present in the experimental investigations [36, 45, 60, 62], but no systematic experimental test of the theoretical (P,σ) diagram is presently available. In Fig. 9 we present a series of profiles of PDMS brushes obtained by neutrons re-

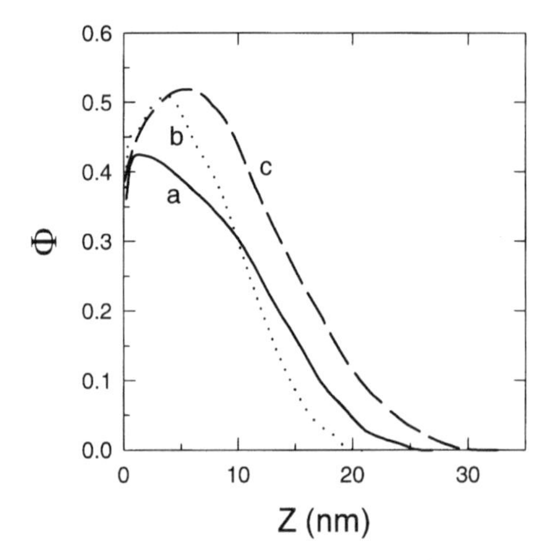

Fig. 9. Concentration profiles determined by neutron reflectivity for three end grafted PDMS layers in contact with PDMS melts. The molecular weight of the grafted chains is mN=92 kg mol^{-1} for all the layers. Curve **a:** surface density in the layer σ= 0.011, molecular weight of the melt mP=90 kg mol^{-1}; curve **b**: σ=0.01, mP=360 kg mol^{-1}; curve c: σ= 0.015, mP=17 kg mol^{-1}. The layer contracts more and more when exposed to a melt of larger molecular weight. In all cases the melt chains penetrate down to the surface, as demonstrated by the volume fraction of end grafted chains which always remains much lower than one

flectivity (the grafted chains are deuterated and the bulk polymer is hydrogenated) for several surface densities and different molecular weights of the bulk polymer. Clearly, increasing the surface density and increasing the molecular weight of the bulk polymer, correspond to a progressive expulsion of the bulk chains from the brush, a fact which is apparent first as a reduction of the extension of the concentration profile of the surface chains, and second as an increase in the steepness of this profile. A remarkable feature, however, is that if the tendencies of the experimental profiles do qualitatively agree with the predictions of the diagram of Fig. 7, the real brushes seem to be more easily interpenetrated by bulk chains than predicted. Indeed, even molecular weights much larger than that of the brush do penetrate a brush with a surface density close to $N^{-1/2}$ (Fig. 7). This might be due either to the polydispersity of the grafted chains (even if small, M_w/M_n <1.1), or more generally to the fact that the (P,σ) diagram of Fig. 7 is based on a scaling theory that assumes a step function for the brush segment density profile. For a detailed comparison with the experimental profiles, one has to use analytical self-consistent calculations [52] which relax the constraint of locating all the chain extremities in the same plane [55, 60].

3.2
Irreversibly Adsorbed Layers in Polymeric Matrices

The behavior of an irreversibly adsorbed layer in contact with a melt of shorter, chemically identical chains was studied theoretically by Aubouy et al. [65, 66]. The pseudo-brush was treated as a polydisperse brush of half loops, characterized by the distribution of Eq. (4). Assuming that the local conformation of a polydisperse polymer brush is similar to the bulk structure of a monodisperse brush with the same grafting density, Aubouy and Raphaël [65] built up a diagram analogous to the diagram of Fig. 7. Again, the global tendency is that a dense layer should expel the chains from the melt, and so more efficiently for bulk chains with a large molecular weight. No systematic experimental investigations are presently available, except a few neutrons reflectivity data [36]. Typical examples of profiles are reported in Fig. 10. These results do show an intrinsic difficulty of such investigations: when comparing data for a layer made with either deuterated or hydrogenated PDMS, in contact with a melt of the corresponding either hydrogenated or deuterated chains, a clear preferential interaction of the hydrogenated specie with the silica surface appears. In order to compare the experimentally determined profiles with a model, such a preferential interaction has to be included. On a qualitative level, however, the data seem to indicate that a pseudo-brush is more easily penetrated by a given polymer melt

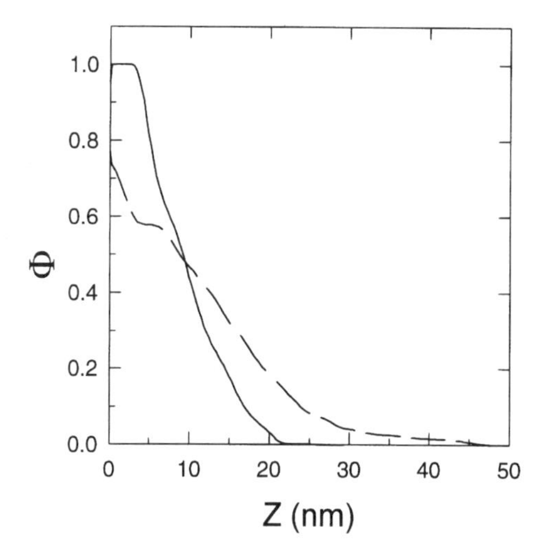

Fig. 10. Concentration profile as determined by neutron reflectivity for two irreversibly adsorbed layers in contact with a PDMS melt. The molecular weight of the surface chains is 92 kg mol^{-1} and is identical to that of the melt. In both cases σ=0.02. The *full line* corresponds to an adsorbed layer made with deuterated chains, in contact with a hydrogenated melt, while the *dotted line* corresponds to the reverse situation (hydrogenated surface chains, deuterated melt). The clear difference between the two profiles is a demonstration of a preferential interaction of the hydrogenated chains with the surface compared to the deuterated one

than the equivalent brush (same surface density and same molecular weight of the surface chains). This might be an indication of the fact mentioned by Laub and Koberstein [67a] that polydispersity may help interdigitation. On the other hand, a pseudo-brush is not really equivalent to a polydisperse brush and one should take the “cactus effect” discussed by Brochard-Wyart et al. [68] into account. The profiles shown inFig. 10 are clearly different from that of a brush: the inner part of the layer is dry (or almost dry in the case of the preferential interactions between the melt and the surface), and tails extend far away from the surface.

4 Role of Surface-Anchored Chains in Adhesion

Understanding what happens when a polymer comes into contact with another material is important for many industrial applications [69, 70]. In the last ten years, there has been a significant improvement in our understanding of the local processes that control the adhesion of polymers [71–73]. Experiments by Hugh Brown and co-workers and by Ed. Kramer and co-workers have shown that the interface between two chemically different glassy polymers can be strengthened by the presence of diblock copolymers at the interface [74–76]. When a fracture propagates at the interface, these coupling chains can either be pulled-out or break. Experiments show that long diblock molecules fail by scission, while shorter diblocks can be pulled out. Random copolymers have also been used recently in this context [77, 78]. Here we shall not discuss adhesion of glassy or semi-crystalline polymers [79–84] but rather focus on the adhesion of elastomers to solids [85–89]. Studies have demonstrated that coupling chains (or “connectors”) that bind to the solid and interdigitate with the elastomer enhance the adhesion energy [90]. The important question is to understand how to optimize practical adhesive joints using connector molecules to enhance adhesion. Early models, considering the connector chains as independent of each other, predict a linear increase of the adhesion energy at zero crack propagation velocity, G_0, with the surface density of connectors [91, 92]. So the question is whether it is interesting to graft as many chains as possible on the solid surface. To try to answer this question, systematic experiments on silica/PDMS systems were undertaken, using the pseudo-brushes presented in Sect. 2.1 (these experiments will be discussed in details below). A typical result for G, the adhesive energy measured at a very low velocity of fracture, as a function of the surface density of connector molecules is reported in Fig. 11. The major feature of these data is the non-monotoneous variation of the adhesion energy with σ. The connector molecules clearly loose their efficiency to promote adhesion when they are too densely packed on the surface. Several attempts have been made very recently to try to rationalize such a behavior. They will be reviewed in Sect. 4.1 for the adhesion between a brush and an elastomer and in Sect. 4.2 for the adhesion between a pseudo-brush and an elastomer. Experimental results will be discussed in Sect. 4.3

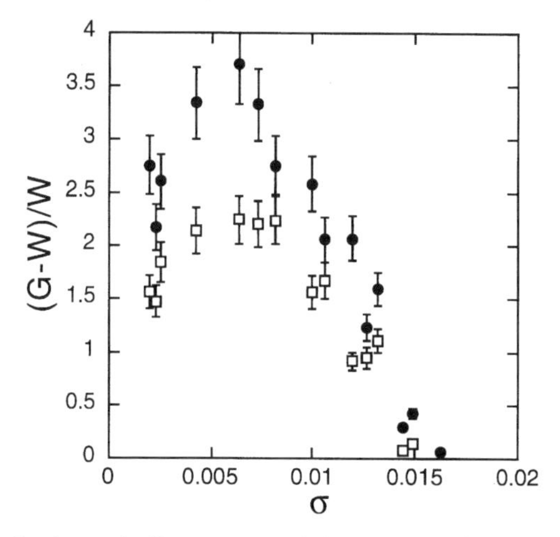

Fig. 11. Normalised enhanced adhesive strength $(G-W)/W$ as a function of the surface density, σ, for two PDMS elastomers in contact with silicon wafers covered with irreversibly adsorbed chains. W is the thermodynamic work of adhesion, $W=2\gamma$, with γ the surface tension of PDMS, $\gamma=21.6\ \mathrm{mN \cdot m^{-1}}$ at 25 °C. The *filled symbols* correspond to a molecular weight between crosslinks in the elastomer $Mc=24.2\ \mathrm{kg\ mol^{-1}}$ while $M_c=10.2\ \mathrm{kg\ mol^{-1}}$ for the *open symbols*. The adhesive strength, G, has been measured by peel tests performed at a very low velocity of the propagation of fracture, 0.17 μm/s. The molecular weight of the surface anchored chains is $M_w=242\ \mathrm{kg\ mol^{-1}}$

4.1
Adhesion Between a Brush and an Elastomer

The interface between a solid and a crosslinked elastomer can be strengthened by the addition of chains (chemically identical to the elastomer) that are tethered by one end to the solid surface as schematically presented in Fig. 12. As the crack grows along the interface, these coupling chains are progressively pulled-out from the elastomer [71, 90, 93] as schematically presented in Fig. 13. This "suction" process is expected to occur in an approximately plane *cohesive zone* directly ahead of the crack tip [94, 95]. A number of models have been proposed to describe the process of chain pull-out and the relation between chain pull-out and interfacial toughness [91, 92, 96–103]. In the model of Raphaël and de Gennes [92, 100], the partially pulled-out chains are assumed to form single-chain fibrils. The minimization of the sum of the surface and stretching energies of these chains shows that there is a minimum force f^* required for a fibril to exist even at zero pull-out rate (such a minimum force has been observed in a very elegant dewetting experiments by Reiter and co-workers [104]). As the force on a chain that is being pulled-out remains finite as the velocity of the fracture line, V, goes to zero, the existence of a threshold toughness G_0 larger than the thermodynamic work of adhesion W due to intermolecular interactions (typically van

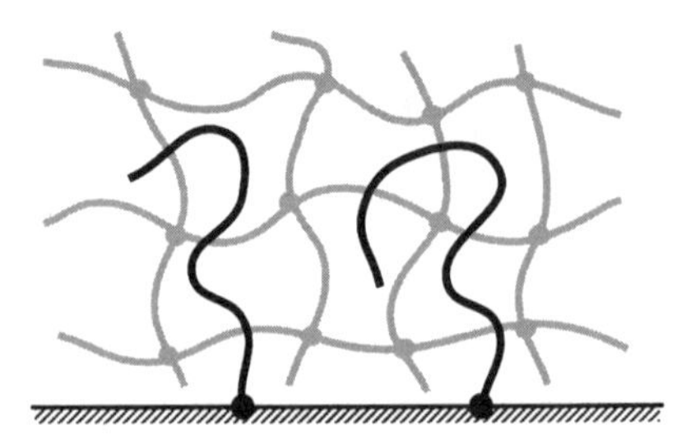

Fig. 12. Schematic representation for an elastomer/solid interface strengthened by the addition of end-grafted chains

der Waals type) is predicted. For many practical cases, the zero-rate fracture energy, G_0, is given by [100, 101]

$$G_0 - W \cong \gamma N \sigma \tag{8}$$

where σ is the dimensionless grafting density of the coupling chains, N their polymerization index, and γ the surface tension of a melt of connector molecules.

Equation (8) is restricted to the limit of low σ, where the connector molecules freely penetrate into the network and have additive contributions. For high σ values, the coupling chains may segregate (at least partially) from the elastomer and G_0 may reduce to W. How to describe the interdigitation between the grafted surface and the network over all the range of σ is a difficult problem which has been recently considered by several authors [3, 68, 105, 106]. Since there are still open questions on the subject, we will here adopt the most simple point of view which is based on the analysis by de Gennes [105] of the equilibrium state between a melt of linear chains (N monomers per chain) and a chemically identical network cross-linked *in the dry state* (N_c monomers between cross-links, with $N_c<N$). Using the fact that the elastic modulus of the network, E, is given by $E \cong kT/a^3 N_c$, de Gennes showed that the behavior of an N-chain inside the net-

Fig. 13. Schematic picture of the pull-out process that takes place as the crack grows along the interface. As the crack advances, each connector is progressively stretched until it reaches its maximal length. At that point the connectors collapses onto the surface

work is very similar to the behavior of the same N-chain when dissolved in a melt of shorter, chemically identical chains, of length N_c. In the later situation, the short chains screen out the excluded volume interaction, and as shown by Edwards, the effective excluded volume parameter is reduced by a factor $1/N_c$. In the network case, a similar screening occurs through the elastic deformation of the network segments. De Gennes further used the above remarkable analogy to show that the behavior of a polymer brush (N monomers per grafted chain) exposed to a chemically identical network cross-linked in the dry state (N_c monomers between cross-links, with $N_c<N$) is identical to the behavior of the same polymer brush when exposed to a melt of shorter, chemically identical chains, of length N_c. The various regimes of interdigitation between a brush and a network are thus identical to those given in Sect. 3.1 for the interdigitation between a brush and a polymer melt (see Fig. 7).

We can now go back to the evaluation of the adhesion energy between the grafted surface and the elastomer. In all what follows we assume that the elastomer was cross-linked in the dry state. As long as $\sigma<N_c/N^{3/2}$, the local volume fraction of grafted chains, Φ, is much smaller than unity. More precisely, $\Phi<<N_c/N$ for $\sigma<<N_c/N^{3/2}$ (see [105]) and we expect Eq. (8) to hold. For $\sigma>N_c/N^{3/2}$, Eq. (8) should be replaced by [107]

$$G_0-W \cong \gamma N\sigma\left(1-\Phi\right) \quad (9)$$

where the factor $(1-\Phi)$ is a phenomenological way of taking into account the fact that the connectors interact less and less with the network as Φ increases [67a, 67b]. For $N_c/N^{3/2}<\sigma<N_c^{-1/2}$, Φ is given by $\sigma^{2/3}N_c^{1/3}$ (see Sect. 3.1) and Eq. (9) can be rewritten as

$$G_0-W \cong \gamma N\sigma\left(1-\sigma^{2/3}N_c^{1/3}\right) \quad (10)$$

For $\sigma>N_c^{-1/2}$, the brush segregates completely from the network and the adhesion energy G_0 reduces to W.

Equation (10) indicates that the adhesion energy passes through an optimum for [3]

$$\sigma_{opt} \approx 0.465 N_c^{-1/2} \quad (11)$$

The corresponding value of G_0 is given by [3]

$$G_{0opt}-W \approx 0.186\gamma \frac{N}{N_c^{1/2}} \quad (12)$$

The full line in Fig. 14 represents the result of Eq. (10) for $N=743$ and $N_c=230$ assuming that $\gamma\approx W/2$ (experimental results with surface chains and elastomers having these characteristics will be presented in Fig. 17).

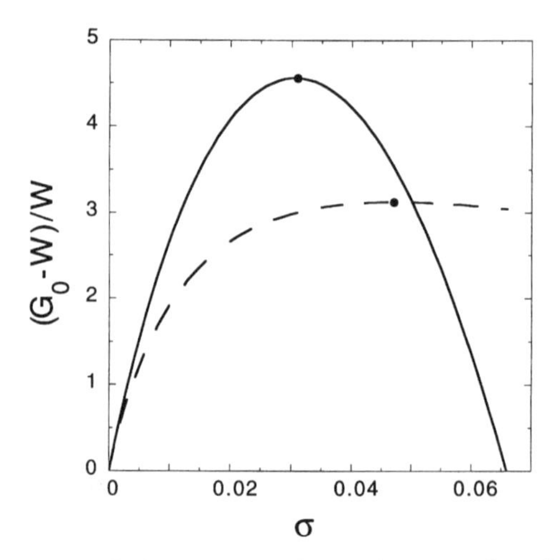

Fig. 14. Normalized enhanced adhesive strength $G_0-W)/W$ vs grafting density σ for a brush (with N=743, N_c=230 and $\gamma\approx W/2$) according to Eq. (10) (*full curve*), and Eqs. (13) and (14) (*dashed curve*). In both cases the optimum is indicated by a *black dot*

Note that if instead of a step-like profile one uses a more realistic profile where Φ is a function of the distance to wall z, Eq. (9) has to be replaced by the more general formula [67a,b,]

$$G_0-W \cong \gamma \int_a^L \frac{dz}{a} \Phi(z)\left(1-\Phi(z)\right) \tag{13}$$

where L is the brush thickness. In particular, if one uses the SCF results of Zhulina et al. [52] for the case of a brush exposed to a melt (N_c momomers per chain) one obtains

$$G_0-W \cong \gamma \int_a^{L_{SCF}} \frac{dz}{a} \exp\left[-3\pi^2 N_c \frac{\left(L_{SCF}{}^2-z^2\right)}{8\,a^2 N^2}\right] \left(1-\exp\left[-3\pi^2 N_c \frac{\left(L_{SCF}{}^2-z^2\right)}{8\,a^2 N^2}\right]\right) \tag{14a}$$

where

$$L_{SCF}=\left(4/\pi^2\right)^{1/3} a N \sigma^{1/3} N_c{}^{-1/3} \tag{14b}$$

One can check numerically that the adhesion energy (Eq. 14a) passes through an optimum. At the level of scaling laws, this optimum has the same characteristics as the optimum predicted by Eq. (10) i.e. $\sigma_{opt} \propto N_c^{-1/2}$ and $G_{0opt}-W \propto \gamma N N_c^{-1/2}$. The dashed line in Fig. 14 presents the result of Eq. (14a) for N=743 and N_c=230 assuming that $\gamma \approx W/2$. The optimum – which is not very visible at the scale of Fig. 14 – is indicated by a black dot.

Let us conclude this section by some remarks [3].

First, the dynamics of interdigitation is expected to be very slow for tethered chains exposed to a network [108]. Thus the above discussion is restricted to long enough times of contact between the brush and the elastomer.

Second, the above discussion is restricted to connectors that are not chemically attached to the elastomer. The opposite situation was analyzed recently by Brochart-Wyart and de Gennes [106] who predicted a switch from adhesive to cohesive rupture.

Third, the above discussion is based on Eq. (8) which is only valid at very low separation velocities. This equation can however be generalized to larger values of V [109].

Finally, the above analysis considers only local crack tip processes. Viscoelastic energy losses during crack propagation [110–113] have been theoretically investigated by de Gennes [114] and by Hui et al. [115], and may also contribute to the sadhesive strength.

4.2 Adhesion Between a Pseudo-Brush and an Elastomer

In order to get an estimate of the adhesion energy between a pseudo-brush and an elastomer (cross-linked in the dry state), we can use the concentration profile calculated [65] for the case of a pseudo-brush exposed to a melt (N_c monomers per chain)

$$\Phi(z) \cong \begin{cases} 1 & a << z << a N_c^{1/2} \phi_0^{7/4} \\ N_c^{1/5} \phi_0^{7/10} (z/a)^{-2/5} & a N_c^{1/2} \phi_0^{7/4} << z << L \end{cases} \tag{15a}$$

with

$$L \cong a N_c^{-1/3} N^{5/6} \phi_0^{7/24} \tag{15b}$$

Inserting Eqs. (15a) and (15b) into Eq. (13), we obtain a quite simple analytical formula for the adhesion energy

$$G_0 - W \cong \gamma N \left\{ \frac{5}{3}\sigma - 5 N_c^{1/3} \sigma^{5/3} + \frac{10}{3} N_c^{1/2} \sigma^2 \right\} \tag{16}$$

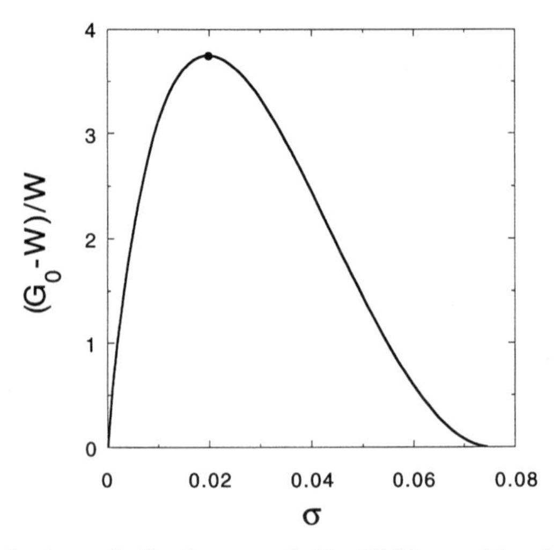

Fig. 15. Normalized enhanced adhesive strength $(G_0-W)/W$ vs grafting density σ for a pseudo-brush (with N=770, N_c=176 and N_c=176) according to Eq. (16) The optimum is indicated by a *black dot*

In Eq. (16), σ is the adimensional surface chain density :$\sigma \approx N^{-1/2}\phi_0^{7/8}$.

Equation (16) indicates that G_0 reduces to W for $\sigma \geq N_c^{-1/2}$. Equation (16) also indicates that the adhesion energy passes through an optimum for

$$\sigma_{opt} \approx 0.263 N_c^{-1/2} \tag{17}$$

The corresponding value of G_0 is given by

$$G_{0opt} - W \approx 0.129\,\gamma \frac{N}{N_c^{1/2}} \tag{18}$$

The full line in Fig. 15 represents the result of Eq. (16) for N=770 and N_c=176 assuming that $\gamma \approx W/2$ (to be compared with the experimental data shown in Fig. 17).

Note that Eq. (16) can be rewritten as

$$\frac{G_0 - W}{G_{0opt} - W} = 3.398\left(\frac{\sigma}{\sigma_{opt}}\right) - 4.184\left(\frac{\sigma}{\sigma_{opt}}\right)^{5/3} + 1.787\left(\frac{\sigma}{\sigma_{opt}}\right)^{2} \tag{19}$$

It is interesting to note that at the level of scaling laws, the optimum in the brush/elastomer adhesion energy predicted by Eq. (10) and the optimum in the pseudo-brush/elastomer adhesion energy predicted by Eq. (16) have the same characteristics [3]:

$$\sigma_{opt} \propto N_c^{-1/2} \tag{20}$$

and

$$G_{0opt} - W \propto \gamma N N_c^{-1/2} \tag{21}$$

These characteristics indicate in particular that one can improve the adhesion between the brush or the pseudo-brush and the elastomer by decreasing the number of monomers between crosslinks, N_c. This result, which is a direct consequence of the de Gennes analogy between a network and a melt (see Sect. 4.1), cannot be correct for too small values of N_c, that is for too highly reticulated networks. For pseudo-brushes, it has been conjectured [107] that Eqs. (20) and (21) are only valid for $N_c > N^{2/3}$ and should be replaced by

$$\sigma_{opt} \propto N_c N^{-1} \tag{22}$$

and

$$G_{0opt} - W \propto \gamma N_c \tag{23}$$

for $N_c < N^{3/2}$.

More theoretical work will be needed to establish the validity of Eqs. (22) and (23), and to derive analogous equations for the brush/elastomer problem.

4.3 Experimental Results

One way to estimate G_0 is to measure the adhesive strengthy G, the energy necessary to open the fracture at finite speed, working at very low speed by using the Johnson-Kendall-Roberts (JKR) technique [116, 117]. In this technique, a lens-shaped elastomer is loaded for a while against a flat solid substrate to allow the adhesion to develop. The load is then removed, and the interfacial crack slowly propagates, driven by the elastic energy stored in the crosslinked elastomer lens [72]. Brown [90] and Creton et al. [93] have used the JKR technique to study chain-pullout process in polyisoprene (PI) where the coupling chains were PS-PI diblock copolymers and the flat substrate was PS (polystyrene). The adhesive strength G was found to increase with diblock copolymer coverage and also with the molecular weight of the PI part of the diblock, in qualitative agreement with Eq. (8).

The adhesion between PDMS chains irreversibly adsorbed on silica and crosslinked PDMS elastomers has been investigated in a systematic way in our group [118–120] over a wider range of surface densities than in the early work by

H.Brown. Both 90° peel test at very low peel velocity (in the range 5 nm/s to 1 μm/s) [118] and JKR test over the same range of fracture velocities were used [121]. The main result of this study is the existence of an optimum surface density to promote the elastomer/solid adhesion, as shown in Fig. 11. The observed loss of efficiency of the connector molecules when increasing their surface density has to be related to the fact that increasing σ leads to a loss of interdigitation (see Sects. 4.1 and 4.2). A remarkable feature of the experimental results is the fact that they display a "universal" behavior: if one plots $G-W)/(G_{opt}-W)$ as a function of σ/σ_{opt} for elastomers with different molecular weights between crosslinks and for layers made with chains having different molecular weights, all the results gather on a single curve (see Fig. 16), in accordance with the prediction of the simple model of Sect. 4.2 (see Eq. (19) and the corresponding full line in Fig. 16).

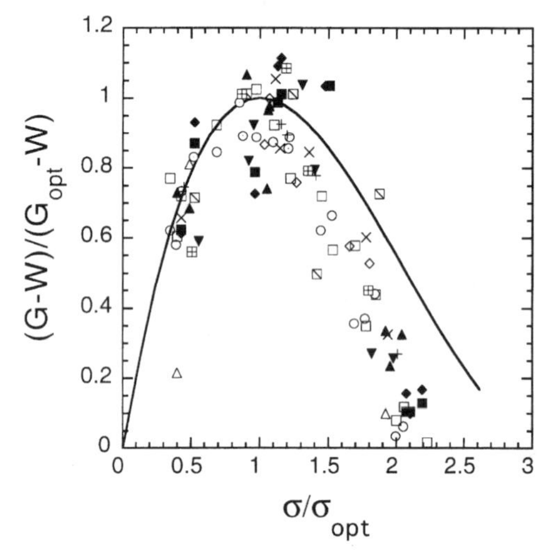

Fig. 16. Universal curve giving the enhanced adhesive strength, G-W, normalized by the optimum enhanced adhesive strength, G_{opt}-W, as a function of the normalized surface density, σ/σ_{opt} (σ_{opt} is the surface density leading to the maximum adhesive strength) for different irreversibly adsorbed PDMS layers in contact with PDMS elastomers. Peel tests with a fracture velocity of 0.017 μm/s have been used to measure G. The molecular weights of the surface chains, M_w, and the molecular weights between crosslinks of the elastomers, M_c, are the following:
(○): M_w=242 kg mol^{-1}, M_c=10.2 kg mol^{-1}; (□): M_w=242 kg mol^{-1}, M_c=24 kg mol^{-1}; (◇): M_w=293 kg mol^{-1}, M_c=24 kg mol^{-1}; (×): M_w=293 kg mol^{-1}, M_c=35.4 kg mol^{-1}; (+): M_w=412.5 kg mol^{-1}, M_c=10.2 kg mol^{-1}; (△): M_w=293 kg mol^{-1}, M_c=77.2 kg mol^{-1}; (■): M_w=412.5 kg mol^{-1}, M_c=24 kg mol^{-1}; (◆): M_w=412.5 kg mol^{-1}, M_c=35.4 kg mol^{-1}; (▲): M_w=412.5 kg mol^{-1}, M_c=77.2 kg mol^{-1}; (▼): M_w=577.5 kg mol^{-1}, M_c=10.2 kg mol^{-1}; (○): M_w=577.5 kg mol^{-1}, M_c=24 kg mol^{-1}; (⧅): M_w=577.5 kg mol^{-1}, M_c=35.4 kg mol^{-1}; (⊞): M_w=577.5 kg mol^{-1}, M_c=77.2 kg mol^{-1}.
The full line is calculated from Eq. (16)

In order to go further in a detailed comparison between experiments and models, it is convenient to work with end-grafted layers rather than irreversibly adsorbed layers because end-grafted layers are somewhat easier to model. After the first experiments with Creton on chain pull-out effects in the adhesion of polyisoprene [90, 93], Hugh Brown [122] used the same JKR technique to study chain-pullout in PDMS elastomers where the coupling chains where PS-PDMS diblock copolymers and the flat substrate was PS. Surprisingly, Brown essentially observed no effect of the end-grafted chains on the adhesive strength of the interface. We have recently repeated these experiments with PDMS brushes formed by using the method of Folkers et al. [43] described at the end of Sect. 2.1, and have observed pull-out. Typical results are shown in Fig. 17, where the enhancement of adhesive strength of the interface (with a fracture velocity equal to 0.1 µm/s) is reported as a function of σ. For comparison, results for irreversibly adsorbed layers (with similar or much larger molecular weights) in contact with crosslinked PDMS elastomers (with a similar molecular weight between crosslinks) are displayed in the same figure. Both the end-grafted layer and the irreversibly adsorbed layers do give adhesion reinforcement, with an optimum surface density, but this optimum appears sooner (when increasing σ) for the brush than for the pseudo-brush. The pseudo-brush thus appears to be somewhat more efficient to promote adhesion than the equivalent brush. A systematic experimental program is presently underway to establish this result firmly. If confirmed, it will become important to understand what makes the

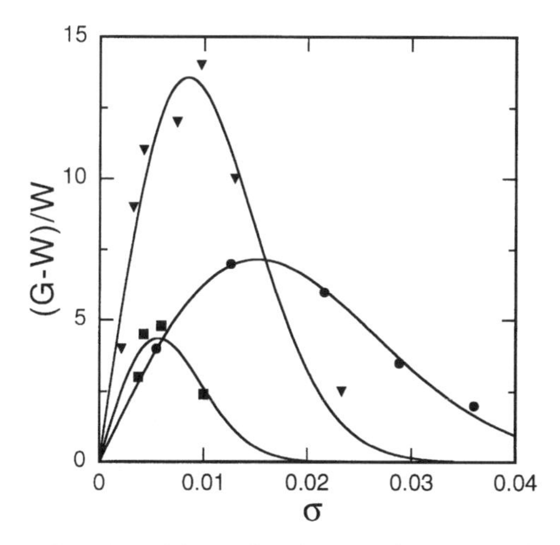

Fig. 17. $(G-W)/W$ as a function of the surface density of connector chains, σ, measured by the JKR test at a velocity of propagation of the fracture of 0.1 µm/s, for 1) (■) a grafted brush with M_w=55 kg mol^{-1} and M_c=17 kg mol^{-1}; 2) (❍) an irreversibly adsorbed PDMS layer with M_w=57 kg mol^{-1} and M_c=13 kg mol^{-1}; 3) (▼) an irreversibly adsorbed layer with M_w=140 kg mol^{-1} and M_c=13 kg mol^{-1}. The *full lines* are only guides for the eye, to help to locate the optimum

pseudo-brush more efficient to promote adhesion than the brush[1]. A natural reason may be the polydispersity of the pseudo-tails. But as mentioned before, a pseudo-brush is not really equivalent to a polydisperse brush and one would have to take into account the "cactus effect" [68].

More theoretical and experimental work is obviously needed for a better understanding of the adhesive properties of surface-anchored polymer chains.

5 Role of Surface-Anchored Chains in Friction

Polymer chains anchored on solid surfaces play a key role on the flow behavior of polymer melts. An important practical example is that of constant speed extrusion processes where various flow instabilities (called "sharkskin", periodic deformation or melt fracture) have been observed to develop above given shear stress thresholds. The origin of these anomalies has long remained poorly understood [123–138]. It is now well admitted that these anomalies are related to the appearance of flow with slip at the wall. It is reasonable to think that the onset of wall slip is related to the strength of the interactions between the solid surface and the melt, and thus should be sensitive to the presence of polymer chains attached to the surface.

Indeed, the shear stress at the solid surface is $\tau_{xz}=\eta_b(\delta V/\delta z)_{z=0}$ (where η_b is the melt viscosity and $(\delta V/\delta z)_{z=0}$ the shear rate at the interface). If there is a finite slip velocity V_s at the interface, the shear stress at the solid surface can also be evaluated as $\tau_{xz}=\beta V_s$, where β is the friction coefficient between the fluid molecules in contact with the surface and the solid surface [139]. Introducing the extrapolation length b of the velocity profile to zero ($b=V_s/(\delta V/\delta z)_{z=0}$, see Fig. 18), one obtains $\beta=\eta_b/b$. Thus, any determination of b will yield β, the friction coefficient between the surface and the fluid. This friction coefficient is a crucial characteristics of the interface: it is obviously directly related to the molecular interactions between the fluid and the solid surface, and it connects these interactions at the molecular level to the rheological properties of the system.

It is important to notice that measuring the shear stress τ_{xz} gives only partial information on the polymer – wall friction, as long as V_s is unknown. Up to 1992, there were no reliable measurements neither of the slip velocity or of the extrapolation length even if direct visualizations of the existence of wall slip had been performed [140]. Migler et al. [141] designed an experimental technique which gives a direct access to the velocity of the fluid very locally at the interface, inside a layer with a typical thickness of 70 nm. This spatial resolution is well adapted to the flows of polymer melts, especially when polymer chains are attached to the wall, the range of the interfacial interactions being of the order of the radius of the polymer chains (between 10 and 100 nm).

We want to review here the main results of these direct measurements of the interfacial velocity, in the case of PDMS melts flowing on silica surface covered

1 The models of Sects. 4.1 and 4.2 are not able to explain this result.

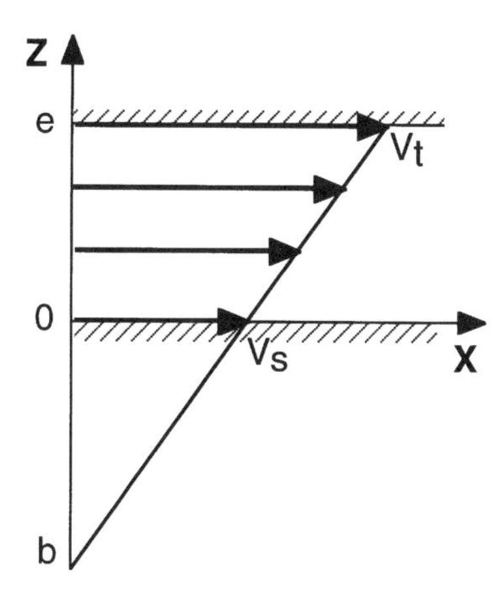

Fig. 18. The simple shear geometry used to characterise the interfacial friction: the fluid thickness is e. The top plate limiting the sample is translated at the velocity V_t and transmits this velocity to the fluid. The bottom plate is immobile, and the local velocity of the fluid at the bottom interface is V_s. The fluid is submitted to a simple shear, with a shear rate $\dot{\gamma} = (V_t - V_s)/e$. The velocity profile extrapolates to zero at a distance b below the interface, with the extrapolation length $b = V_s / \left(\partial V / \partial z\right)_{z=0}$

with various anchored polymer layers, along with the models which have been developed to describe the different friction regimes put into evidence by these experiments.

5.1 The Three Friction Regimes

The polymer system which has been extensively studied was PDMS flowing (simple shear) on smooth silica surfaces bearing either irreversibly adsorbed or end grafted PDMS chains. In the case of adsorbed layers with low densities, it was necessary to protect the surface in order to avoid any further adsorption during the flow experiment (since the surface was put into contact with a PDMS melt). To do so, the silica surface was previously treated by grafting a monolayer of octadecyltrichlorosilane, the adsorption of the surface PDMS chains taking place in the holes of this protective layer which had an adjusted surface density [142]. End grafted layers were produced following the procedure described in Sect. 2.1. Over the range of shear rates explored (from $10^{-2}\,s^{-1}$ to $100\,s^{-1}$), and for all the samples investigated, three different regimes have been observed for the evolution of both the slip velocity and the extrapolation length of the velocity profile as shown in Fig. 19. In Fig. 19a, the average velocity measured in a layer from the interface with a thickness 70 nm, V_s, is displayed as a function of the velocity imposed at the upper limiting surface of the sample, V_t. First, wall slip

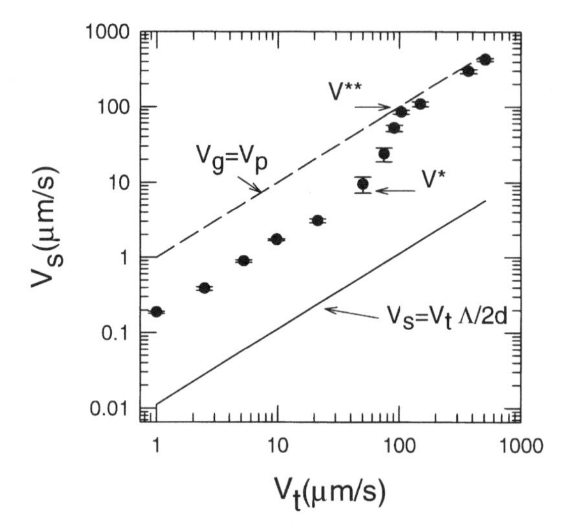

Fig. 19a

exists at all V_t, the measured average local velocity being always at least ten times larger than the average velocity one would obtain in the same layer of fluid, in the case of a zero boundary condition for the velocity at the interface (full line in Fig. 19a). The second clear result is the existence of a transition from low slip to high slip, when the applied shear rate becomes larger than a critical shear rate $\dot{\gamma}^*$. The slip velocity at $\dot{\gamma}^*$ is V^*. In Fig. 19b, the extrapolation length (or slip length) b is reported as a function of the slip velocity V_s. For V_s smaller than V^*, b admits a constant value b_0 (of the order of 1 μm): this is a linear friction regime characterized by a friction coefficient independent of V_s. Between V^* and V^{**}, b becomes proportional to V_s, implying a friction coefficient β inversely propor-

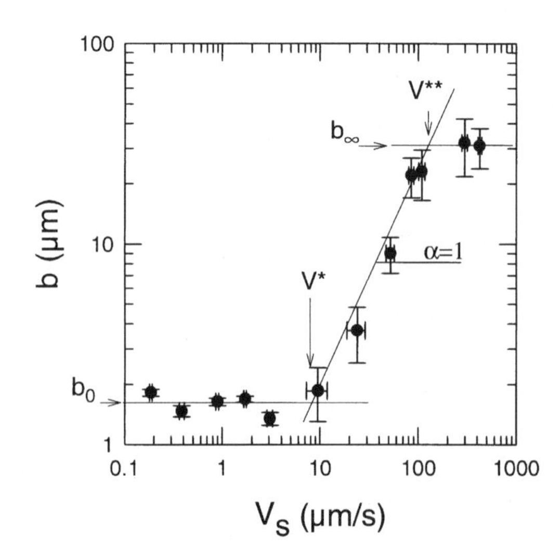

Fig. 19b

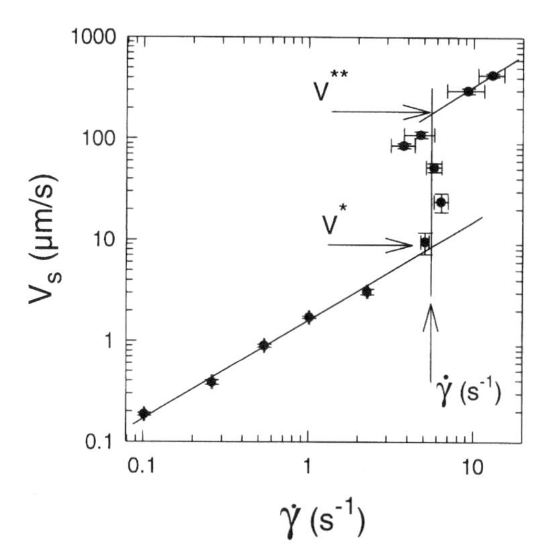

Fig. 19c

Fig. 19a–c. Typical results obtained at the interface between a PDMS melt mP= 970 kg mol-$^{-1}$, with m the molar mass of the monomer m = 0.074 kg/mol) and a grafted PDMS layer (mN=96 kg mol^{-1} and σ=0.0055): **a** the local velocity at the interface V_s (measured in a layer from the interface with a thickness 70 nm) as a function of the imposed top plate velocity V_t; **b** the same data reported in terms of the extrapolation length b as a function of the slip velocity V_s, in log scales. Three different friction regimes are clearly seen: at low shear rates, b is low and constant. Above V^*, b increases linearly with V_s, over more than one decade, and above V^{**}, b becomes constant again with a value b_∞ more than ten times larger than the value at low V_s, b_0; **c** the same data reported in terms of the slip velocity V_s as a function of the shear rate experienced by the polymer melt, $\dot{\gamma}$. The three friction regimes are also clearly visible in this representation. In the marginal regime where b increases linearly with V_s, the shear rate appears locked at the threshold value $\dot{\gamma}^*$

tional to V_S. This non-linear friction regime (the shear stress at the wall is not proportional to the slip velocity) is called the "marginal regime". Finally, for slip velocities larger than V^{**}, b levels off to a value b_∞, about 20 times larger than b_0. In this high slip regime, the friction coefficient between the polymer and the solid is strongly decreased compared to what happens at low velocity, and the polymer melt and the surface are almost decoupled. In Fig. 19c, the variations of the slip velocity vs the shear rate close to the surface, $\dot{\gamma}$ exhibit the same three regimes. The striking feature is that in the whole "marginal regime" $\dot{\gamma}$ remains constant and equal to $\dot{\gamma}^*$.

Such behavior has been interpreted in terms of a molecular model proposed by Brochard-Wyart and de Gennes [143] and further refined [145, 146]. The first version of these models considers a solid surface bearing a few end grafted polymer chains, with a surface density, σ, below the onset of the mushroom regime $\sigma N<1$, with N the polymerization index of the anchored chains). The melt chains have a polymerization index P. Both N and P are assumed to be much larger than N_e, the average number of monomers needed to form an entanglement. Thus the

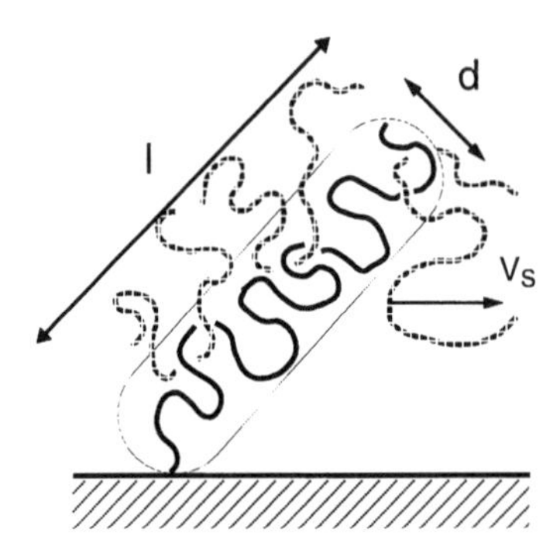

Fig. 20. Schematic representation of one grafted chain entangled with melt chains and deformed under the effect of the friction forces

surface chains behave independently of each other and are entangled with the melt chains which fully penetrate the surface layer. When flowing, the bulk chains exert a friction force on the immobilized surface chains. If large enough, this friction force can deform the surface chains, which in turn develop a restoring elastic force. In a steady state regime these two forces counterbalance. In the first version of the models, the resulting equilibrium shape of the deformed grafted chains was assumed to be a cylinder with a length l and a diameter d, as shown in Fig. 20. In order to evaluate the friction force, let us consider one surface chain and one bulk chain entangled with it. The process allowing the relative motion of these two chains is schematically depicted in Fig. 21: to allow the bulk P chain to flow, the entanglement has to be released in a time shorter than the time taken by the P chain to move parallel to the surface over a distance comparable to the average distance between entanglements, $d^* \cong aN_e^{1/2}$. The velocity of the P chain along its own tube (in the reptation picture [147]) needs thus to be larger than the surface velocity, V_s, by a factor P/N_e. Counting the number of entanglements between the two chains, the friction force f_f between one surface N chain and one bulk P chain can be evaluated as $f_f \cong a\eta_b V_s$. The total friction force between one N chain and the melt is then $F_f \cong X f_f$, where X is the number of bulk chains entangled with one surface chain. F_f increases with V_s, and, consequently, the elongation of the grafted chain increases with V_s. The diameter of the grafted chain is thus a decreasing function of V_s. This description assuming entangled bulk and surface chains, cannot remain valid at large slip velocities: when the diameter of the elongated N chains reaches d^* (that is for $V_s \cong V^*$), one enters into a new friction regime. The cornerstone of the model is that when d reaches d^*, the elongation of the surface chain no longer increases with V_s but remains locked to d^* over a wide range of V_s values. Indeed, if increasing V_s over V^* the diameter of the grafted chain would decrease below d^*, the surface and bulk chains would disentangle, and the friction force would decrease. It would then not be able to balance the large elastic force associated with the large elongation of the N chain which would then recoil and re-entangle. This is the "mar-

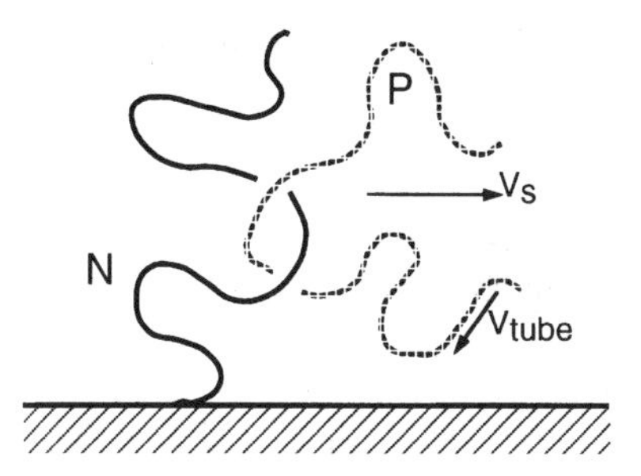

Fig. 21. Schematic representation of the process allowing one melt chain, P, entangled with one grafted chain N to move parallel to the surface at the velocity V_s: the P chain has to release the entanglement fast enough to allow for the relative motion of the two chains. The velocity of the P chain along its tube (in the reptation picture) is thus $V_{tube}=V_s P/N_e$ (N_e is the average number of monomers between entanglements)

ginal regime" where the progressive transition from low slip to high slip takes place. It is remarkable that the transition is progressive in terms of the evolution of the extrapolation length with the slip velocity (in Fig. 19, the marginal regime extends over more than one decade in V_s), while it appears sudden in terms of the evolution of the surface stress as a function of the slip velocity: all along the marginal regime, the surface stress is locked at the value τ^*,

$$\tau^* \cong \frac{\sigma kT}{d^* a^2} \cong \eta_b \dot{\gamma}^* \tag{24}$$

while b increases with V_s as

$$b \cong \frac{\eta_b V_s N_e^{1/2} a^3}{\sigma kT} \tag{25}$$

The onset of the marginal regime is reached for $V_s \cong V^*$ with:

$$V^* \cong \frac{kT}{\eta_b N a^2} \tag{26}$$

When the slip velocity is further increased, the Rouse friction [148] finally becomes dominant, for $V_s > V^{**} \propto N^{-1}$. A linear friction regime is then recovered, with a constant extrapolation length, b_∞, much larger than b_0 and comparable to what would be observed on an ideal surface without anchored chains [139].

The experimental data obtained at low surface densities, for end grafted chains, are in very good agreement with these theoretical predictions, not only for the overall evolution of the slip velocity vs the shear rate or of the slip length vs the slip velocity as shown in Fig. 19, but also for the molecular weight dependence of the critical velocity V^* which do follow exactly the laws implied by

Eq. (26): $V^{*} \propto P^{-3.3} N^{-1}$ [144, 149–152]. The extension of the nonlinear friction regime over a wide range of slip velocities and the reproducibility of the location of the onset of the marginal regime can be taken as clear evidence that the deformation of the surface anchored chains under the effect of the friction force exerted on them by the flowing melt is the correct framework to analyze the dynamic decoupling between the bulk and surface chains. A decoupling mechanism between the melt and the surface based on the breaking or the desorption of the surface chains [153] would lead to a sudden onset of strong slip, the width of the transition being governed by the polydispersity of the surface anchored chains. It would also lead, because the adsorption process is not a rapid one [34], to an evolution of the parameters of the transition (value of the critical shear rate) with the time elapsed under shear, evolution which has never been observed for the PDMS-silica system discussed here.

5.2 Modulation of the Interfacial Friction

From the above discussion, the friction β is given by

$$\beta \cong \frac{\sigma kT}{N_e^{1/2} a V_s} \tag{27}$$

and can be controlled by modulating σ, the surface density of the surface anchored chains. This has been investigated in details by Durliat et al. [151, 152] for end grafted chains. Figure 22 displays the variations of $\dot{\gamma}^{*}$ vs the grafting density, σ. From $\sigma^{*}=0.003$ to $\sigma^{*}=0.01$, $\dot{\gamma}^{*}$ increases linearly with σ as predicted by Eq. 27. This is surprising, because the model has been developed for surface chains well below the overlapping density, $\sigma^{*}>N^{-1}$ while the experiments presented in Fig. 22 all correspond to surface densities above the overlapping density: $N \approx 1000$ gives $N^{-1} \approx 0.001$. In fact, σ remains low enough to allow a good penetration of the bulk chains into the surface layer, screening the interactions between the surface chains, so that they still behave independently of each other [6, 65] and the surface stress increases linearly with σ. At the same time, as shown in Fig. 23, the extrapolation length at low slip velocity, b_0, appears to be independent of σ (151). The explanation is due to Gay [144, 145]: the friction between the surface and the melt is fixed by the number of bulk chains entangled with the surface chains. Starting from very low surface densities of grafted chains, and increasing progressively σ, it is natural to think that the number of bulk chains able to entangle with the surface layer first increases linearly with σ. However, the total number of bulk chains trapped by a unit area of the grafted layer cannot increase more than the surface density of bulk chains, $\sigma_b \cong P^{-1/2}$. With X the number of entanglements between one surface chain and one bulk chain, this means that above the grafting density $\sigma_c \cong X^{-1} P^{-1/2}$, the surface layer is saturated with bulk chains and the number of entanglements no longer increases when σ increases. The friction coefficient between the surface chains and the

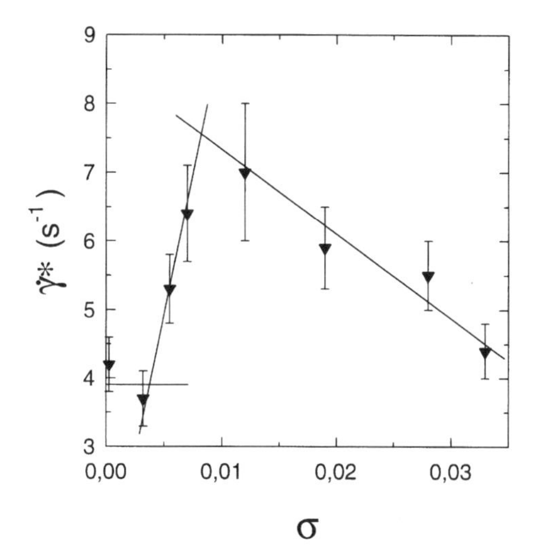

Fig. 22. Evolution of the critical shear rate $\dot{\gamma}^*$ at the onset of the marginal regime as a function of the surface density in the grafted surface layer, σ, for mP=970 kg mol^{-1}, and mN= 96 kg mol^{-1}. For 0.003≤σ≤0.01, $\dot{\gamma}^*$ increases linearly with σ, indicating an additive effect of the surface chains to the friction, i.e. surface chains contributing to the surface friction independently of each other. Above a surface density of 0.01, a collective behavior of the surface chains shows up

fluid is then independent of σ, while the onset of the marginal regime, which corresponds to a given elongation of each surface chain, is characterized by a critical shear rate increasing linearly with σ.

For larger σ values, σ>0.01 in Fig. 22, $\dot{\gamma}^*$ becomes a decreasing function of σ. The friction between the bulk and surface chains is no longer additive as it was at lower surface densities, meaning that a "collective behavior" of the grafted chains shows up. Such a decrease of the critical surface stress for the onset of strong slip means that it becomes easier to disentangle the bulk and surface chains compared to the situation at low surface densities. This is coherent with the fact that the bulk chains are progressively expelled from the surface layer when σ is increased (see Sect. 3.1). Additionally, it is also possible that the mechanical response of the dense grafted layer is different from that at low densities. Up to now there is no quantitative model to describe this situation.

Similar trends have been observed qualitatively for a melt in contact with irreversibly adsorbed layers. Three friction regimes also show up, with a wide marginal regime in which the extrapolation length b follows a power law dependence vs the slip velocity V_s^{α}. Quantitatively, the exponent α of this power law is always smaller than 1 at low grafting density [149] and varies from 0.7 to 1.4 at high grafting densities [151]. At the same time, the shear rate is not exactly locked at a constant value inside the marginal regime, but rather follows a S shape curve as a function of the slip velocity, a fact that can be attributed to the polydispersity of the loops and tails forming the surface anchored layer.

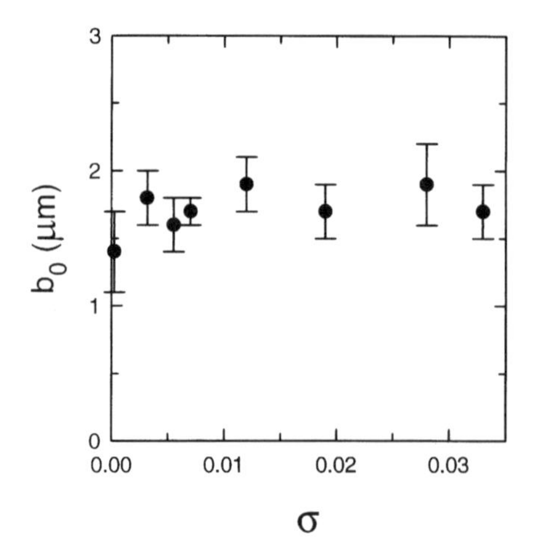

Fig. 23. Evolution of the extrapolation length at low slip velocity, b_0, as a function of the surface density of grafted chains for the experiments reported in Fig. 22. The fact that b_0 appears independent of σ when $\dot{\gamma}^*$ increases linearly with σ indicates that in this range of surface densities, the surface layer has saturated the number of melt chains it can capture

Besides the essential possibility of controlling the surface friction by changing the surface density of anchored chains all other parameters being kept constant, the value of V^* (Eq. 26) can also be adjusted by changing the molecular weight of the anchored chains [150, 152, 154].

The preceding analysis of the progressive decoupling between the surface and bulk chains assumes that the anchored and the bulk chains are entangled, i.e. that N and P are both much larger than N_e. What does happen if this is not the case? An interesting situation should be that corresponding to $N<N_e$, $P>>N_e$ and a high σ value. According to [6, 65] and to the discussion of Sect. 3.1, the bulk chains are then not allowed to penetrate into the dense surface layer of short chains. The surface should thus behave like an ideal one, with a very low friction coefficient, comparable to that observed in the high slip regime previously described. This has indeed been shown by Durliat [151] with small PDMS grafted chains (M_w= kg mol^{-1}, $N<N_e$) and long PDMS bulk chains (M_w=970 kg mol^{-1}): high slip was always present, even at the lower shear rates experimentally attainable.

All the above results show clearly that adsorbing or grafting polymer chains on a solid surface facing a flowing polymer melt can change drastically the friction coefficient and lead to non trivial non linear friction regimes. For the PDMS/silica system, where the surface chains are strongly anchored to the surface, the dynamic decoupling between the surface and bulk chains occurs through a coil to stretch transition of the surface chains. This leads to a friction governed both by the molecular parameters of the system (surface density of the surface chains, polymerization indices of both the surface and bulk chains) and the level of shear. For weaker anchoring of the surface chains, however, an alternative mechanism of decoupling is the desorption of these chains, also leading to an onset of strong wall slip above a critical shear stress and to an abrupt transition [153].

6 Conclusions

There have been considerable advances in understanding adhesion and friction in polymeric systems in the last few years. In this work we have examined in detail the behavior of surface-anchored polydimethylsiloxane chains attached to a plane silica surface. Depending on the fact that either all the monomers in the macromolecule or only one chain extremity are able to bind to the surface, two different categories of layers can be formed (both with controlled molecular characteristics) in a rather wide range of surface density of anchored chains. End-grafted chains form brushes, well described by the SCF parabolic profile of Milner et al. [10, 11] and Zhulina et al. [12] when swollen by a good solvent, while irreversibly adsorbed chains lead to concentration profiles which agree quite well with Guiselin' s description of pseudo-brushes [35].

These surface-anchored layers can be used as adhesion promoters to enhance elastomer-solid adhesive strength. We have shown that there is then an optimum surface density of surface anchored chains to do so. At high coverage of the surface, the layers lose their efficiency, a tendency which can be rationalized in terms of interdigitation between the elastomer and the surface chains. A full characterization of the different regimes of adhesion enhancement associated with the different regimes of interdigitation between the surface chains and the elastomer is a difficult experimental task, not fully accomplished to date.

We have demonstrated that these surface-anchored layers were also quite efficient to adjust the friction between a polymer melt and a surface. This friction is governed by the ability of the surface chains to interdigitate into the melt and to entangle with the bulk chains. However, because the surface anchored chains are flexible and deformable objects, they can elongate under the effect of the friction forces, thus leading to a non-trivial friction law when the local velocity at the interface is progressively increased. Indeed, at sufficiently high shear rates the surface chains become elongated enough so that they can disentangle from the bulk chains, leading to an onset of flow with high slip at the wall, i.e. to a dynamic decoupling between the layer and the bulk polymer. We have shown that it is possible to account for the different observed friction regimes in terms of a molecular model based on this notion of shear induced disentanglement. These investigations thus open the way to the design of surfaces with adjusted friction properties.

An important question is now to understand how these two effects, adhesion enhancement and adjusted friction, can interplay. For example, when peel tests are used to estimate the adhesive strength, the curvature of the peeled ribbon implies a fracture advancing through both modes I and II of opening [72, 155]. This means that in the presence of surface anchored chains, the connector chains are solicited both in elongation and shear deformation. The total energy necessary to open the fracture, and its velocity dependence should thus be affected by the non-linearity of the interfacial friction. Systematic analysis of such effects just start to appear [156, 157] and should be of great fundamental and practical importance in the near future.

Acknowledgements. We are much indebted to P.-G. de Gennes for stimulating our interest in the behavior of surface-anchored polymer chains; we would like to thank him for his constant and active interest in the work presented here. We have benefited greatly from discussions and collaborations with H. Brown, F. BrochardWyart, M. Tirrell, J. Folkers and K. Migler. We have also benefited from discussions with M. Rubinstein, A. Ajdari, G. Schorsch and L. Vovelle. An important part of the work discussed in the present paper relies on the PhD theses of M. Aubouy, M. Deruelle, E. Durliat, C. Gay, Y. Marciano, C. Marzolin, and G. Massey. We thank them for their essential contribution to this work. We thank C. Creton and M. Aubouy for a critical reading of the manuscript.

7
References

1. Fleer GJ, Cohen Stuart M, Scheutjens J, Cosgrove T, Vincent B (1993) Polymers at interfaces. Chapman & Hall
2. Edwards SF (1994) Faraday Discuss 98:1
3) de Gennes PG (1997) Soft Interfaces (Cambridge University Press).
4. de Gennes PG (1987) Adv Coll Int Sci 27:189
5. Alexander S (1977) J Phys (Paris) 38:977
6. de Gennes PG (1980) Macromolecules 13:1069
7. Hirz SJ (1986) Modeling of interactions between adsorbed block copolymer. M. S. Thesis, University of Minnesota, Minneapolis, MN
8. Cosgrove T, Heath T, van Lent B, Leermakers F, Scheutjens J (1987) Macromolecules 20:1692
9. Skvortsov AM, Pavlushkov IV, Gorbunov AA, Zhulina EB, Borisov OV, Pryamitsin VA (1988) Polymer Sci USSR 30:1706
10. Milner ST, Witten TA, Cates ME (1988) Europhys Lett 5:413
11. Milner ST, Witten TA, Cates ME (1988) Macromolecules 21:2610; Milner ST, Witten TA, Cates ME (1989) Macromolecules 22:853
12. Zhulina EB, Pryamitsyn VA, Borisov OV (1989) Vysokomol Soedin Ser A 31:185
13. Zhulina EB, Borisov OV, Pryamitsyn VA (1990) J Colloid Interface Sci 137:495
14. Murat M, Grest GS (1989) Macromolecules 22:4054
15. Chakrabarti A, Toral R (1990) Macromolecules 23:2016
16. Lai PY, Binder K (1991) J Chem Phys 95:9299
17 Milner ST (1991) Science 251:905
18. Halperin A, Tirrell M, Lodge TP (1991) Adv Polym Sci 100:31
19. Grest GS, Murat M (1995) In: Binder K (ed) Monte Carlo and molecular simulations in polymer science. Oxford University Press, New York, p476
20. Szleifer I, Carignano MA (1996) In: Prigogine I, Rice SA (eds) Advances in chemical physics, vol 94. Wiley, New York, p165
21. Vincent B (1974) Adv Colloid Interface Sci 4:193
22) Napper D (1983) Polymeric stabilisation of colloidal dispersions. Academic Press
23. Klein J (1989) Physics World, 2 June
24. Taunton HJ, Toprakcioglu C, Fetters LJ, Klein J (1988) Nature 332:712
25. Auroy P, Auvray L, Léger L (1991) Macromolecules 24:2523
26. Auroy P, Auvray L, Léger L (1991) Macromolecules 24:5158

27. Auroy P, Auvray L, Léger L (1991) Phys Rev Letters 66:719
28. Hadziioannou G, Patel S, Granick S, Tirrell M (1986) J Amer Chem Soc 108:2869
29. Tirrell M, Patel S, Hadziioannou G (1987) Proc Nat Acad Sci 84:4725
30. Granick S, Herz J (1985) Macromolecules 18:460
31. Factor BJ, Lee LT, Kent MS, Rondelez F (1993) Phys Rev E 48:2354
32. Kent MS, Lee LT, Factor BJ, Rondelez F, Smith GS (1995) J Chem Phys 103:2320
33. Bates FS (1991) Science 251:898
34. Cohen Addad JP, Viallat AM, Pouchelon A (1986) Polymer 27:843
35. Guiselin O (1992) Europhys Lett 17:225
36. Marzolin C (1995) PhD Thesis, University Paris VI, France
37. Deruelle M, Ober R, Vovelle L, Hervet H, Léger L (to be published)
38) Deruelle M, PhD Thesis (1995) University Paris VI, France
39. de Gennes PG (1981) Macromolecules 14:1637
40. de Gennes PG (1982) Macromolecules 15:492
41. Auvray L, Cotton JP (1987) Macromolecules 20:202
42. de Gennes PG (1985) Scaling concepts in polymer physics. Cornell Univ Press, Ithaca
43. Folkers JP, Deruelle M, Durliat E, Marzolin C, Hervet H, Léger L (submitted to Macromolecules)
44. Pincus P (1991) Macromolecules 24:2912
45. Marzolin C, Menelle A, Hervet H, Léger L (submitted to Macromolecules)
46. Leibler L (1988) Makromol Chem, Macromol Symp 16:1
47. Shull KR (1991) J Chem Phys 94:5723
48. Zhulina EB, Borisov OV (1990) J Colloid Interface Sci 144:507
49. Zhulina EB, Borisov OV, Brombacher L (1991) Macromolecules 24:4679
50. Wijmans CM, Scheutjens JMHM, Zhulina EB (1992) Macromolecules 25:2657
51. Raphaël E, Pincus P, Fredrickson GH (1993) Macromolecules 26:1996
52. Wijmans CM, Zhulina EB, Fleer GJ (1994) Macromolecules 27:3238
53. Aubouy M, Raphaël E (1993) J Phys II France 3:443
54. Aubouy M, Fredrickson GH, Pincus P, Raphaël E (1995) Macromolecules 28:2979
55. Wijmans CM, Factor BJ (1996) Macromolecules 29:4406
56. Shull KR (1996) Macromolecules 29:2659
57. Budkowski A, Steiner U, Klein J, Fetters LJ (1992) Europhys Lett 20:499
58. Budkowski A, Klein J, Steiner U, Fetters LJ (1993) Macromolecules 26:2470
59. Jones RAL, Norton LJ, Shull KR, Kramer EJ, Felcher GP, Karim A, Fetters LJ (1992) Macromolecules 25:2359
60. Clark CJ, Jones RAL, Kramer EJ, Shull KR, Penfold J (1995) Macromolecules 28:2042
61. Auroy P, Auvray L (1996) Macromolecules 29:337
62. Lee LT, Factor BJ, Rondelez F, Kent M (1994) Faraday Discuss 98:139
63. Zhao X, Zhao W, Zheng X, Rafailovich MH, Sokolov J, Schwarz SA, Pudensi MAA, Russell TP, Kumar SK, Fetters LJ (1992) Phys Rev Lett 69:776
64. Grest GS (1996) J Chem Phys 105:5532
65. Aubouy M, Raphaël E (1994) Macromolecules 27:5182
66. Aubouy M (1995) Ph.D Thesis, Pierre and Marie Curie University, Paris; Aubouy M, Guiselin O, Raphaël E (1996) Macromolecules 29:7261
67. (a) Laub CF, Koberstein JT (1994) Macromolecules 27:5016; (b) Creton C, Kramer EJ, Hui CY, Brown HR (1992) Macromolecules 25:3075
68. BrochardWyart F, de Gennes PG, Léger L, Marciano Y, Raphaël E (1994) J Chem Phys 98:9405
69. Wool RP (1995) Structure and strength of polymer interfaces. Hanser/Gardner Publications, NY
70. Brown HR (1996) Physics World, January:38
71. Brown HR (1991) Ann Rev Mat Sci 21:463
72. Brown HR (1994) IBM J Res Develop 38:379
73. Baljon ARC, Robbins MO (1996) Science 271:483

74. Brown HR, Deline VR, Green PF (1989) Nature 341:221
75. Creton C, Kramer EJ, Hadziioanou (1991) Macromolecules 24:1846
76. Creton C, Kramer EJ, Hui CY, Brown HR (1992) Macromolecules 25:3075
77. Brown HR, Char K, Deline VR, Grenn PF (1993) Macromolecules 26:4155
78. Dai CA, Dair BJ, Dai KH, Ober CK, Kramer EJ, Hui CY, Jelinski LW (1994) Phys Rev Lett 73:2472
79. Duchet J, Chapel JP, Chabert B, Spitz R, Gerard JF (1997) J Appl Polym Sci 65:2480
80. Gérard JF, Chabert B (1996) Macromol Symp 108:137
81. Lin R, Wang H, Kalika DS, Penn LS (1996) J Adhesion Sci Tech 10:327
82. Lin R, Quirk RP, Kuang J, Penn LS (1996) J Adhesion Sci Tech 10:431
83. Boucher E, Folkers JP, Hervet H, Léger L, Creton C (1996) Macromolecules 29:774
84. Boucher E, Folkers JP, Creton C, Hervet H, Léger L (1997) Macromolecules 30:2102
85. Shanahan MER, Schreck P, Schultz J (1988) C R Acad Sci (Paris) II 306:1325
86. Vallat MF, Ziegler P, Pasquet V, Schultz J (1990) C R Acad Sci (Paris) II 310:477
87. Shanahan MER, Michel F (1991) Int J Adhes Adhes 11:170
88. Maugis D (1991) J Colloid Interface Science 150:243
89. Nardin M, Alloun A, Schultz J (1993) Proceedings of the 16th Annual Meeting of the Adhesion Society, Williamsburg
90. Brown HR (1993) Macromolecules 23:1666
91. de Gennes PG (1989) J Phys France 50:2551
92. Raphaël E, de Gennes PG (1992) J Phys Chem 96:4002
93. Creton C, Brown HR, Shull KR (1994) Macromolecules 27:3174
94. de Gennes PG (1990) Canadian J of Phys 68:1049
95. Fager LO, Bassani JL, Hui CY, Xu DB (1991) Int J Fracture Mech 52:119
96. de Gennes PG (1989) C R Acad Sci (Paris) II 309:1125
97. Xu DB, Hui CY, Kramer EJ, Creton C (1991) Mech Mater 11:257. This model is aimed mainly at the pullout process in glassy polymers
98. Hong JI, de Gennes PG (1993) Macromolecules 26:520
99. Rubinstein M, Ajdari A, Leibler L, BrochardWyart F, de Gennes PG (1993) C R Acad Sci (Paris) II 316:317
100. Raphaël E, de Gennes PG (1994) In: Bruisma R, Rabin Y (eds) Soft order in physical systems. Plenum Press, NY
101. Brown HR, Hui CY, Raphaël E (1994) Macromolecules 27:607
102. Ligoure C (1996) Macromolecules 29:5459
103. Ligoure C, Harden JL (1997) J Phys Chem B 101:4613
104. Reiter G, Schultz J, Auroy P, Auvray (1996) Europhys Lett 33:1
105. de Gennes PG (1994) C R Acad Sci (Paris) II 318:165
106. BrochardWyart F, de Gennes PG (1996) J of Adhesion 57:21
107. Léger L, Raphaël E (1995) In: Daillant J, Guenoun P, Marques C, Muller P, Tran Than Van J (eds) Short and long chains at interfaces. Editions Frontières, GifsurYvette
108. O'Connor KP, McLeish TCB (1993) Macromolecules 26:7322
109. Marciano Y, Raphaël E (1994) Int Journ of Fracture 67:R23
110. Gent AN, Petrich R (1969) Proc R Soc London A 310:433
111. Gent AN, Schultz J (1972) J Adhes 3:281
112. Andrews EH, Kinloch AJ (1973) Proc R Soc London 332:385
113. Carré A, Schultz J (1984) J Adhes 18:171
114. de Gennes PG (1988) C R Acad Sci (Paris) II 307:1949
115. Hui CY, Xu DB, Kramer EJ (1992) J Appl Phys 72:3294
116. Johnson KL, Kendall K, Roberts AD (1971) Proc R Soc London A 324:301
117. Chaudhury MK, Whitesides GM (1991) Langmuir 7:1013
118. Marciano Y (1994) PhD Thesis, University Paris XI France
119. Deruelle M, Tirrell M, Marciano Y, Hervet H, Léger L (1994) Faraday Discuss 98:55
120. Deruelle M, Léger L, Tirrell M (1995) Macromolecules 28:7419
121. Deruelle M, Hervet H, Jandeau G, Léger L (1998) Adhesion Sci. Technol. 12:225

122. Brown HR (1995) (unpublished results)
123. ShiQuing W, Drda PA (1996) Macromolecules 29:2627
124. Benbow JJ, Lamb P (1963) SPE Trans 3:7
125. de Smet, Nam S (1987) Plastic Rubber Process Appl 8:11
126. Denn MM (1990) Ann Rev Fluid Mech 22:13
127. El Kissi N, Léger L, Piau JM (1994) J Non Newtonian Fluid Mech 52:249
128. El Kissi N, Piau JM (1990) J Non Newtonian Fluid Mech 37:55
129. Hatzikiriakos SG, Dealy JM (1991) J Rheol 35:497
130. Hatzikiriakos SG, Dealy JM (1992) J Rheol 36:703
131. Hatzikiriakos SG, Dealy JM (1992) J Rheol 36:845
132. Kalika DS, Denn MM (1987) J Rheol 31:815
133. Larson RG (1992) Rheol. Acta 31:213
134. Piau JM, El Kissi N (1994) J Non Newtonian Fluid Mech 54:121
135. Piau JM, El Kissi N (1994) J Rheol 38:1447
136. Piau JM, El Kissi N, Tremblay BJ (1990) J Non Newtonian Fluid Mech 34:145
137. Ramamurthy AV (1986) J Rheol 30:337
138. Vinogradov GV, Protasov VP, Dreval VE (1984) Rheol Acta 23:46
139. de Gennes PG (1979) CR Acad Paris 288:219
140. Atwood BT, Schowalter WR (1989) Rheol Acta 28:134
141. Migler KB, Hervet H, Léger L. (1993) Phys Rev Lett 70:287
142. Silberzan P, Léger L, Ausserré D, Benattar JJ (1991) Langmuir 7:1647
143. BrochardWyart F, de Gennes PG (1992) Langmuir 8:3033
144. Leger L, Hervet H, Marciano Y, Deruelle M, Massey G (1995) Israel J Chem 35:65
145. BrochardWyart F, Gay C, de Gennes PG (1996) Macromolecules 29:1992
146. Gay C (1997) PhD Thesis, University Paris VI (France)
147. (a) de Gennes PG (1971) J Chem Phys 55:572; (b) Ajdari A, BrochardWyart F, Gay C, de Gennes PG, Viovy JL (1995) J. Phys. II France 5:491
148. Rouse PE (1953) J Chem Phys 21:1273
149. Massey G (1996) PhD Thesis, University Paris VI (France)
150. Léger L, Hervet H, Massey G (1996) In: Piau JM, Agassant JF (eds) Rheology for polymer melt processing. Elsevier, p 337
151. Durliat E (1997) PhD Thesis, University Paris VI (France)
152. Durliat E, Hervet H, Leger L (1997) Europhys Lett 38:383
153. Yongwoo I, ShiQuing W (1996) Phys Rev Lett 76:467
154. Massey G, Hervet H, Léger L (1998) Europhys Lett 43:83
155. Kanninen M, Popelar C (1985) Advanced Fracture Mechanics. Oxford University Press
156. Brown H (1994) Science 263:1411
157. Newby BZ, Chaudhury MK (1997) Langmuir 13:1805

Received: February 1998

Molecular Transitions and Dynamics at Polymer / Wall Interfaces: Origins of Flow Instabilities and Wall Slip

Shi-Qing Wang

Departments of Macromolecular Science and Physics, Case Western Reserve University, Cleveland, Ohio 44106-7202
(e-mail: sxw13@po.cwru.edu)

This article reviews recent results on capillary melt flow anomalies. Long standing controversies and debates in this field are illustrated by summarizing previous results and clarified with an extensive discussion of the most recent results. Explicit molecular mechanisms for flow instabilities are presented in contrast to a background of 40 years' continuous and far ranging research. New experiments show that the widely observed extrusion anomalies (including oscillating flow, discontinuous flow transition and sharkskin) of linear polyethylenes (LPE) originate from interfacial molecular transitions, which may or may not be stable depending the specific flow conditions. A global flow instability (commonly known as oscillating capillary flow) evidently arises from a time-dependent oscillation of the global hydrodynamic boundary condition (HBC) between no-slip and slip limits at the capillary die wall. Other convincing observations show that sharkskin originates from a local instability of HBC at the die exit wall. The global and local interfacial instabilities both originate from a reversible coil-stretch transition involving interfacial unbound chains that are entangled with the adsorbed chains. In other words, local and global stress oscillations result in the observed macroscopic sharkskin-like and bamboo-like extrudate distortions respectively. A second molecular mechanism for wall slip is also clearly identified, involving stress-induced chain desorption off low surface energy walls. An organic coating of capillary die walls produces massive chain desorption and a large magnitude wall slip at rather low stresses, whereas bare metallic and inorganic surfaces (e.g., steel, aluminum, and glass) usually retain sufficient chain adsorption and prevent catastrophic slip up to the critical stress for the coil-stretch transition. The intricate interfacial flow instabilities exhibited by LPE are also shared by other highly entangled melts such as polybutadienes. In contrast, monodisperse melts with high critical entanglement molecular weight (M_e) such as polystyrene of $M_w=10^6$ show massive wall slip on low energy surfaces but no measurable interfacial stick-slip transition before reaching the plateau around 0.2 MPa. Tasks for future work include (i) direct molecular probe of melt chain adsorption and desorption processes at a melt/wall interface as a function of the surface condition, (ii) new theoretical studies of chain dynamics in an entangling melt/wall interfacial region as well as in bulk at high stresses, (iii) test of universality of the established physical laws governing melt/wall interfacial behavior and flow for all polymers, and (iv) development of tractable experimental and theoretical methods to study boundary discontinuities and stress singularities.

Keywords. Polymer interfaces, Melt flow instabilities, Interfacial wall slip, Chain dynamics in fast flow

1 **Introduction** 229

2 **Navier-de Gennes Slip Boundary Condition** 230

3 **Molecular Behavior of Entangled Polymers at Solid Surfaces** 232

3.1 Stress Induced Chain Desorption. 232
3.2 Coil-Stretch Transition of Interfacial Chains 234

4 **Boundary Discontinuities and Stress Singularities** 238

5 **Interfacial Slip in Various Flow Geometries**. 240

5.1 Wall Slip in Drag Flows 240
5.1.1 Slip-Like Behavior at Relatively Low Steady Shear Stresses 240
5.1.2 Edge Failure vs "Dynamic Slip" 243
5.1.3 Ordering in Suspensions and Interfacial Slip 244
5.2 Wall Slip in Pressure Driven Flows 245

6 **Phenomenology of Melt Flow Anomalies** 247

6.1 Flow Discontinuity Transition. 247
6.2 Rheological Behavior of Other Polymers 248
6.3 Capillary Extrusion Instability 249

7 **Interfacial Chain Disentanglement at Flow Discontinuity** 251

7.1 Discontinuity Upon Molecular Transition at Melt/Wall Interfaces . . 251
7.2 Mooney vs Navier Treatment of Wall Slip 254
7.3 Critical Condition and Molecular Characteristics of Stick-Slip Transition 255
7.3.1 Critical Stress and Interfacial Adsorption 256
7.3.2 Molecular Weight Dependence of Constant-Stress Viscosity and Extrapolation Length 258

8 **Molecular Mechanism for Oscillating Capillary Melt Flow** 259

9 **Interfacial Origin of Anomalous Capillary Flow Behavior** 261

10 **Molecular Instabilities – Origin of Sharkskin** 263

11 **Constitutive Instabilities, Extrudate Distortions and "Melt Fracture"** 267

11.1 Constitutive Instabilities. 267
11.2 Gross Extrudate Distortion and "Melt Fracture" 269

12 **Summary** 270

13 **References**. 272

1 Introduction

Polymers can be confined one-dimensionally by an impenetrable surface besides the more familiar confinements of higher dimensions. Introduction of a planar surface to a bulk polymer breaks the translational symmetry and produces a polymer/wall interface. Interfacial chain behavior of polymer *solutions* has been extensively studied both experimentally and theoretically [1–6]. In contrast, polymer *melt*/solid interfaces are one of the least understood subjects in polymer science. Many recent interfacial studies have begun to investigate effects of surface confinement on chain mobility and glass transition [7]. *Melt* adsorption on and desorption off a solid surface pertain to dispersion and preparation of filled polymers containing a great deal of particle/matrix interfaces [8]. The state of chain adsorption also determine the hydrodynamic boundary condition (HBC) at the interface between an extruded melt and wall of an extrusion die, where the HBC can directly influence the flow behavior in polymer processing.

The need to understand the nature of hydrodynamic boundary conditions near melt/wall interfaces is underscored by striking melt flow anomalies in capillary extrusion of various engineering polymers, especially linear polyethylenes (LPE). Characteristic flow instabilities in melt extrusion can be best described in terms of a typical flow curve for LPE. The first type of melt flow anomalies occurs when the capillary extrudate develops surface roughness commonly known as *sharkskin*. This peculiar flow phenomenon is followed by a second more striking anomaly whose features depend on the driving mode of capillary flow. Under a constant piston speed that yields a nominal shear rate inside the shaded region, symptoms of the *flow instability* include temporal pressure oscillation and quasi-periodic extrudate distortion [9, 10]. The alternating variation of the extrudate diameter along the extrudate has been categorically termed "melt fracture" [9, 11]. Under controlled pressure, the capillary flow "spurts" discontinuously to a higher rate at a critical pressure [9]. The extrudate is often grossly distorted upon the transition to the upper "slip branch." This type of extrudate distortion also carries the phrase "melt fracture" [9, 12]. During a continuous variation of the applied pressure a notable hysteresis loop may develop in the shaded region in Fig. 1 [13]. Appearance of a hysteresis loop and gross extrudate distortion has often been taken as evidence that LPE exhibit a constitutive instability or bulk failure (i.e., melt fracture) in the shaded region.

One frontier in polymer science has been to explore plausible molecular mechanisms for the flow anomalies indicated in Fig. 1. The task to uncover molecular origins of these important flow phenomena can be regarded as one of the few remaining experimental challenges in the field of polymer melt rheology of single-component systems. The subject of melt flow instabilities has invited publication of over three thousand papers since the late 1950s. Existing review articles [10, 11, 14] still represent an accurate summary of the history and progress from the 1950s up to the 1990s. The present article is devoted to a thorough review of our current understanding of the flow anomalies, which occur

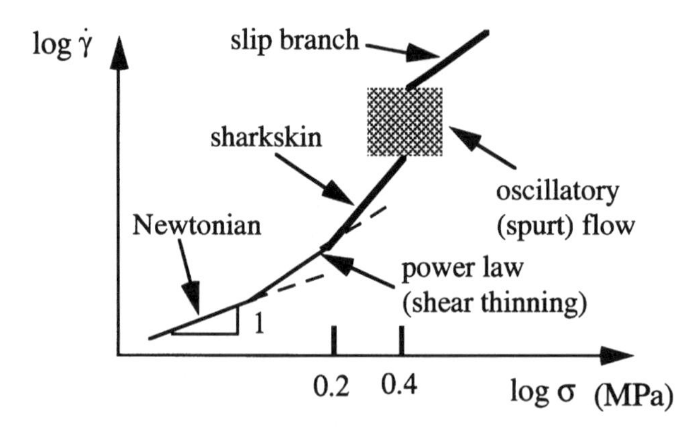

Fig. 1. Typical flow curve of commercial LPE. There are five characteristic flow regimes: (i) Newtonian; (ii) shear thinning; (iii) sharkskin; (iv) flow discontinuity or stick-slip transition in controlled stress, and oscillating flow in controlled rate; (v) slip flow. There are three leading types of extrudate distortion: (a) *sharkskin like*, (b) alternating *bamboo like* in the shaded region, and (c) *spiral like* on the slip branch. Industrial extrusion of polyethylenes is most concerned with flow instabilities occurring in regimes (iii) to (v) where the three kinds of extrudate distortion must be dealt with. The unit shows the approximate levels of stress where the sharkskin and flow discontinuity occur respectively. There is appreciable molecular weight and temperature dependence of the critical stress for the discontinuity. Other highly entangled melts such as 1,4 polybutadienes also exhibit most of the features illustrated herein

beyond the Newtonian and power law (shear thinning) regimes. It begins in the first three sections by providing a historical background and summarizing our up-to-date knowledge. Major results from our, as well as other, research groups will be reviewed and discussed in the subsequent sections against the background provided in Sects. 2–4. The paper will end with a summary.

2 Navier-de Gennes Slip Boundary Condition

The subject of hydrodynamic boundary condition at a liquid/solid interface is a fundamental element of fluid mechanics and is particularly pertinent to numerical simulations of polymer flow and processing. The possible existence of a slip hydrodynamic boundary condition (where a fluid attains a finite velocity at its interface with a stationary wall) is conceptually intriguing and has been explored since the time of Navier [15] and Stokes [16]. For over a century, the molecular origin of wall slip has been unclear to both the founding fathers (Navier, Stokes, Hagen, Poiseuille, Darcy, Couette, etc.) of fluid mechanics and leading experts of this century including Goldstein [17]. To one's disappointment, low molar mass liquids have never been found to exhibit any measurable degree of slip on *macroscopic* scales [18]. Despite the intrinsic difficulty to find measurable slip corrections in *moderately* entangled melts, efforts have continued around the world in this direction [19–22].

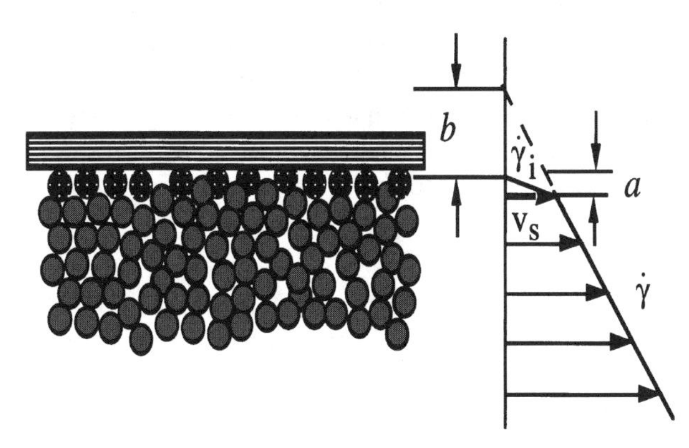

Fig. 2. Schematic illustration of "slip" at an interface between a wall and a low molar mass liquid, where the *black dots* denote either adsorbed molecules of the same species, or adsorbed molecules of a different species, or atoms/molecules of the wall material

Today it is generally understood that the no-slip condition is valid because "the interaction between a fluid particle and a wall is similar to that between neighboring fluid particles" [23]. This is clearly true for ordinary fluids made of small or short chain molecules. Thus in formulation of hydrodynamics for non-polymeric liquids it is not necessary to consider any measurable violation of the no-slip boundary condition on macroscopic length scales.

It is imperative to describe first a microscopic picture of fluid/wall interfacial interactions involving ordinary fluids. Figure 2 shows why it is impossible to observe any macroscopic magnitude of interfacial slip regardless of the surface condition (wetting or non-wetting). Furthermore, it illustrates a first principles derivation of the Navier's slip boundary condition: $v_s=\sigma/\beta$, where v_s is the slip velocity, β is an interfacial friction coefficient, and σ is the shear stress at the boundary. Let us assign an interfacial viscosity η_i to account for the intermolecular interactions between the dark and light dots. Let us also assume that the intermolecular interactions among light dots produce a fluid viscosity η. Then the stress continuity condition yields, in steady state, the dynamic boundary condition

$$\begin{aligned}\sigma &= \eta\dot{\gamma} = \eta_i\dot{\gamma}_i \\ &= \beta v_s,\end{aligned} \tag{1}$$

where $\dot{\gamma}_i$ is the interfacial shear rate present between the adsorbed layer and the neighboring unbound layer, and $\dot{\gamma}$ is the shear rate in the unbound layers near the wall. The last equality of Eq. (1) followed from the definitions:

$$v_s \equiv a_i\dot{\gamma}_i \tag{2a}$$

and

$$\beta \equiv \eta_i / a, \tag{2b}$$

where a is the layer thickness on the order of molecular dimensions. Thus the preceding discussion constitutes a first-principles derivation of Eq. (2b) proposed by de Gennes in 1979 based on a dimensional argument.

The Navier relation in Eq. (1), i.e., $\eta\,\dot{\gamma}=\beta v_s$, naturally introduces a length scale known as the extrapolation length b

$$b = v_s / \dot{\gamma} \equiv \eta / \beta = (\eta / \eta_i)a, \tag{3}$$

where the third equality follows from use of Eq. (2b) for β. The slip behavior is described by the ratio of b to characteristic dimensions of a flow apparatus, such as the capillary die diameter D in a capillary rheometer or the gap distance H in the flow geometry of two parallel plates. For instance, at a given pressure gradient G, capillary flow of a Newtonian fluid in a die of diameter D would have a volumetric flow rate Q

$$Q / Q_0 = 1 + 8b / D \tag{4a}$$

in presence of a slip HBC [24], where

$$Q_0 = (\pi G / 128\eta) D^4 \tag{4b}$$

is the flow rate at G in absence of slip. Historically, search for violation of the no-slip HBC has been to look for any deviation of the measured Q from the D^4 dependence, i.e., to search for a finite D^3 dependence in Q given by Eqs. (4a) and (4b) [24].

Since for ordinary fluids the bulk intermolecular interactions are not much stronger than those between the liquid molecules (gray dots) and molecules/atoms (black dots) on the surface, the viscosity ratio η/η_i in Eq. (3) is on the order of unity. Thus according to Eq. (3), the extrapolation length b is not much larger than the molecular size a, and the slip correction measured by b/D (e.g., the second term in Eq. (4a)) is infinitesimally small for a macroscopically large D. Clearly, non-polymeric liquids would not exhibit any measurable slip on *macroscopic* length scales and the no-slip HBC is virtually exact. In 1979, De Gennes rediscovered the concept of b and found the true meaning of Eq. (3) for polymeric liquids [25].

3 Molecular Behavior of Entangled Melts at Solid Surfaces

3.1 Stress-Induced Chain Desorption

The same Navier dynamic boundary condition Eq. (1) and the subsequent expression Eq. 3 for the extrapolation length b can also be written down for non-Newtonian and polymeric fluids, where η is the shear viscosity and η_i is the local viscosity at the interface. The expression Eq. (2b) for β is equally valid for polymeric fluids, as first proposed by de Gennes on the basis of a dimensional analysis [25]. Equation (3) predicts that a macroscopically large *b* may be attained when an exceedingly large viscosity ratio η/η_i arises. It is important to remark that the relation between stress σ and slip velocity v_s is implicitly nonlinear since β is generally a function of σ. Similarly, the extrapolation length *b* is generally a highly nonlinear function of σ.

Polymeric fluids may actually violate the widely accepted law of the no-slip boundary condition. This would occur when the viscosity ratio in Eq. (3) becomes enormously large. However, the physical origin of a large interfacial slip in an entangled polymer melt remained unclear until de Gennes first envisioned a situation of vanishing polymer adsorption [25]. Suppose there is no polymer adsorption as depicted in Fig. 3 above a *certain level* of stress at a polymer/wall interface [26]. Then the first monolayer (denoted by thick chains) does not meet much resistance from the wall, and the interfacial interactions are merely frictional between the polymer segments and the molecules on the wall. On the other hand, the first monolayer fully entangles with the bulk chains away from the wall. This chain-chain entanglement interaction can be many orders of magnitude stronger than the monomer/wall friction. In this case, the viscosity ratio in Eq. (3) can be immense. Here the large η originates from chain entanglements, whereas the small η_i simply accounts for monomer/wall frictional interactions and is independent of the chain length N. As a consequence, the extrapolation length *b* of Eq. (3) can be hundreds of thousands times larger than the molecular dimensions such as the interfacial thickness *a*. Within the framework of the reptation theory, one can show [25] that the extrapolation length *b* rapidly increases with the chain length N in absence of *interfacial* chain entanglement interactions as

$$b_0 = (\eta / \eta_i)a = \mathrm{N}(\mathrm{N}/\mathrm{N_e})^2 a, \tag{5}$$

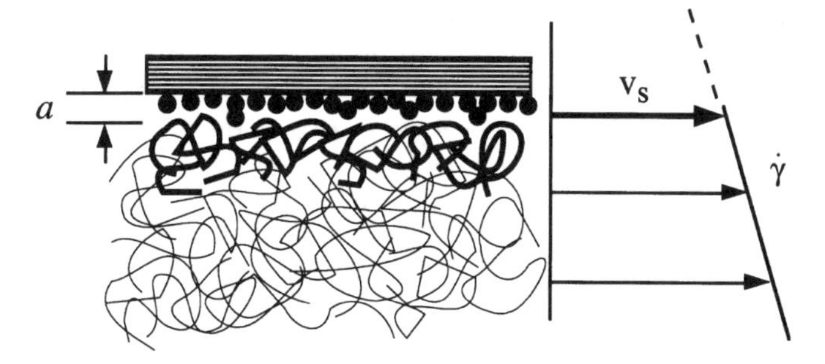

Fig. 3. Interfacial slip of an entangled melt at a non-adsorbing perfectly smooth surface, where the dots represent an organic surface (e.g., obtained by a fluoropolymer coating), which invites little chain adsorption. Lack of polymer adsorption produces an enormous shear rate $\dot{\gamma}_i$ at the entanglement-free interface between the dots and the first layer of (thick) chains. $\dot{\gamma}_i = v_s/a$ is much greater than the shear rate $\dot{\gamma}$ present in the entangled bulk. This yields an extrapolation length *b*, which is too large in comparison to the chain dimensions to be depicted here

where N_e is the number of segments between neighboring entanglement points and the subscript 0 on b indicates the zero-shear limit. It is implicitly assumed in the derivation of Eq. (5) that the interfacial viscosity η_i is related to the Rouse friction coefficient ζ as $\zeta=\eta_i a$, where a is the size of a Rouse bead. The necessary condition for realization of a large b is existence of an enormously large ratio η/η_i, which may arise from absence of chain adsorption as depicted in Fig. 3. It is important to emphasize that although Eq. (5) is derived under the condition of zero shear the same molecular weight scaling may also hold in the strong flow shear thinning regime [27].

Recent experimental data [28] indeed show that a weak interface can be created by lowering the surface energy with an fluorocarbon elastomer coating. On such a weakly adsorbing die wall, a macroscopic slip occurs in linear polyethylenes during capillary die extrusion. However, the same surface fails to produce any observable wall slip at low stresses that can be reliably generated in a parallel-plate flow cell. This contrast emphasizes that massive polymer desorption and interfacial slip occur only beyond a critical wall stress.

Polymer melt adsorption may be quantified in terms of the lifetime of chain adsorption. Melt adsorption can be regarded as non-existent if each adsorbed chain only spends a duration on a surface that is significantly shorter than its disentanglement relaxation time τ_d. Sufficient interfacial stress can lower the energy barrier for desorption and shorten the residence time of adsorbed chains on a surface. The chain adsorption state is critically dependent on the surface energy. At present, little is *quantitatively* known about how the critical interfacial stress for polymer melt desorption varies with the surface energy.

3.2 Coil-Stretch Transition of Interfacial Chains

Most inorganic solid surfaces are of sufficiently high surface energy and usually allow sufficient polymer melt adsorption, in the sense that chains remain adsorbed for a long time to assure no-slip boundary condition. When enough chain adsorption exists and the first "tethered" monolayer is in complete entanglement with the second layer of free polymer chains, the viscosity ratio η/η_i in Eq. (3) should be around unity, and no macroscopic slip is possible. Figure 4a illustrates this situation. Thus, under low stresses, we do not expect any appreciable slip behavior on inorganic surfaces.

It is possible to perceive a situation of strong melt flow when the corresponding wall stress exceeds a critical value and causes the adsorbed chains to undergo a coil-stretch transition. Recent experimental results suggest [29] that such a strong flow can be readily produced in a capillary die by pressure-extrusion of a linear polyethylene (LPE) melt. A coil-stretch transition occurs when each adsorbed PE chain experiences a sufficient force, which overcomes its entropic force and causes complete chain stretch [29, 30]. Figure 4b shows that the adsorbed chains take a dimension of D_{tube} in the velocity gradient direction. The entanglement-free interface has a thickness a, which is comparable to a segmen-

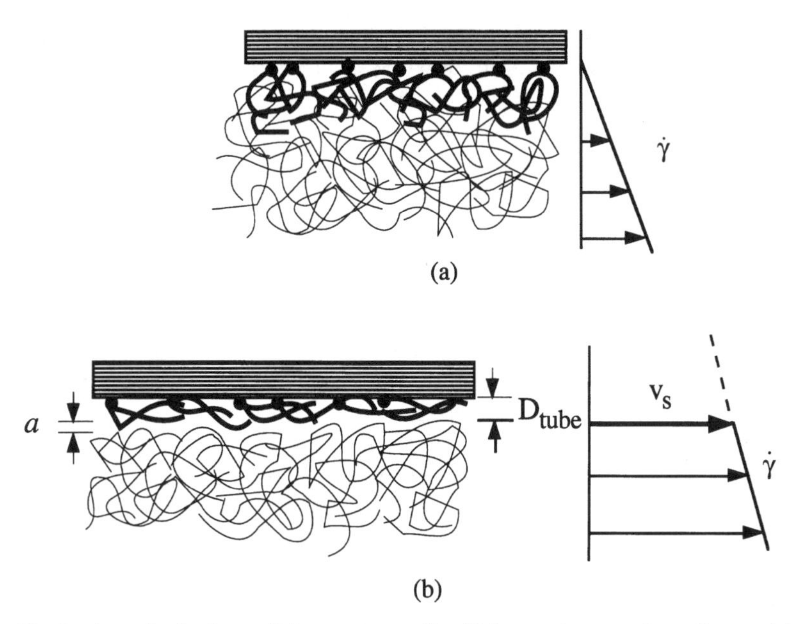

Fig. 4.a A perfectly flat wall in presence of sufficient polymer adsorption and interfacial chain entanglements. **b** An entangling melt under high stresses ($\sigma>\sigma_c$) in contact with a molecularly smooth wall, where the adsorbed (thick) chains undergo a coil-stretch transition and the unbound chains are no longer in entanglement with the "tethered" chains. Here the first layer of adsorbed chains is stagnant, as the unbound chains flow by.

tal length and is certainly not larger than D_{tube}. To the best of our knowledge, a hypothetical idea close to that described in Fig. 4b was first perceived and mentioned by Bergem in a 1976 conference proceeding [31]. Bergem's idea says that "the temporary network of entangled chain molecules yields through sudden disentanglement when the shear stress is increased to a definite value, thus providing a thin layer of low-viscosity polymer where this value first is reached, i.e., at the periphery of the die entrance, and subsequently over the whole inner surface of the die" [31]. Figure 4a,b gives an idealistic view of a coil-stretch transition beyond a critical wall stress σ_c. It is assumed that the die wall surface is molecularly smooth, so that only the first molecular layer is involved in the entanglement-disentanglement transition through the coil-stretch transition of the adsorbed chains.

It is important to point out that this coil-stretch transition mechanism for interfacial slip does not require the adsorbed chains to stay permanently anchored to the wall. The polymer melt adsorption only needs to be strong enough to keep adsorbed chains tethered onto the surface for as long as it takes for them to undergo the stress-induced coil-stretch transition. Approximately, this "residence time" only needs to be comparable to the overall chain relaxation time τ_d. Thus, any surface that provides such a minimum degree of polymer adsorption would allow a disentanglement transition at a critical wall stress in entangled melts. Such surfaces may include most inorganic surfaces. Thus, there should not be any qualitative changes of the transition characteristics as the wall material varies from steel to aluminum to glass.

A coil-stretch transition, corresponding to the interfacial transition from Fig. 4a to Fig. 4b, may produce a dramatic *stick-slip* transition as the extrapolation length *b* jumps from a molecular to a macroscopic scale. The amplitude of this stick-slip transition is determined by the ratio of the extrapolation length *b* to the capillary die diameter D, where *b* scales with the chain length N according to Eq. (5) even in the shear thinning limit [27]. Thus, when *b* is comparable to D, such a stick-slip transition will be observed. Typically, D is 0.5 mm or larger in rheological apparatus and industrial processing machinery. This means only highly entangled melts may display a noticeable stick-slip transition *when* the adsorbed chains undergo a coil-stretch transition to produce a sufficiently large viscosity ratio η/η_i in Eq. (3). It is most appropriate in this context to term the extrapolation length *b* the *Navier-de Gennes length* because it was de Gennes [25] who first recognized the possibility of a macroscopically large Navier extrapolation length *b* in flow of highly entangled polymer melts.

The critical interfacial condition (CIC) for a coil-stretch transition has not been fully explored. Even in the ideal case of a molecularly smooth surface, we do not know how the CIC changes from one polymer to another, e.g., from polyethylene (PE) to polybutadiene (PB) because we do not have enough knowledge about the state of chain adsorption for a given pair of polymer and surface. For a given surface, a polymer A may effectively have only one adsorption point per chain and a polymer B may establish several adsorption sites per chain at any moment, as depicted in Fig. 5a,b respectively. A case of strong adsorption is depicted in Fig. 5c, where polymer C may form many small loops on the surface. Conversely, as the surface energy changes, the interfacial condition may vary from Fig. 5a to Fig. 5c for a given polymer melt. For a given polymer/wall interface, we know little about which of the three adsorption states prevails. In a case of strong chain adsorption, we would replace the light dots in Fig. 5a,b with dark dots. Then the state in Fig. 5a of one contact per chain and multiple contact points per area of R_g^2 is statistically improbable, and the state in Fig. 5b becomes that in Fig. 5c. In this case, the first layer of unbound chains (not drawn in Fig. 5) that are entrapped and entangled with the loops can undergo a coil-stretch transition beyond a critical wall stress σ_c to disentangle with the bulk chains. It is possible that the state of chain adsorption is not that of Fig. 4a and is more accurately described by Fig. 5b or Fig. 5c before the transition. Therefore Fig. 4a,b only serves an illustrative purpose. The experimental results to be discussed in Sect. 7.3 however show that there are many chains per R_g^2 that are involved in a coil-stretch transition, implying Fig. 4a is approximate accurate for polyethylenes.

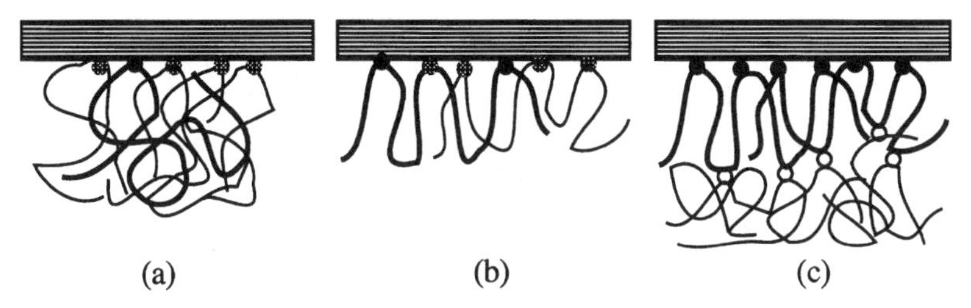

Fig. 5a–c. Polymer adsorption on a smooth surface, where the state of chain adsorption varies between: **a** weak; **b** intermediate **c** strong with the *light and dark dots* denoting weak and strong adsorption sites respectively, and the *thin lines* denoting either weakly adsorbed or non-adsorbed chains. These three adsorption states may refer to increasingly favorable interfacial interactions between a polymer and three walls of increasing surface energy. They may also refer to adsorption states on a given wall of three different polymers. The state (c) has a saturated number of loops, and there are the entanglements between the loops and the first layer of unbound chains as symbolized by the *open circles*. An average loop size may or may not exceed the tube diameter D_t depending on the chain length N and N_e between entanglements

Another parameter influencing melt flow behavior in a die is the roughness of the wall surface. In reality, the interfacial transition may occur at an effective interface as shown in Fig. 6. The unbound polymer chains in the upside down valleys may be stagnant to provide an "entangling surface" for the free chains (thick) on the inner side of the interface. This hypothesis of Tordella [9c], Bergem [31], and Blyler and Hart [32] indicate that besides the adsorbed chains at the effective interface, the coil-stretch transition may occur to a layer of unbound chains trapped by the stagnant chains. Consequently, Tordella speculated [9b]: "Slip appears to be the ultimate result of (die) land fracture. The lack of dependence of land fracture and slip on material of construction of the capillary indicates that the phenomena are predominantly dependent on polymer properties rather than polymer-solid interfacial forces." Thus the idea that a *bulk failure* produced the observed flow anomalies dates back to the observations of Tordella [9] and Blyler and Hart [32] who first visualized the plausible picture of Fig. 6. They thought that the flow discontinuity and oscillatory flow reflect bulk properties. Bergem subsequently concluded that "polymer slip at a surface is not necessary for the occurrence of a discontinuity of the flow curve and a corresponding oscillation region" [31]. It remains to be shown whether surface roughness is large enough for us to perceive the coil-stretch disentanglement transition to occur at an effective interface.

Evidently, desorption and coil-stretch transition are two interfacial molecular processes, both of which may produce massive interfacial slip between a highly entangled melt and a solid surface. The former can be viewed as a true *adhesive* failure whereas the latter may not be regarded as such and can be viewed as *cohesive*. However, it is misleading and inappropriate to link the coil-stretch transition of *adsorbed* or *entrapped* chains to any bulk behavior in shear because these chains are either confined or trapped at the surface and should be distinguished

Fig. 6. A rough surface of die wall, where the stagnant (*thin*) chains allow the unbound chains (the *middle thick chain without dot*) to entangle and "adsorb" at the effective interface (*dashed line*). Adsorbed chains (*thick chains with dots*) are also present. Such an interface can also produce a stick-slip transition upon a coil-stretch transition involving the thick chains. See Ref. [27]. For clarity, we only draw three "tethered" chains (two adsorbed and one entangled) besides the stagnant chains in one valley. The chains not drawn here, of course, fill up all the space away from the rough wall

from any other *unbound* layers in the bulk. Both kinds of interfacial chain behavior depend critically on the stress field near the wall. Upon increasing the applied stress, which process occurs first depends on the surface energy. On fairly weakly adsorbing surfaces, chain desorption is usually the dominant process. On high energy surfaces, chain adsorption prevails and a coil-stretch transition may occur abruptly above a critical wall stress σ_c. At *sufficiently* low stresses where most polymer adsorption survives and adsorbed chains fully entangle with unbound chains, the melt/wall interfacial region may be depicted by Fig. 4a or Fig. 5c, and wall slip is difficult to observe on macroscopic length scales.

4 Boundary Discontinuities and Stress Singularities

Polymer melt rheologists constantly applied the notion of wall slip in describing a variety of polymer melt flow anomalies [10, 11, 14]. However, no explicit knowledge of a molecular mechanism for and microscopic origin of wall slip had been available. Therefore, for 40 years no explanation was given about why wall slip was pertinent and how it produced such melt flow anomalies as flow oscillations and sharkskin-like extrudate distortion. One objective of current research around the world is to explore the molecular nature of wall slip and establish a correlation between wall slip and some of the anomalous melt flow phenomena.

Usually, for metallic surfaces such as steel and aluminum, most polymers adhere well due to adequate chain adsorption. Even polyethylenes apparently achieve sufficient chain adsorption on bare metallic walls for stresses up to 0.5 MPa. This means that only a coil-stretch transition may occur to produce disentanglement and to cause *complete* interfacial slip on metallic walls. Since the coil-stretch transition requires exertion of *sufficient* forces on each adsorbed or stagnant chain at the interface, it is not possible to observe on clean metallic surfaces in any geometry other than capillary (or slit) dies. It is well known [14] that the meniscus of a torsional parallel-plate flow cell suffers from severe edge instabilities at relatively moderate stresses. Thus it may not be reliable to measure wall slip on high energy surfaces with a conventional drag-flow apparatus.

We may recognize three classes of flow geometry where a sharp discontinuity in the boundary condition may produce singularly strong stresses around the discontinuity. The first two known examples of flow singularities are that of the channel (e.g., slit or capillary die) flow and coating flow. In *channel flow*, the HBC is discontinuous at the exit rim and the local stress tends to grow much larger than the stress inside the channel [34, 35]. Flow geometry involving sharp corners also leads to singularly large stresses. In *coating flow*, the HBC is discontinuous at the contact line [36]. In both cases, the local stress is inherently large, and often a slip HBC is applied even for Newtonian fluids to remedy the singularity in numerical computations [36]. One of the great theoretical challenges is to derive a finite, physically meaningful stress distribution around a boundary discontinuity. The expected stress distribution at a die end will be discussed for the case of capillary flow in Sect. 10 where the origin of sharkskin is discussed. The second case of coating flow does not concern us here. There might be a third type of boundary discontinuity and stress singularity involving *parallel plate geometry* that has never been suggested in the literature: We *speculate* that the torsional parallel plate cell may possess a similar stress build-up due to the boundary discontinuity at the contact lines between the free meniscus surface and the two plates.

Suppose the upper plate is rotating with an angular speed of Ω. Conventionally, one measures the stress by assuming the velocity field $V_\theta(z, r)=\Omega r(z/H)$, as shown in Fig. 7a throughout the sample in between the parallel plates. The free meniscus surface has the stress-free condition: $\mathbf{n}\cdot\vec{\sigma}=0$ where $\mathbf{n}$ is the unit vector normal to the meniscus surface. To assure $\mathbf{n}\cdot\vec{\sigma}=0, \frac{\partial V_\theta}{\partial z}=0$ must be satisfied at r=R. This is depicted in Fig. 7b, where the velocity V_θ on the free surface is uniform across the gap. This means there must be a strong discontinuity in the

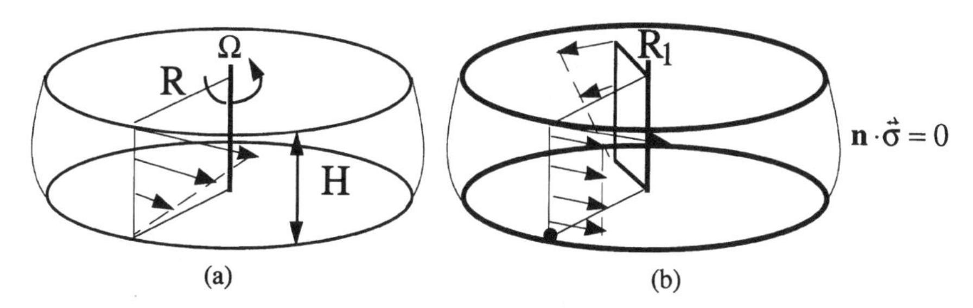

Fig. 7.a Conventional view of the flow field in a parallel plate flow cell. **b** The actual flow fields on the stress-free meniscus and inside the cell. There appear to be contact (thick) lines along the rims of both plates where the boundary condition is discontinuous

velocity field at the rims (at r=R) of both plates, i.e., the contact lines. Sufficiently far inward from the meniscus, we may expect a familiar velocity gradient as indicated at the radius R_1 in Fig. 7b.

The existence of a stress-free meniscus introduces a boundary discontinuity on both upper and lower rims. This discontinuity is bound to result in sharply enhanced stress build-up at the upper and lower contact lines. It may produce an effective "slip" layer on the sample/plate interfaces at the meniscus, yielding an overall torque, which is less than anticipated on the basis of Fig. 7a, since a large portion of the measured torque on the plate arises from the stress contribution at the rim $r \approx R$. This correction may increase with lowering the gap distance. Without a proper analytical treatment, it remains unknown how the magnitude of such an apparent slip depends on the rheological properties of the sample and whether the reported slip like behavior [19, 33] is a manifestation of such a plausible edge effect.

5 Interfacial Slip in Various Flow Geometries

5.1 Wall Slip in Drag Flows

In this subsection we review experimental reports of *weak* wall slip in drag flows, i.e., non-pressure-driven flows, involving parallel plate or sliding plate geometry. Such studies are particularly challenging because one has to identify a very small correction as due to wall slip and not due to other artifacts.

5.1.1 *Slip-Like Behavior at Relatively Low Steady Shear Stresses*

The first suggestion of an apparent wall slip in a parallel plate cell was the 1983 report of Burton et al. [19] on gap dependent measurements of polystyrene (PS). Two explanations were offered. The first hinted at an idea of reduced viscous interactions at the melt/wall interface when the anchored (i.e., adsorbed) chains become oriented. This is somewhat close to the earlier idea of Bergem [31]. However, unlike polyethylene capillary extrusion considered by Bergem, the parallel plate measurements of Burton et al. involved rather low stresses and shear rates at which interfacial chain disentanglement is not expected to occur on bare metallic plates. A second explanation went as far as to suggest that there could be a stagnant adsorption layer of thickness around 20 μm, which reduced the effective gap separation [19]. The physical mechanism is rather unclear for such a layer formation on a scale three orders of magnitude larger than the chain dimension at an experimental temperature around 200 °C, well above the glass transition temperature of PS.

More extensive and quantitative experiments along this line have been carried out recently by Henson and Mackay (HM) [33]. Their results agree with the earlier observations of Burton et al. [19]. An apparent wall slip was again inferred from the observed gap dependence of the measured stress at a fixed nominal shear rate. HM arrived at a finite extrapolation length b ranging from 2.4 to 20 μm for four polystyrenes with molecular weight M_w ranging from 49,550 to 392,000, respectively. In order to determine such a small b, HM had to adopt gap separations between two parallel plates as small as 10 μm while avoiding misalignment. HM believed that their measurements could not be interpreted in any other way but as a result of a true wall slip. They imagined that polystyrene chains would slide along the steel plates as if there was little chain adsorption at stresses as low as 4 KPa. It was perceived that insufficient polystyrene adsorption on bare steel surfaces would allow partial slip at the melt/wall interface [33]. The reported linear dependence of b on M_w cannot be explained by the current theoretical picture given in Eqs. (3) or (5). The absence of b's dependence on the applied stress σ is also puzzling and unexplained since chain desorption should increase with σ. More experiments will be needed to verify these unexpected results.

Hatzikiriakos and Dealy (HD) performed an earlier study in 1991 on the weak wall slip behavior of high density polyethylene using Dealy's innovative sliding plate rheometer [37]. On the basis of the observed gap dependence, they also deduced that the HDPE slips on *clean* steel surfaces. From their data, one can compute an extrapolation length b as a function of stress and finds that b is immeasurably small below a certain stress level, becomes non-negligible for $\sigma \geq 0.1$ MPa, and gradually increases with the wall stress σ. Figure 8 shows the result calculated from the data in [37]. The magnitude of wall slip as characterized by b is sufficiently large. It should be possible to confirm such kind of wall slip

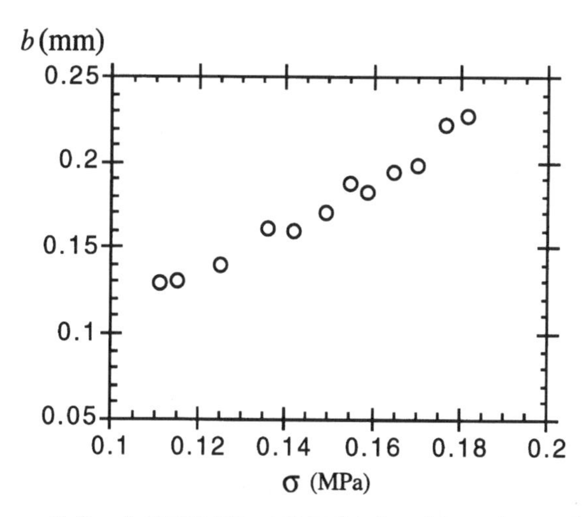

Fig. 8. Apparent wall slip of a HDPE (M_w=177,800) inferred from the gap dependence of the sliding plate rheometry measurements involving bare steel surfaces, calculated based on the data from HD's paper [37]

with other instruments. Using a commercial controlled-pressure capillary rheometer we have never been able to observe any wall slip of any HDPE in clean steel dies. The apparent wall slip as described in Fig. 8 may actually be related to unstable flow in the sliding rheometer and not interfacial in nature. Presumably, a true wall slip would arise from insufficient chain adsorption. It is key to be able to characterize the state of chain adsorption in melt flow. Currently no experimental technique is available to enable in situ characterization of state of chain adsorption at melt/wall interfaces exposed to strong flow.

A new experimental method has been invented in the laboratory of Léger et al. to enable direct observation of wall slip in a simple shear cell under the condition of controlled rate [38]. The technique, involving a combination of evanescent-wave-induced fluorescence and fringe pattern fluorescence recovery after photo bleaching, directly measures the melt velocity in the vicinity of a melt/wall interface (up to 70 nm into the bulk). For a weak interface such as that between a silane-treated glass plate and high molecular weight poly(dimethyl siloxane), wall slip is found and observed to increase with the applied shear rate above a certain threshold, thus confirming existence of wall slip phenomenon in melt flow.

This optical study of interfacial slip also shows that there is no detectable slip for poly(dimethyl siloxane) (PDMS) melts on *bare* glass plates. This [38], along with the rheological studies of PS [33], would suggest that PS adsorption on steel would be weaker than PDMS adsorption on glass. As a consequence, PS would display wall slip even in a parallel plate flow cell under normal measurement conditions according to [33]. Such a conclusion would have profound implications. This would mean that it may not be reliable to make any rheological measurements on certain *high* molecular weight polymers (e.g., PS) without properly accounting for a possible wall slip. A direct method such as the optical techniques pioneered in the laboratory of Léger et al. [38] would be a good way to verify the reported slip behavior of PS.

In the open geometry of torsional parallel plates, it is possible that the source responsible for the reported apparent gap dependence [19, 33] actually exists elsewhere. Besides the proposed idea of a global wall slip between the PS and the parallel plates, and apart from the possibility of misalignment, the peculiar gap dependence found by HM [33] could partially or completely originate from the boundary discontinuity and stress singularity described in Sect. 4. The effect of the boundary discontinuity in torsional parallel plates, as illustrated in Fig. 7b, would produce a larger correction at a smaller gap separation and affect the measurements in such a way as observed by HM. At sufficiently high stresses, the boundary discontinuity at the meniscus may produce a coil-stretch transition of the adsorbed chains on the rims of the two plates. It seems that new work needs to be done to determine the influence of the free meniscus surface on the *small gap* rheological measurements.

The only observation of a stick-slip transition in a simple shear geometry is the unique experimental study of Laun [39]. This controlled stress experiment not only observed a stick-slip transition but also explicitly recorded the time scale (a few milliseconds) on which the boundary condition (BC) evolved from no-slip to slip. Since the simple shear flow does not involve melt compression and decompression, the flow field is capable of responding to the BC change on molecular time scales. The essence of this study is that it was able to capture the short time scales of the stress relaxation in a HDPE.

5.1.2
Edge Failure vs "Dynamic Slip"

It is well known in polymer rheology that a torsional parallel-plate flow cell develops certain secondary flow and meniscus distortion beyond some stress level [14]. For viscoelastic melts, this can happen at an embarrassingly low stress. The critical condition for these instabilities has not been clearly identified in terms of the shear stress, normal stress, and surface tension. It is very plausible that the boundary discontinuity and stress intensification discussed in Sect. 4 is the primary source for the meniscus instability. On the other hand, it is well documented that the first indication of an unstable flow in parallel plates is not a visually observable meniscus distortion or "edge fracture," but a measurable *decay of stress* at a given shear rate [40]. The decay of the average stress can occur in both steady shear and frequency-dependent dynamic shear.

Recently several reports have related some observed stress decay behavior in parallel plate and sliding plate flow to a dynamic interfacial wall slip instead of edge instability. Using a sliding plate rheometer, Hatzikiriakos and Dealy [37] first suggested that decay of the stress maximum and distortion of the sinusoidal stress wave form in an oscillatory shear of HDPE are an indication of dynamic slip. Similar experiments have been carried out for other polymer melts [41]. Such a correlation between stress decay and a time dependent "dynamic slip" would have broad implications. It would indicate that a slip hydrodynamic boundary condition would develop *slowly* over time even as the generated stress decays at a constant shear rate. Chen et al. reported a very slow recovery of the measured stress in polystyrene after a stress decay at a higher overall stress level in oscillatory shear [21]. Regarding this stress growth as a recovery from slip to no-slip boundary condition, the long recovery was taken as an indication that chain dynamics at a wall were glassy [21, 42]. In other words, tethered chains at a wall would possess relaxation times that are three or four orders of magnitude longer than those of unbound chains in the bulk. An alternative explanation for the observed stress decay is that development of secondary flow and meniscus failure gradually caused detachment of the sample from the plates. This was indeed observed by Chen et al. [21].

Several theoretical studies have been carried out on a phenomenological basis to justify the idea that a dynamic slip can indeed occur [43–45]. New experimental techniques and theoretical treatments must be developed to reveal the molecular origin of dynamic slip and characteristic time scales governing the dynamic processes in the melt/wall interfacial region. Our own experiments on polyethylenes [27–29] have persistently indicated that slip, either through chain desorption on weakly interacting surfaces or interfacial chain disentanglement transition on clean metallic surface, occurs instantaneously, i.e., on time scales too short to determine, which is consistent with the observation of Laun [39]. More experimental studies must be conducted to verify that the observed stress decay would indeed originate from a dynamic interfacial wall slip and not from development of a secondary flow or other unknown flow instabilities. A particularly convincing confirmation for existence of dynamic slip may require application of the optical measurements at the melt/wall interface [38].

5.1.3
Ordering in Suspensions and Interfacial Slip

Several years ago we were led to believe that the only reliable way to observe true interfacial slip in drag flow was to consider "internal" slip between suspended spheres and a highly entangled polymeric matrix. A model system was established, consisting of glass spheres with diameters around 10 μm suspended in high molecular weight linear poly(dimethyl siloxane) (PDMS). Lowering the surface energy of glass spheres with silanation reaction to eliminate polymer adsorption, the idea was to measure the suspension viscosity as a function of the surface condition. A totally erroneous interpretation of the observed stress decay as interfacial slip was first reported in an earlier paper [42]. Such a conclusion was drawn to obtain a favorable agreement with the previously published papers [21, 37] reporting observations of stress decay due to "dynamic slip" of PS and PE respectively on high energy surfaces such as steel. Actually the stress decay of the glass sphere suspensions had a completely different origin [46, 47]. The homogeneously suspended spheres tend to order in oscillatory shear even at a strain amplitude of 1% as long as the maximum stress is sufficiently high. It is important to emphasize that this peculiar ordering of particles in a viscoelastic medium only occurs in oscillatory shear and does not occur in steady shear.

The ordering of glass spheres in PDMS can be readily avoided by carrying out steady shear measurements or taking initial values in oscillatory shear measurements before an appreciable development of ordering causes the suspension viscosity to drop. The interfacial condition depicted in Fig. 3 can be readily realized by using bare glass spheres. To achieve the interfacial state of Fig. 4a required a delicate silane treatment of the glass spheres. Figure 9 suggests that the lower suspension viscosity involving the silane-treated spheres arises from interfacial slip between the weakly adsorbing silane-treated spheres and PDMS matrix. Since both samples have the same concentration and uniform dispersion, an internal slip is the only sensible explanation. This conclusion was deduced from an extrapolation of the only known theoretical result for a dilute suspension of spheres of radius R exhibiting interfacial slip at the sphere/matrix interfaces: $\eta/\eta_m=[1+5(1+2b/R)\phi/2(1+5b/R)]$, which reduces to $\eta/\eta_m=1+\phi$ in the complete slip limit of $b/R>>1$. The slip behavior was seen to be stress and molecular weight dependent, as expected. Furthermore the onset stress for slip was found to be higher at a lower temperature [48]. This would be consistent with a chain desorption mechanism corresponding to a transformation of the interfacial

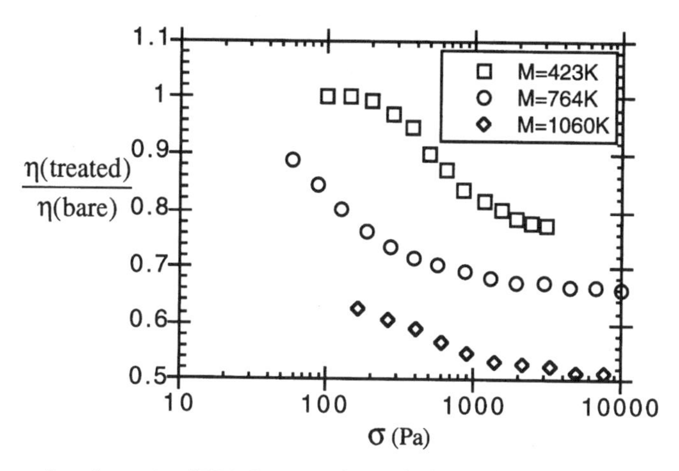

Fig. 9. Shear viscosity ratio of PDMS suspensions of silane-treated glass spheres (ca. 10 mm in diameter) to that of identical bare glass spheres at the same volume fraction ϕ=0.28, measured at T=25 °C in steady state for three PDMS molecular weights, reproduced from [48]

condition from Fig. 4a to Fig. 3. At sufficiently low stresses, e.g., for the overall stress σ<200 Pa, the PDMS suspension (M_w=423 K) shows no slip, and the sphere/PDMS interface can be depicted by Fig. 4a. It is important to comment that the slip was visualized to occur between the spheres and PDMS matrix, and slip between the sample and parallel plates has a negligible effect on the measurements. The advantage of this approach is that the viscosity measurement could detect a rather small amount of slip as long as b and R are comparable, where R≈5 μm is over two orders of magnitude smaller than the characteristic dimension of most rheological apparatuses.

5.2
Wall Slip in Pressure Driven Flows

Pressure-driven flows in capillary and slit dies do not have such complicating factors as a free meniscus surface. Consequently, much higher interfacial wall stresses can be generated without encountering potential instabilities related to a *free* surface. Capillary and slit die flows are especially suited for studies of interfacial melt/wall slip since breakdown of interfacial interactions often requires a stress level that is difficult to attain reliably in a parallel plate rheometer. The notion of wall slip has been around since Navier although an experimental signature of wall slip is much more recent [13]. Over the past fifteen years a large number of papers on wall slip have surged into the literature [49–56], and earlier papers have been discussed in review articles [10, 14]. For example, following the same line of their sliding plate measurements, Hatzikiriakos and Dealy also characterized a pre-transitional wall slip of a high density polyethylene (HDPE) using a controlled-speed capillary rheometer equipped with clean steel dies of different die diameters [54]. In our opinion, a controlled-speed capillary rheometer may be inaccurate to determine whether a small amount of wall slip occurs on clean metallic dies. Using a *controlled-pressure* capillary rheometer, we never find any measurable wall slip of any HDPE in clean steel dies. The only time a small wall slip was observed at stresses lower than the critical stress for a stick-slip transition was when a fluorocarbon elastomer (Dynamar, a 3M product) was coated onto the die wall [28].

Schowalter and coworkers [49–51] developed a hot film anemometer to measure "wall slip" [49]. This particular experimental technique attempted to correlate the pressure and hot film oscillation with the alternation of boundary condition in slit capillary flow of monodisperse polybutadiene (PB).

Many polymers exhibit neither a measurable stick-slip transition nor flow oscillation. For example, commercial polystyrene (PS), polypropylene (PP), and low density polyethylene (LDPE) usually do not undergo a flow discontinuity transition nor oscillating flow. This does not mean that their extrudate would remain smooth. The often observed spiral-like extrudate distortion of PS, LDPE and PP, among other polymer melts, normally arises from a secondary (vortex) flow in the barrel due to a sharp die entry and is unrelated to interfacial slip. Section 11 discusses this type of extrudate distortion in some detail. Here we focus on the question of why polymers such as PS often do not exhibit interfacial flow instabilities and flow discontinuity. The answer is contained in the celebrated formula Eqs. (3) or (5). For a polymer to show an observable wall slip on a length scale of 1 mm requires a viscosity ratio η/η_i equal to 10^5 or larger. In other words, there should be a sufficient level of bulk chain entanglement at the critical stress for an interfacial breakdown (i.e., disentanglement transition between adsorbed and unbound chains). The above-mentioned commercial polymers do not meet this criterion.

To see whether a given polymer melt would display a macroscopically measurable wall slip at various stresses, one merely needs to find out whether the flow behavior is sensitive to the surface condition of die walls. The simplest strategy is to carry out measurements under the same flow condition (i.e., at the same pressure) using two identical dies that only differ in their surface energy. Such experiments concern two different kinds of melt/wall interfaces corresponding to those depicted in Figs 4a and 3 respectively. In other words, on bare metallic walls sufficient chain adsorption is expected as illustrated in Fig. 4a up to fairly high stresses, whereas on an organic (such as Dynamar) surface little chain adsorption may remain in the same stress range.

Figure 10 shows the capillary flow behavior of a polystyrene (M_w=280,000). A noticeable magnitude of interfacial slip is observed by a comparison between the two flow curves. It is of interest to note that in the shear thinning region (described by the power law exponent n=3.2) the slip diminishes with increasing stress causing the two flow curves to converge. This means that the shear thinning has an adverse effect on slip. This is expected since the effective degree of *chain entanglement* is lower due to a greater level of chain deformation at a higher stress. The more profound implication of this portion of data states that there

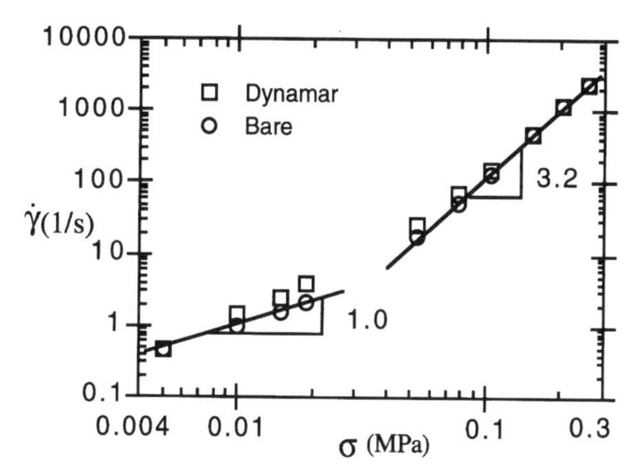

Fig. 10. Flow curves of polystyrene (M_w=280,000) measured at T=200 °C in capillary flow using dies of L/D=15 and 80 respectively with the same D=1.04 mm. The wall surface condition is either bare aluminum or Dynamar (a fluorocarbon elastomer from 3 M)

cannot be any measurable flow discontinuity transition or flow oscillation in the stress range of 0.1 MPa and beyond since the two curves overlap. The two flow curves also converge in the Newtonian region with lowering stress. This shows that the no-slip condition on the Dynamar coated die wall is recovered at stresses below 0.005 MPa where *sufficient* PS chain adsorption survives to *prevent* any measurable interfacial slip. Influence of the surface condition on capillary flow behavior of other polymers such as LDPE has also been examined. A similar phenomenon, i.e., convergence of slip and no-slip flow curves in the stress range where a stick-slip transition would be expected, is found [57]. This study has begun to answer the fundamental question of why LDPE resins do not undergo the discontinuous flow transition or flow oscillation.

6 Phenomenology of Melt Flow Anomalies

6.1 Flow Discontinuity Transition

Since the Navier's slip hypothesis of the last century, most experiments have failed to obtain positive evidence for a slip boundary condition on *macroscopic* scales in *low molar mass* liquids. However, Navier's notion of slip turns out to be extremely useful and convenient for the latest description of flow anomalies of highly *entangled* polymer *melts* including linear polyethylenes (LPE). The ability of a melt/solid interface to possess two profoundly different states as shown by Fig. 4a,b clearly reveals the potential role of interfacial slip in governing various melt flow phenomena in high pressure extrusion. Before reviewing recent experimental studies that have elucidated the molecular origins of different flow

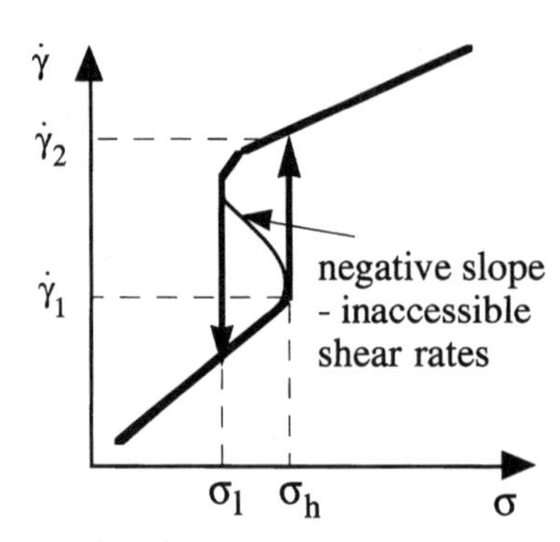

Fig. 11. Hysteresis behavior of the flow discontinuity in LPE, indicating that the shear rate range covered by the negative-slope line is not experimentally accessible under controlled pressure. The separation (σ_h-σ_l) depends on the rate of the pressure variation

instabilities and molecular mechanisms for wall slip, it is necessary to provide a historical background of 40 years of efforts to uncover the phenomenology of melt flow anomalies.

Bagley et al. [13] and Tordella [9] first observed in the 1950s that PE exhibited a discontinuous flow transition (i.e., a jump in the apparent flow rate) when the pressure applied to a PE-containing barrel was beyond a critical value. More often than not, the extrudate from the capillary extrusion was found to be grossly distorted upon the transition [32]. This led Tordella to believe that the flow discontinuity phenomenon was a genuine "melt fracture," or a form of instability, although Tordella [9c] bulk flow as well as others have also noted smooth extrudates [58, 59]. The observed hysteresis [9b, 13, 60] as illustrated in Fig. 11 never led previous workers to realize that the transition is interfacial in origin..

It was thought in the past that the only mechanism for wall slip would be polymer desorption, i.e., an adhesive breakdown [25, 53]. However, lack of a strong temperature dependence would be inconsistent with an activation process of chain desorption. Since the onset of the flow discontinuity (i.e., stick-slip) transition was found to occur at about the same stress over a range of experimental temperatures, it was concluded from the outset [9] that the phenomena could not possibly have an interfacial origin. Thus, the idea of regarding the flow discontinuity as *interfacial* did not receive sufficient and convincing theoretical and experimental support in the past, not only because the transition was often accompanied by severe extrudate distortion and hysteresis, but also because the molecular mechanism for such an interfacial transition involving wall slip was elusive.

6.2 Rheological Behavior of Other Polymers

It has been a profound enigma that polymers other than linear polyethylene rarely exhibit any flow instabilities such as flow oscillation under controlled piston speed. As to be discussed in later sections, we can generally understand this situation in terms of the theoretical prediction given by Eq. (5). In short, engineering polymers such as polystyrene, polypropylene, and polymethylmethacrylate do not possess sufficient chain entanglement and the level of interfacial slip may not be measurable during extrusion or under other processing conditions. Figure 10 clearly shows that a PS of M_w=280,000, which is among the highest molecular weight of any commercially available PS, is incapable of exhibiting a measurable amount of slip beyond 0.1 MPa even on an organic surface.

Polybutadiene (PB) and polyisoprene (PI) show similar flow behavior, including a flow discontinuity transition first observed in capillary flow of LPE [61]. In their pioneering experimental study [61], Vinogradov and coworkers reported the *transition-like* sharp increase of the apparent shear rate in monodisperse PB and PI beyond a certain level of applied stress. They proposed, based on the linear viscoelastic properties, that the flow transition corresponds to a bulk transition from a fluid to an elastic state. Calling the phenomenon "spurting" by borrowing the terminology used for describing the flow discontinuity of LPE, the Russian school suggested that this bulk transition mechanism also applied to explain the flow behavior of LPE.

Poly(dimethyl siloxane) (PDMS), has been extensively studied by the French group of Piau and coworkers [20, 22, 52, 56]. The PDMS samples studied by Piau et al. exhibit a flow discontinuity transition. Wall slip has been suggested to be responsible for the observed flow behavior. However, there was no revelation of any explicit molecular mechanisms for the interfacial slip.

Highly entangled *monodisperse* melts display rather special flow behavior reminiscent of flow discontinuity (stick-slip) transition characteristic of LPE. After the Newtonian regime, such melts abruptly enter the regime of extreme "shear thinning" beyond a certain shear rate comparable to the relaxation (reptation) rate where the flow rate begins to increase over several order of magnitude within a very narrow range of stress. Our most recent studies [62] on monodisperse polystyrene (PS) and polybutadiene (PB) show that the "plateau" (i.e., the extreme shear thinning) regime is characterized by a stress level σ^* that coincides with the plateau modulus G_N^0 measured in oscillatory shear. Because of the high entanglement molecular weight M_e for PS, G_N^0 is only 0.2 MPa, and σ^* is so low that no interfacial coil-stretch transition could occur without first entering the plateau regime where no slip would be observable due to the "extreme shear thinning." In contrast, PB shows a clear interfacial stick-slip transition at σ_c significantly below σ^*, i.e., prior to the "plateau" regime. These new data clearly indicate that an interfacial flow discontinuity transition *precedes* transition-like "plateau" behavior of highly entangled monodisperse melts in contrast to the conclusions from Vinogradov and coworkers [61].

6.3 Capillary Extrusion Instability

Ignoring the fundamental difference between constant speed and constant pressure conditions for capillary flow of LPE, most workers (except for the early work of Blyler and Hart [32]) abandoned the approach of controlled pressure used by Tordella and went on to study capillary flow of polyethylene with *controlled piston speed* rheometers. When the prescribed shear rate falls in the range of "inaccessible" rates approximately between $\dot{\gamma}_1$ and $\dot{\gamma}_2$ as indicated in Fig. 11, the flow becomes "unstable" and undergoes pulsation. The phenomenon is characterized by a pressure oscillation, corresponding to cycles of compression and decompression of LPE in the barrel reservoir prior to entry into the die, and by the large scale bamboo-like extrudate irregularity. This phenomenon is typically known as oscillatory flow and sometimes also termed interchangeably slip-stick behavior, melt flow instability, melt fracture and so on. Since the instability occurs under the condition of a constant piston speed, i.e., at a constant volumetric throughput, the appearance of acceleration ("slip") and deceleration ("stick") arises merely from the compressibility of PE under high pressure, which oscillates between two limiting values. Lupton and Regester [63] and Pearson [64] showed that a combination of melt compressibility and presence of two disconnected branches in the capillary flow curve would cause pressure oscillation with a period proportional to the melt volume in the barrel and inversely proportional to the shear rate. However, it was unclear to the previous workers [55, 63, 64] why there existed both no-slip and slip branches in the flow curve.

Evidence for wall slip was suggested over thirty years ago [9, 32, 63]. One of the first attempts at a slip mechanism was the performance of a Mooney analysis by Blyler and Hart [32]. Working in the condition of constant pressure, they explicitly pointed out "melt slip at or near the wall of the capillary" as the cause of flow discontinuity. On the other hand, they continued to insist that "bulk elastic properties of the polymer melt are responsible for the flow breakdown" on the basis that the critical stress for the flow discontinuity transition was found to be "quite insensitive to molecular weight." Lack of an explicit interfacial mechanism for slip prevented Blyler and Hart from generating a satisfactory explanation for the flow oscillation observed under a constant piston speed.

One leading explanation attributes the anomalous melt flow behavior (i.e., flow discontinuity and oscillation) to constitutive instabilities [65]. In other words, the anomalies would be constitutive in nature and non-interfacial in origin. Such an opinion has not only been expressed phenomenologically by Tordella [9b] and many other rheologists but found support from several theoretical studies [65–67]. However, these theories only attempt to describe inherent bulk flow behavior. Thus, a connection between the anomalous flow phenomena and constitutive instabilities was often explored without any account for possible molecular processes in the melt/wall interfacial region.

In the vast literature on melt flow instabilities in capillary extrusion, the most misleading information is the report that "the material of construction of the capillary die has no effect on the flow curve of linear polyethylene, or, in particular, on the instability region" [32, 68] – see a quotation by Tordella cited in Sect. 3. Experiments using screw-threaded dies have further led people to believe that "the slip (at the flow discontinuity transition) with linear polyethylene therefore appears not to result from adhesive breakdown at the polymer-die interface" [32]. The fact that a processing additive had no effect on the *upper slip branch* was also used as evidence for absence of interfacial failure during the transition [55, 59]. Blyler and Hart [32], followed Tordella's suggestion [9c], conjectured that the die surface may be rough as illustrated in Fig. 6, and perceived that slip may occur at an interface of maximum stress "between the stagnant layer and the main body of flowing melt." Since such an interface is primarily a polymer/polymer interface, a slip on this interface was regarded to be *cohesive*.

Extrudate distortion has been viewed as explicit evidence for melt flow instabilities or melt fracture. This is another calamitously misleading assertion in the massive literature of over 3000 papers on the subject. Because extrudate distortion was also observed even without any signature of slip, it was concluded by Blyler and Hart [32] that "the slip process is not an essential part of the flow instability." This dilemma, that the flow anomalies including flow oscillation cannot be accounted for in terms of either a constitutive instability or interfacial slip mechanism, has persisted until very recently. Denn coined this plight "the paradox" [10b].

7 Interfacial Chain Disentanglement at Flow Discontinuity

7.1 Discontinuity Upon Molecular Transition at Melt/Wall Interfaces

New results began to pour in on melt flow anomalies in extrusion of linear polyethylenes 2 years ago against the background of 40 years of predicament and frustration described in review articles of Denn [10b], Larson [14], and Hatzikiriakos and Dealy [55], a background that has been summarized in the preceding Sect. 6. On the basis of Navier's notion for interfacial slip [15] and de Gennes' insight of the magnitude of slip [25], the molecular origins of melt flow instabilities were explored from a fresh outlook [27–29, 69]. In this section we focus on a description of the current view of spurt flow phenomenon (i.e., flow discontinuity transition or stick-slip transition).

Each of the previously encountered difficulties can be overcome to unambiguously unravel the nature of the flow discontinuity transition in capillary flow of linear polyethylenes. The first step is to eliminate any secondary flow and associated extrudate distortion originating from a die entry instability by selecting a capillary die with a converging (e.g., 60°) entry angle. It turns out that such a small change in the die design removes spiral-like extrudate distortions even beyond the flow transition [57], proving that the commonly observed spiral like distortion due to a flat 180° entry is merely a propagation of the entry disturbance.

Besides eliminating the cause for extrudate distortion, it is important to remove the obscuring hysteresis around the transition so that the critical condition for the onset of spurt flow can be precisely determined. This can be accomplished by applying a discrete fixed pressure within each load (filling) of the barrel to insure that a *steady* state is established in the melt/wall interfacial region. We found that no hysteresis would arise unless the applied pressure was being varied during flow.

The ultimate proof or disproof that the flow discontinuity has an interfacial origin relies on an experiment to determine whether the surface condition of a metallic wall affects the transition characteristics. A change from a steel to aluminum die is inadequate because these metallic surfaces are sufficiently adsorbing for polyethylenes and chain adsorption is saturated on most metallic surfaces. Metzger and Hamilton [68] failed to produce a weakly adsorbing surface to remove the transition with a *glass-reinforced* Teflon capillary because glass is of sufficiently high surface energy. In order to have an entirely organic surface, we adopted a well-established procedure [37] to coat a steel die inner wall with a fluorocarbon elastomer (Dynamar). Such a surface modification indeed completely altered the basic features of linear PE in capillary flow [28], proving that the flow discontinuity is non-constitutive in nature and interfacial in origin. We will discuss these new features in Sect. 9.

The molecular process accompanying the interfacial transition is expected to show an explicit temperature dependence. In particular, the stick-slip transition occurs either because of an adsorption-desorption transition of the interfacial state from Fig. 4a to Fig. 3 or as a result of an interfacial entanglement-disentanglement transition from Fig. 4a to Fig. 4b. The temperature dependence of these two distinctly different processes is opposite of each other. The former involves polymer desorption. The critical stress for such an activation process would be higher at a lower temperature. The latter involves a coil-stretch transition. The condition for this conformational transition is a critical external force overcoming the entropic force F_e of each disentangling chain. Assuming the area number density of chains undergoing the disentanglement is ν, the critical stress σ_c for such a coil-stretch transition at a surface should be given by

$$\sigma_c = \nu F_e, \text{ with } F_e = k_B T / D_{tube} \quad (6)$$

where D_{tube} is the tube diameter in the reptation theory.

In the theory of Brochard and de Gennes [30], $\sigma_c = \nu F_e$ is taken as the stress in a marginal state where *no* disentanglement occurs yet. In the simplified picture given in [30], ν was assumed to represent the density of adsorbed chains that have only one adsorption site per chain. In reality for an adsorbing surface, there should be $N^{1/2}$ loops per chain as depicted in Fig. 5c, and the number ν_{loop} of loops per unit area is on the order of $\nu_{loop} \approx 1/a'^2$, independent of the chain length N where a' is the Kuhn length. There is a distribution of loop sizes, ranging from a minimum size smaller than the tube diameter $D_t = N_e^{1/2} a'$ to a maximum size comparable to the coil size R_g. An average loop may contain $N^{1/2}$ segments. When N is not large, i.e., $N < N_e^2$, many loops have a size smaller than the tube diameter where N_e is related to the critical molecular weight M_e for entanglement as defined in Eq. (5). When $N > N_e^2$, more loops are capable of a coil-stretch transition. The wall stress required for such a transition of loops would be as

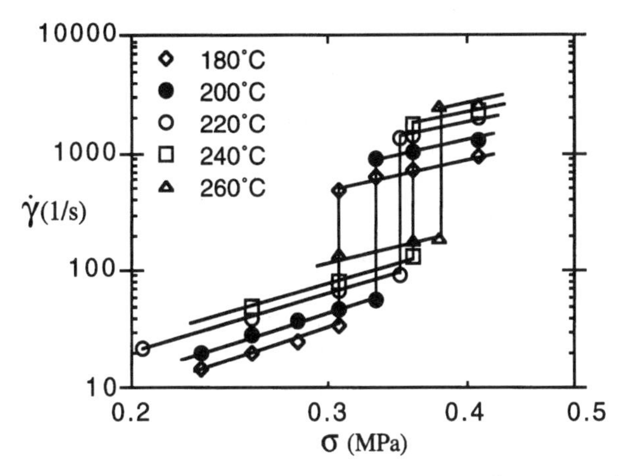

Fig. 12. Temperature dependence of a stick-slip transition of an HDPE (MH20 of M_w= 316,600) from BP Chemicals in the controlled pressure mode using a capillary die of L/D= 15 and D=1.0 mm, reproduced from [29]. The identical amplitude observed at different temperatures is an impressive feature of the interfacial transition

high as $\nu_{loop}F_e$. This value is estimated to be much larger than σ_c, observed experimentally to be around 0.3–0.5 MPa. For a strongly adsorbing wall, the stick-slip transition may not involve the loops undergoing a coil-stretch transition. Thus the meaning of ν in Eq. (6) remains to be explored. In any event, Eq. (6) does indicate that σ_c would increase linearly with temperature T, provided that ν does not change with T.

Plagued by the difficulty of hysteresis and other factors such as entry instability, most previous efforts failed to observe a *systematic* temperature dependence of the stick-slip transition under the controlled pressure operation condition. Despite the small variations of σ_c with temperature over a relatively narrow range from 180 to 260° a clear linear relationship between σ_c and T can be established by discretely applying the piston pressure and employing a die with a converging entry angle. Figure 12 shows a monotonically increasing σ_c with respect to T and thus unambiguously rules out polymer desorption as the molecular mechanism for the spurt flow discontinuity. Figure 12 forcefully suggests that the stress-induced coil-stretch transition and the resulting disentanglement between adsorbed and unbound chains are the origin of the interfacial stick-slip transition.

The data in Fig. 12 actually collapse onto a *master* curve when the wall stress σ is rescaled by temperature T and the nominal shear rate $\dot{\gamma}$ is normalized by a WLF factor a_T [29]. Thus Eq. (6) for the critical stress σ_c is supported by the data in Fig. 12, where ν does not change with T. Another feature of the transition is that the amplitude of the flow discontinuity does not vary with T. In other words, the extrapolation length b_c, which is evaluated according to Eq. (4a) *at the transition*, is a constant with respect to T. Thus for a given surface, b_c is more than just a material property such as the melt viscosity η. It essentially depends only on the molecular weight, its distribution and molecular architecture. The temperature independence of b_c is not universal. At lower temperatures, b_c is found to be larger for some high density polyethylenes. The ability of linear PE to have ordered (i.e., nematic) mesophases under flow [70] may be responsible for the low temperature anomaly [27].

7.2 Mooney vs Navier Treatment of Wall Slip

Mooney [71], a century after Navier's notion for interfacial slip [15], ignored the essence of the Navier extrapolation length *b*, and produced a different formula, which employs the slip velocity v_s as the *essential quantity* to describe wall slip. Instead of evaluating *b* from Eq. (4a), which has been available since the last century [15, 24], Mooney wrote down

$$Q = Q_0 + (\pi D^2 / 4) v_s \tag{7a}$$

which differs from Eq. (4a) just by a multiplicative factor Q_0, and suggested to compute v_s from

$$\dot{\gamma} = \dot{\gamma}_0 + 8 v_s / D \tag{7b}$$

as the slope in a plot of the apparent shear rate $\dot{\gamma}$ vs 8/D. Note that Eq. (7b) holds true only for Newtonian fluids for which $Q=(\pi/32)D^3 \dot{\gamma}$.

Conventionally, polymer rheologists uniformly apply the Mooney analysis (Eq. 7b) to quantify the degree of melt/wall interfacial slip. Their task of evaluating wall slip centers on measuring the apparent shear rate $\dot{\gamma}$ or flow rate Q as a function of D at the same stress and computing v_s from Eqs. (7b) or (7a) by fitting to the data. Table 1 lists several parameters including the slip velocity v_s^c evaluated at the stick-slip transition in the temperature range covered in Fig. 12. It is striking but not surprising to see that v_s^c increases rapidly with temperature, as the critical shear rate $\dot{\gamma}_c$ does at the transition, in sharp contrast to b_c, which remains essentially constant within the experimental uncertainty. Clearly, v_s is a kinematic quantity rather than a material property and cannot itself provide any meaningful quantification of wall slip.

Table 1. Temperature dependence for polyethylene MH20 (L/D=15)

T (°C)	$\dot{\gamma}_c$ (s^{-1})	v_s^c (cm/s)	σ_c (MPa)	b_c (mm)
180	34	5.9	0.307	1.74
200	57	11.1	0.332	1.95
220	92	16.3	0.348	1.77
240	132	21.4	0.358	1.62
260	186	29.9	0.373	1.61

In view of the more effective treatment of wall slip known to fluid mechanicians including Navier and Stokes since the last century, the Mooney analysis is indeed a setback in the development of this entire field. *One particular drawback of the Mooney analysis is that it is incapable of comparing the magnitude of wall slip at different temperatures or for different polymers.* The absolute value of v_s reveals little information about the magnitude of slip. In other words, if v_s of polymer A is larger than v_s of polymer B, it does not mean the polymer A exhibits a larger degree of wall slip, because v_s must be evaluated relative to the corresponding wall shear rate $\dot{\gamma}$. The slip velocity v_s itself is not a material property. This can be illustrated explicitly and quantitatively. *b* in Eq. (3) depends only on the ratio of two material parameters, the shear viscosity η and friction coefficient β. The slip velocity v_s, given by $v_s=b\,\dot{\gamma}$, directly depends on the shear rate $\dot{\gamma}$ and therefore only provides a *kinematic* description of the slip behavior. Since most workers in the field continue to use v_s to examine slip phenomena, it is the purpose of this subsection to publicize the more effective century-old method of using the extrapolation length *b* for quantification of interfacial slip.

The difference between *b* and v_s is profound yet simple: To properly evaluate wall slip, we simply need to use Eq. (4a) instead of Eq. (7a), i.e., to normalize both sides of Eq. (7a) by Q_0. Only when the relative change $(Q\text{-}Q_0)/Q_0=8b/D$ is beyond the experimental uncertainty, can wall slip be reliably measured rheologically. In other words, evidence for wall slip would be questionable if the magnitude of slip correction, measured by $8b/D$, is much smaller than unity, and the difference $(Q\text{-}Q_0)/Q_0$ is experimentally vanishingly small.

7.3 Critical Condition and Molecular Characteristics of Stick-Slip Transition

Linear polyethylenes (PE) are one polymer that possess an important ingredient necessary for a display of interfacial stick-slip transition. In the past, the coincidence that PE is both the most widely used polymer and most prone to suffer from melt flow instabilities has challenged the PE industry. Today we still face the task of how to effectively remove instabilities that result in various types of extrudate distortions.

To achieve the goal of improving polymer processing efficiency, it is essential first to determine the crucial parameters controlling the onset and magnitude of the stick-slip transition. The onset stress σ_c for a disentanglement transition in the melt/wall interfacial region is given by Eq. (6); the magnitude of slip resulting from this interfacial transition is given by Eq. (3). Because PS is much less entangled than PE, Eq. (3) indicates that an extrapolation length for PS would be two orders of magnitude smaller than that of a linear PE for the same molecular weight, taking the critical entanglement molecular weights $M_e(PS)=13{,}300$ and $M_e(PE)=830$ from Fetters et al. [72]. This explains why many engineering thermoplastics such as PS behave differently in capillary rheology and commercial processing: they are not of sufficiently high molecular weight to produce a measurable interfacial slip.

We have little idea about the strength and extent of polyethylene adhesion with a given solid surface. We do know that the chain adsorption is sufficiently strong, i.e., unbound chains well entangle with the adsorbed chains to prevent a massive interfacial slip at stresses below the critical stress for the stick-slip transition. It is entirely reasonable that other polymers also establish strong adsorption with such high energy surfaces as steel and aluminum. Their effective chain entanglement density at a critical stress for interfacial chain disentanglement determines whether a measurable stick-slip transition takes place or not.

7.3.1
Critical Stress and Interfacial Adsorption

The linear relation $\sigma_c \propto T$ observed in Fig. 12 is not sufficient evidence that would unambiguously support Eq. (6) and reveal the interfacial nature of the transition, because a bulk phenomenon may also produce such a temperature dependence. For instance, one might think of "melt fracture" and write down: $\sigma_c = G\gamma_c$ that would be independent of M_w, where γ_c would correspond to the critical effective strain for cohesive failure and modulus G would be proportional to $k_B T$. Previous experimental studies [9, 32] lack the required accuracy to detect any systematic dependence of σ_c on M_w and T. This has led to pioneers such as Tordella [9] to overlook the interfacial origin of spurt flow of LPE. It is in this sense that our discovery of an explicit molecular weight and temperature dependence of σ_c and of the extrapolation length b_c is critical. The temperature dependence has been discussed in Sect. 7.1. We will focus on the M_w dependence of the transition characteristics.

We found the critical stress σ_c for the stick-slip (coil-stretch) transition scales with the molecular weight M_w as [27]

$$\sigma_c \propto M_w^{-0.5} \tag{8}$$

This scaling perhaps reveals a rather simple situation: $\sigma_c \propto \nu \propto M_w^{-0.5}$, where ν is the number of the *entanglement-effective* chains per unit area, with each chain sharing a force F_e, which is independent of M_w according to Eq. (6). Inserting the experimental value of $\approx$0.3 MPa from Fig. 12 into Eq. (6), we infer that $\nu >> 1/R_g^2$ where R_g is the radius of gyration of a PE chain. If there are many adsorbed chains per area of R_g^2, as the value of σ_c indicates, then it is unlikely that these chains would each only have one adsorption site without forming any loops. This consideration clearly rejects the state of adsorption depicted in Fig. 5a where no loops are present. Thus the molecular picture of Fig. 4a,b can only be regarded to be illustrative rather than quantitative. Equation 8 appears to favor Fig. 5b where a few loops [13] are present per adsorbed chain since Fig. 5c would not give any molecular weight dependence.

For chains of uniform length N, the number ν of chains penetrating a unit area inside a melt can be estimated by $\nu \propto \rho/R_g^2$, where $\rho \propto R_g^3/Na'^3$ is the number of chains in a volume of R_g^3. Recalling $R_g \propto N^{1/2}a'$, we have

$$\nu \approx 1/R_g a' \propto M_w^{-1/2}. \tag{9}$$

Thus Eq. (9) shows that the number density ν of chains entangled with the loops in Fig. 5c would indeed have an $M_w^{-0.5}$ dependence. When each of these adsorbed chains experiences a critical force of F_e they become disentangled and a stick-slip transition follows as observed in Fig. 12. The molecular weight dependence of the critical stress given by the combination of Eqs. (6) and (9) explains the experimental finding [27] for linear polyethylenes .

Substituting Eq. (9) into Eq. (6), we obtain the following estimate of the critical stress σ_c for the stick-slip transition:

$$\sigma_c \approx k_B T / R_g a' D_{tub} \approx 0.31 \text{ MPa}, \tag{10}$$

if we take T=473 K, $k_B=1.38\times10^{-16}$ erg/K, D_{tube}=33 Å, estimate R_g=629 Å for M_w= 316,600 from [72], and assume a'=1 Å. This value is close to the experimental value of $\sigma_c \approx 0.32$ MPa for MH20 at 200 °C. This agreement supports the proposed interfacial mechanism for the transition. The validity of Eq. (10) can only be explained by the notion that the stick-slip transition is produced by disentanglement of the interfacial chains from the bulk chains. The full implication of this comparison between experiment and theory remains to be further explored in the future.

A sufficiently rough die surface (perhaps larger than microns) could also yield such behavior as described by Eqs. (8) and (10). In this case, the coil-stretch transition takes place on an effective interface between the stagnant chains in the "valleys" and flowing chains on the inner side of the interface, as sketched in Fig. 6 and simulated by using threaded dies in [28]. At the effective penetrable interface there are $\nu \approx 1/R_g a'$ chains per unit area that are entangled into the stagnant chains. At the critical stress $\sigma_c = \nu F_e$, each of the entrapped chains at the "interface" experiences a critical force F_e given in Eq. (6), which is sufficient to cause full chain stretching.

It remains a formidable task to develop a new surface sensitive spectroscopic method, which can differentiate the adsorption states shown in Fig. 5a–c. Moreover, it remains to verify whether the scaling given by Eq. (8) originates from the surface roughness, i.e., whether Eq. (8) applies to a molecularly smooth die wall. Throughout this article, we will simply regard the stick-slip transition as due to a coil-stretch transition that results in removal of all chain entanglement in the melt/wall interfacial region. We have also taken the terms "interfacial" and "interface" to mean either a melt/wall boundary or an effective boundary where the chain disentanglement occurs. The pre-transitional wall slip phenomena [29] occur perhaps due to stress-induced detachment of weakly adsorbed chains on low energy walls.

7.3.2
Molecular Weight Dependence of Constant-Stress Viscosity and Extrapolation Length

We have noted in Sect. 7.1 that the extrapolation length b is essentially independent of temperature in the range 180–260 °C and is only a function of molecular weight, its distribution and architecture. This makes b a very special material property of linear polyethylenes, as basic as the molecular weight itself. The physical meaning of b is depicted pictorially in Fig. 2 in the case of small molecule liquids and in Figs. 3 and 4b in the case of highly entangled melts where the size of b is too large to be depicted on that scale. The kinematic definition of b is given by the first equality of Eq. (3). It is merely an imaginary distance away from the sample/boundary interface where the fluid would attain zero velocity.

The molecular meaning of b is best seen from the second or third equality of Eq. (3). In other words, b is explicitly related to the steady shear melt viscosity η and depends on the chain-chain interactions near the melt/wall interface as quantified by the friction coefficient β. In the limit of no polymer adsorption or in absence of interfacial chain entanglements due to the coil-stretch transition, β involves an interfacial viscosity η_i, which is as small as the viscosity of a monomeric liquid and independent of the molecular weight M_w: $\beta=\eta_i/a$, where a is a molecular length. Thus at the stick-slip transition, the molecular weight dependence of b arises entirely from η in Eq. (3).

In the existing literature, the molecular weight dependence of steady shear melt viscosity η has rarely been adequately discussed in the *shear thinning regime*. To reveal a possible *universal* scaling form of $\eta(M_w)$ in the shear thinning limit, one must consider η_σ, i.e., measure η of different molecular weights at the same stress level σ. Recent constant-stress steady shear viscosity measurements of a series of four different M_w revealed that η_σ scales with M_w as $\eta_\sigma \propto M_w^{3.4}$ in the shear thinning (power law) regime [27]. This may imply that in strong shear the overall melt relaxation time $\tau(\sigma, M_w)$ still varies with M_w as $\tau(\sigma, M_w) \propto M_w^{3.4}$, as the zero shear disentanglement relaxation time τ_d does. The familiar exponent 3.4 arises because the four samples happened to share a sufficiently similar molecular weight distribution. The molecular meaning of this kind of experimental data remains to be explored.

The molecular weight dependence of the extrapolation length b originates from the same molecular weight dependence of the constant stress viscosity η_σ. A recent experimental study shows

$$b_c \propto M_w^{3.4}, \tag{12}$$

which measures the amount of interfacial slip at the transition [27]. We observe such an apparent scaling law for b_c because σ_c only weakly varies with M_w as $\sigma_c \propto M_w^{-0.5}$ and can be almost regarded as independent of M_w. Thus, $b_c(\sigma_c) \propto \eta(\sigma_c) \approx \eta_\sigma \propto M_w^{3.4}$. Equations (8) and (12) depict the salient features of the interfacial stick-slip transition and emphasize that an *proper extent of polymer adsorption* and a *sufficient degree of chain entanglement* combine to dictate the transition characteristics. As we remarked above, it appears that linear polyethylenes (LPE) possess these two basic ingredients for a stick-slip transition in capillary flow. It is this interfacial transition that causes flow instabilities such as oscillating flow in controlled speed extrusion. Many other engineering polymers are insufficiently entangled. As a consequence, the breakdown of interfacial interactions i.e., disappearance of chain entanglement at the polymer/wall interface, does not lead to significant modification of the melt flow and thus does not produce extrusion instabilities. Gross extrudate distortion is a separate issue and should not be confused with such interfacial flow instabilities as oscillating flow due to the unsteady boundary condition. In closing, it is important to add that the influence of molecular weight on the magnitude of the flow discontinuity has been qualitatively investigated previously by other groups for HDPE [32, 52].

8
Molecular Mechanism for Oscillating Capillary Melt Flow

Over the last four decades, a vast amount of research has been performed in an attempt to uncover the physical origins of flow anomalies such as flow oscillation and extrudate distortion. Figure 1 shows the stress level, at which the flow anomalies occur in linear polyethylenes (LPE). There were at least two competing mechanisms proposed for the melt flow anomalies (i.e., an oscillatory flow in controlled speed extrusion and flow discontinuity observed under controlled pressure). One links the phenomena to a constitutive instability and another relates to wall slip behavior. The debate has continued until very recently. Wall slip has been advertised as playing an important role in producing the flow anomalies (flow oscillation and spurt – flow discontinuity) [11, 32]. No previous work proposed an explicit account of just how wall slip would be involved in the melt flow instabilities. This long standing predicament or "paradox" [10b] arises partially from poor understanding of polymer melt/wall interfacial interactions, i.e., lack of an explicit molecular mechanism, and partially from unclear knowledge of constitutive flow properties of entangled melts. As a result, it has not been obvious whether and when an interfacial flow discontinuity transition or constitutive flow behavior would prevail in capillary extrusion of highly entangled polymers including LPE, PDMS, PB, PI, and PS.

As discussed in more detail below, recent experiments convincingly showed that the flow oscillation in capillary extrusion of LPE is interfacial in nature due to a reversible coil-stretch transition at the melt/die wall boundary. Pressure oscillation phenomenon has also been reported in extrusion of other polymer melts. In particular, there are well-defined oscillations in controlled-rate capillary flow of PB that were found to arise from the same interfacial molecular instability [62].

Having unraveled the specific characteristics of the stick-slip transition, it is rather straightforward to describe the physical origin of the oscillating flow observed under a constant piston speed or extrusion rate. A die diameter dependence study of the stick-slip transition shows [74] that there is insignificant wall slip below the critical stress σ_c, implying that there is plenty of polyethylene melt adsorption at bare metallic walls. The temperature dependence shown in Fig. 12 in Sect. 7.1 indicates that there must be sufficient polymer adsorption on bare die walls. The chain adsorption provides the necessary condition for a stress-induced stick-slip transition to occur over a range of wall stress corresponding to the finite hysteresis loop indicated in Figs. 1 and 11. Given such a remarkable property of polymer/wall interfaces where a reversible coil-stretch transition can take place, oscillatory flow occurs inevitably under controlled piston speed. The striking symptoms of the oscillating flow (i.e., pressure oscillation and quasi periodic variation of extrudate – bamboo like distortion) can be clearly explained in terms of the interfacial stick-slip (coil-stretch) transition. Such an explanation has previously appeared; however, the phenomenon of flow oscillation was inappropriately termed spurt flow in [29, 57].

Specifically, when the volumetric flow rate Q produced by a constant piston speed exceeds a certain value, the nominal wall shear stress $\sigma=\eta\,\dot{\gamma}$ becomes larger than the critical stress σ_c and induces wall slip resulting from a coil-stretch transition. The velocity field profile changes discontinuously and Q is now given by Eq. (4a), where Q_0 is related to the actual shear rate $\dot{\gamma}_{slip}$ as $\dot{\gamma}_{slip}=32Q_0/\pi D^3$. Due to the wall slip, the true wall shear rate drops from $\dot{\gamma}=32Q/\pi D^3$ to $\dot{\gamma}_{slip}=\dot{\gamma}/(1+8b/D)$ according to Eq. (4a).

If b/D is very large so that the actual shear rate $\dot{\gamma}_{slip}$ in this slip state is very small, the actual wall stress $\sigma=\eta\,\dot{\gamma}_{slip}$ can drop considerably below the critical stress σ_c for the stick-slip transition. The stretched chains will recoil, i.e., undergo a reverse stretch-to-coil transition since there is insufficient wall stress to counterbalance the internal entropic force within each stretched chain. Upon the recoil, the unbound flowing chains become re-entangled with the adsorbed chains in a short period, presumably on the order of the chain relaxation time. At this moment, the stick hydrodynamic boundary condition (HBC) is restored, and the wall stress level returns to the no-slip value $\sigma=\eta\,\dot{\gamma}>\sigma_c$. This completes one cycle of oscillation in the measured wall stress and extrudate swell. The cycle will repeat itself, with the period dependent *not* on the molecular time scales, on which the adsorbed chains undergo the coil-stretch and its reverse, but on the melt compressibility and amount of PE in the barrel [63, 64]. The above analysis indicates that the oscillating flow would have a diminishing magnitude for a sufficiently large D, such that $b/D<<1$ and $Q\approx Q_0$. This means wall slip is inconsequential and the flow oscillation is barely measurable. In other words, when $b/D<<1$, the width of the instability window, given in Fig. 11 as $\dot{\gamma}_2/\dot{\gamma}_1=1+8b/D$, diminishes and no flow oscillations would be visible.

In summary, the origin of the oscillating flow (sometimes termed slip-stick regime) observed in the constant piston speed mode is the oscillation of the HBC between the no-slip and slip states due to a reversible coil-stretch transition of either adsorbed chains or the first layer of unbound chains entrapped with the adsorbed chains. The experimental demonstration of an abrupt large stick-slip transition in pressure-controlled capillary flow of linear polyethylenes makes it clear that the symptoms of the oscillating flow would occur in extrusion at a constant piston speed. In particular, the stress oscillation caused by the unstable boundary condition unequivocally elucidates that the oft-observed bamboo like extrudate distortion merely originates from the temporal fluctuation of extrudate swell and should not be regarded as a signature of any form of cohesive bulk failure, bulk instability, or melt fracture, as having been mistakenly called ever since its first reports [9, 13] in the late 1950s. Polyethylene chain adsorption on the bare metallic die walls is necessary for the occurrence of this flow oscillation phenomenon. Absence of chain adsorption results in interfacial wall slip and in elimination of the flow oscillation along with the quasi-periodic extrudate distortion [28].

9 Interfacial Origin of Anomalous Capillary Flow Behavior

Polymer/wall interfacial interactions depend strongly on the extent and strength of chain adsorption. As the surface condition varies from strongly adsorbing to weakly adsorbing, the boundary condition may change from no-slip, to partial slip, to complete slip at a high stress. It depends on the extent of effective chain entanglement upon slip whether such a slip produces any observable correction to rheological measurements. If the applied stress could cause complete chain desorption off a given weakly adsorbing die wall, we would have an effective method to evaluate the maximum wall slip of a given polymer. The magnitude of an interfacial slip measured as a function of the wall stress is *not only a profound indicator of the degree of effective chain entanglement in flow but also a barometer for predicting whether a given polymer could undergo a flow instability such as flow oscillation.*

Figure 13 shows the influence of a Dynamar coating on the capillary flow behavior of MH20 at stresses well below the critical stress for the stick-slip transition. In sharp contrast to the small level of PS slip revealed in Fig. 10, the ability of MH20 to exhibit a huge interfacial slip is shown by the large separation between circles and squares. The extrapolation length b for PE is a huge factor p times larger than that for PS, given the same molecular weight: $p=b(\text{PE})/b(\text{PS})$ $\approx(100/28)(13309/828)^2=922$, taking $M_e=13{,}309$ for PS and $M_e=828$ for PE from Fetters et al. [72], and assuming b can be estimated from Eq. (5). This estimate is not too far from the actual ratio of the two measured extrapolation lengths. One finds $b(\text{PS280})=0.06$ mm at $\sigma=0.05$ MPa from Fig. 10 and $b(\text{MH20})=$ 9.5 mm at $\sigma=0.03$ MPa from Fig. 13. This agreement is remarkably good, knowing that the two polymers may not necessarily experience the same level of shear thinning at the respective stresses and that MH20 has a much broader molecular weight distribution. This huge difference in slip between PE and PS clearly originates from the large difference in their molecular chain entanglement. We also notice that $b(\text{MH20})$ at such a low stress is much larger than b_c measured at the

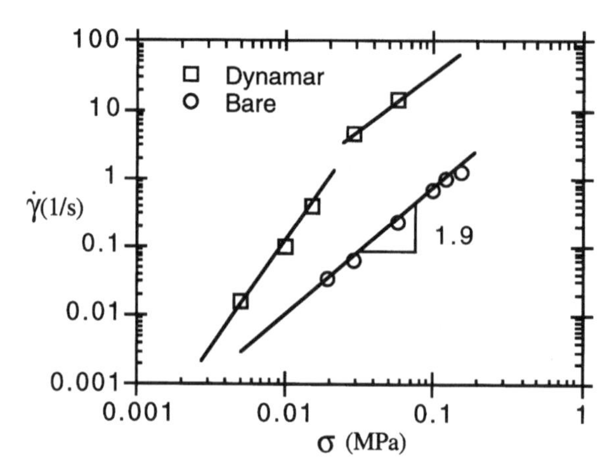

Fig. 13. The influence of the Dynamar surface coating on flow behavior of HDPE (MH20) shown by the comparison between the circles and squares obtained at T =150 °C with the same dies that are used to obtain Fig. 10

stick-slip transition and listed in Table 1, demonstrating once again the adverse effect of shear thinning on the magnitude of interfacial slip. In other words, b is larger because chain entanglement is significantly higher at a lower stress.

To illustrate further the interfacial origin of the spurt flow behavior of linear polyethylenes, it is important to modify the chemical composition of the die wall. In theory, one would like to prepare surfaces with well-defined surface energy and systematically vary the extent and strength of chain adsorption. In practice, this is a rather challenging task. We have attempted to modify the die wall surface with various degrees of fluorocarbon elastomer coating. Figure 14 captures the essence of the capillary flow phenomena of linear polyethylenes and reveals their interfacial origin. The flow behavior depends critically on the nature of chain/wall interfacial interactions characterized by the amount of polyethylene adsorption. The behavior observed in bare aluminum dies (full circles) changes as the surface condition is modified with poor (3M1) and good Dynamar coating (3M2) respectively as shown in Fig. 14. It can be shown that the two slip-slip transitions in open squares and circles involve chain desorption rather than coil-stretch transition associated with the remaining adsorbed chains on the imperfectly coated die walls [28]. Finally a fluorocarbon elastomer coating using the du Pont silverstone technology was able to eliminate thoroughly direct aluminum-PE contact at the PE/coated-wall interface and to produce complete wall slip throughout the explored stress range. If the spurt flow observed in the bare die was non-interfacial, it would have also occurred in the coated dies. The open symbols indicate otherwise. In other words, the surface condition can completely alter the observed capillary flow behavior over the *same* stress range and eliminate the flow discontinuity (stick-slip transition) by causing PE melt to slip throughout a broad range of stresses. Clearly no bulk

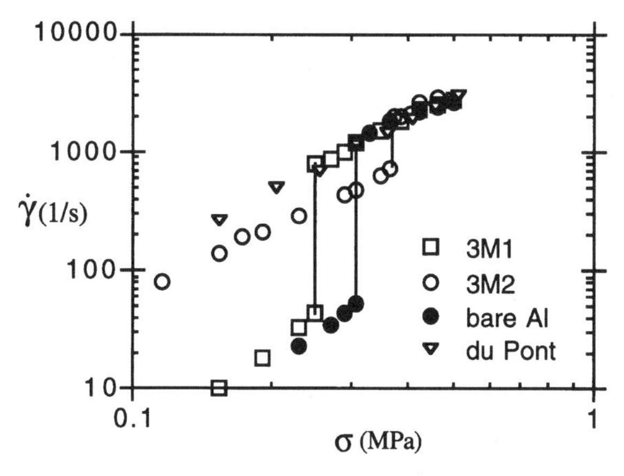

Fig. 14. Flow behavior of MH20 (M_w=316,600) HDPE at T=200 °C as influenced by the surface condition of the inner die walls (L/D=15 and D=1.0 mm)

constitutive mechanism for the flow discontinuity transition is consistent with this experimental observation.

10
Molecular Instabilities – Origin of Sharkskin

There are at least three types of extrudate distortion that have created difficulties in polyethylene processing. Taking linear low density polyethylene (LLDPE) as an example: (a) we observe *sharkskin* as the first kind of extrudate roughness regardless of the extrusion mode; (b) with increasing flow rate we encounter the second kind – *bamboo-like or alternating* extrudate distortion accompanying a pressure oscillation during a constant piston speed extrusion; (c) a third more severe often *spiral-like* distortion may occur at a similar (or higher) flow rate or applied pressure. The main scientific objective of the last 40 year research has been to unravel the physical origins of these three leading types of extrudate distortion. Upon accomplishing this goal, one hopes to develop higher processing efficiency in film blowing, blow molding, wire coating and even injection molding of all grades of polyethylenes.

We have presented a thorough description and discussion about the molecular origin of the second kind (b) – bamboo like extrudate distortion – in the preceding Sect. 8. The present section is devoted to a specific illustration of the molecular origin of the type (a) distortion, i.e., *sharkskin*, which occurs in a range of stress/rate below the oscillatory flow or stick-slip transition, as indicated in Fig. 1. The next section will provide a brief discussion of the origin of the type (c), often spiral-like distortion. The macroscopic nature of the type (c) distortion was first discussed at least over 20 years ago [75]. Note that when the type (c) spiral distortion occurs on very fine length scales on the extrudate it can be and has sometimes been mistaken as sharkskin.

Previous literature on the subject of melt flow anomalies has often neglected the need to distinguish and separate these three types of extrudate distortions. Along with the term "melt fracture" for bamboo-like distortion, the sharkskin is often called "surface melt fracture" [10b]. During a constant speed extrusion in the oscillating flow regime, one may actually observe all three types of distortions occurring simultaneously on the extrudate. In the past, the type (c) distortion has frequently interfered with observation and investigation of the type (b) distortion. It is crucial that the type (c) extrudate distortion be *eliminated* when studying the flow discontinuity (i.e., spurt) and flow oscillations. Failing to do so has generated confusion and made an unambiguous investigation of the oscillating flow and spurt phenomena an impossible task. Apparently Tordella [76] and Bagley and Birks [77] understood the origin of the type (c) distortion. Somehow they did not separate type (c) distortion from the other types by eliminating the spiral-like distortion and did not carry through their pioneering work to achieve a satisfactory understanding of polyethylene extrusion instabilities. The type (c) distortion usually does not occur in the sharkskin regime.

Sharkskin is itself a sufficiently intricate phenomenon. In particular, there is no strong global rheological or rheo-optical signature associated with the extrudate behavior. In other words, there is virtually no clue from rheological measurements regarding the origin of sharkskin. The little slope change of the flow curve has only beguiled previous workers to correlate sharkskin with a *non-local* wall slip [11, 78]. Despite the apparent complexity, research efforts [11, 52, 78–82] on sharkskin phenomenon have intensified over the past decade, driven by an industrial urge to find a satisfactory solution for such polyethylene extrusion processes as film blowing of linear low density and metallocene-catalyzed polyethylenes. Many workers have measured the time scale associated with the sharkskin characteristics but none arrived at a molecular origin of the instability. Thus, no explicit *molecular mechanism* existed until our recent finding of a correlation between sharkskin dynamics and overall chain relaxation time [69].

In a recent work [69], we were able to reveal that the sharkskin merely originates from a local interfacial instability of the boundary condition near the die exit wall. Specifically, the oscillation of adsorbed chains between their coil and stretch states produces a small scale periodic perturbation on the overall die swell and makes the extrudate surface appear rough or sharkskin like.

Two critical pieces of information made it possible for us to arrive at this proposed molecular mechanism for sharkskin. *First*, previous work using birefringence [61b, 82] and theoretical modeling [34, 35, 83] have suggested that due to the boundary discontinuity at the die exit the local stress is significantly higher than the wall stress in the die land. In other words, the actual stress σ_{ex} in the exit region well exceeds the nominal wall stress σ in the die land, which is usually taken as related to the applied pressure P at the die entry through the aspect ratio L/D, i.e., $\sigma_{ex}>\sigma=(D/4L)P$. Thus we can have a case where σ is below σ_c, which is the critical wall stress for the interfacial coil-stretch transition, and yet σ_{ex} is already above σ_c. The local stress build-up due to the boundary discontinuity look

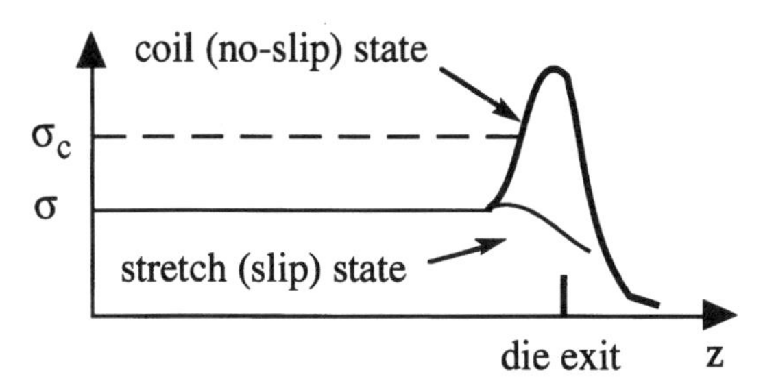

Fig. 15. Expected stress distribution at the die exit both in the no-slip and slip states corresponding to the coil (entangled) and stretch (disentangled) states (a) and (b) of the adsorbed chains on the exit wall (see Fig. 16)

qualitatively like the thick line in Fig. 15. The existing theories [34, 35] indicate a strong divergence of the exit stress, which is obviously unphysical. This predicted divergence stems from the limitation of applying a *continuum* description to a sharply discontinuous geometry. A coarse-grained renormalized theory remains to be developed to deal with such mathematical difficulties.

Second, we have learned from our investigation of the molecular origin of the stick-slip transition that adsorbed linear chains undergo a coil-stretch transition and produce a slip boundary condition above a critical stress σ_c [27–29]. The adsorbed chains at the die exit rim are expected to undergo a coil-stretch transition at nominal stresses $\sigma<\sigma_c$ as long as the peak stress σ_{ex} exceeds σ_c. At the present, we do not know much about how σ_{ex} increases with σ, i.e., we do not know the explicit shape of the stress distribution depicted by the thick curve in Fig. 15.

Upon the localized coil-stretch transition, the boundary discontinuity at the exit rim of a circular die is greatly reduced, and so is the local exit stress as sketched by the thin line in Fig. 15. The actual stress upon the local slip is now substantially below σ_c. This allows the stretched chains at the exit wall to recoil, assuming that the adsorbed chains do not suffer a stress induced desorption at the transition. If the desorption would occur *after* the local coil-stretch transition, then there must be immediate re-adsorption of new chains because absence of chain adsorption would have led to complete interfacial wall slip and disappearance of the sharkskin [69], which is clearly not the case in bare metallic dies.

It can be experimentally elucidated that the sharkskin arises from a molecular oscillation between coil and stretch states of the adsorbed or entrapped chains in the interfacial region of the PE/wall boundary at the die exit. This is illustratively shown in Fig. 16. To explore this molecular mechanism, we first need to characterize the regular sharkskin period τ at a given stress. A well-defined τ reveals presence of an internal clock in the system. τ can be evaluated as a function of the applied stress σ and temperature T according to $\tau=\lambda/V$, where λ is the measured sharkskin wavelength and V is the measured linear velocity of the extrudate exiting the die. If formation of the sharkskin-like surface roughness is

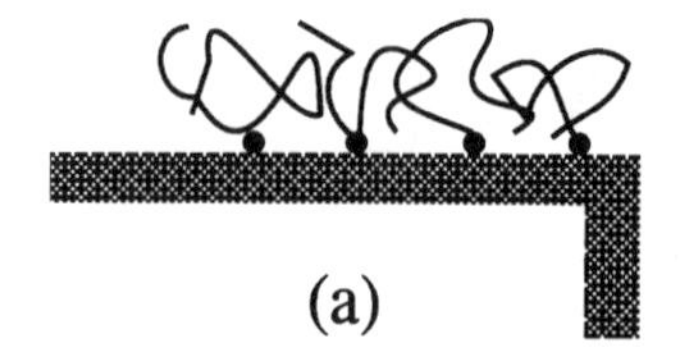

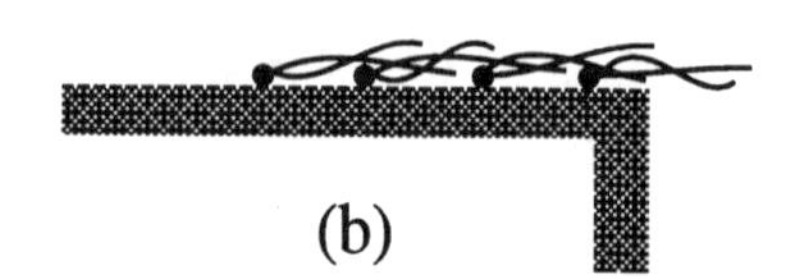

Fig. 16a,b. Reversible transitions between: **a** coil; **b** stretch states of adsorbed chains at the die *exit* wall, which cause the local stress to oscillate between the thick and thin curves in Fig. 15

controlled by a molecular relaxation process triggered by a reversible coil-stretch transition at the exit wall, then the internal time scale must be related to the melt relaxation dynamics. To verify this idea, we determine how τ varies with the experimental temperature T, and find out whether the melt relaxation time τ^* would depend on T in the same manner. A standard method, widely used by rheologists and polymer scientists to evaluate the overall chain relaxation time τ^*, is to perform oscillatory shear measurements using parallel plates or cone/plate rheometers. The crossover frequency ω^* at which the storage modulus G' and loss modulus G" intercept each other represents an overall chain relaxation rate and $\tau^*=1/\omega^*$ is regarded as the characteristic molecular relaxation time.

Figure 17 discloses a fundamental correlation between the dynamics of the sharkskin instability depicted in Fig. 16 and the molecular relaxation process. To further verify the origin of the sharkskin formation, we attempt to remove the source of the instability by a fluoropolymer (Dynamar) coating of the exit portion of the die wall. Such a surface treatment should result in massive chain desorption at the exit wall according to Fig. 14 and eliminate sharkskin roughness on the extrudate surface. This indeed occurred as we expected [69]. Thus the *source* of sharkskin instability is apparently the *presence* of adsorbed chains at the *exit wall* and their conformational instability. In other words, the origin of sharkskin is *interfacial.* It has also been suggested previously that the sharkskin distortion would arise from cycles of chain desorption and re-adsorption at the die exit wall [53, 78]. Unfortunately this desorption mechanism would not produce such a correlation as we find experimentally in Fig. 17. In particular, Fig. 17 shows that the sharkskin dynamics follow the same WLF law as the overall chain relaxation time does, whereas an activation process such as desorption would show an opposing trend to the experimental observations.

It is simplest to think of sharkskin as a result of a quasi-periodic perturbation on the overall extrudate swell. This small amplitude fluctuation of extrudate swell arises from the oscillation of the boundary condition at the exit wall that produces an oscillation of the local stress level as the interfacial chains suffer a conformational instability. The local boundary condition oscillates between no-slip and slip, resulting in the fluctuation of the stress level at the die exit. To determine whether some sort of melt fracture occurs, we need to know not only the

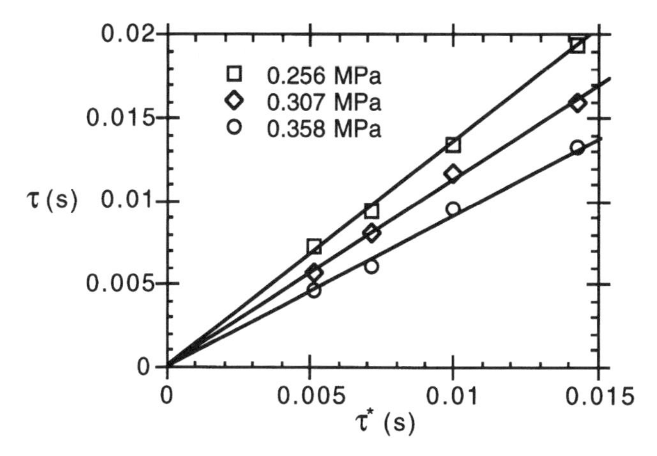

Fig. 17. Explicit correlation between the time scale τ of sharkskin formation and the overall chain relaxation time τ^* as a function of the experimental temperatures at 160, 180, 200 and 220 °C, reproduced from [69]

stress distribution and its maximum at the exit but the criterion for melt fracture. Unfortunately, such knowledge does not exist. In any event, the remarkable correlation between τ and τ^* supports the idea that sharkskin is produced by oscillation of the local boundary condition arising from reversible coil-stretch transitions in the *interfacial* region of the exit wall/melt boundary.

11 Constitutive Instabilities, Extrudate Distortions and Melt Fracture

11.1 Constitutive Instabilities

Most rheological theories for constitutive equations consider deformation of *bulk* materials and evaluate the stress generated by a specified deformation or flow field without accounting for presence of a boundary or an interface. The preceding sections show that global and localized *interfacial* processes on *molecular* length scales are responsible for the observed *macroscopic* phenomena of various flow anomalies in linear polyethylenes (LPE). In other words, the nature of the polymer/wall interface and degree of chain entanglement determine their capillary flow behavior. Thus, theoretical predictions based on *boundary-less constitutive* models (either molecular or continuum) have severe limitations and cannot be applied to describe such phenomena as sharkskin, oscillating flow and flow discontinuity or stick-slip transition. Our recent studies unambiguously demonstrate that the interfacial transitions *precede*, i.e., occur at lower stresses/rates than an onset stress for, any constitutive instabilities if the latter would ever exist. The *interfacial nature* of the stick-slip transition is most clearly elucidated by Fig. 14.

How does a constitutive instability occur in flow of polymer melts? What are the prerequisites for constitutive instabilities? To answer such questions, we need to give a clear definition. Theoretical speaking, the constitutive instabilities might be associated with a maximum or *double-valued* plateau region in the stress vs shear rate flow curve. Traditionally, vortex formation and secondary flow are not considered a constitutive instability although they are obviously constitutive in nature and viscoelastic in origin. From an experimental viewpoint, a constitutive instability might be associated with a spurt like jump of the observed flow rate upon reaching a critical level of stress. In the molecular theory of McLeish and Ball [67], the plateau width, defined as the ratio of the two shear rates on both ends of the plateau, is argued to scale with the molecular weight M_w with an exponent equal to 1.4. In the "double-valued region," stress oscillation would be expected at a constant flow rate. This may not be observable in capillary flow because the sample is continously be renewed as it is extruded out. Being a bulk property, a constitutive instability should show features that are invariant with respect to the dimensions of the apparatus, e.g., the die diameter D.

Monodisperse melts appear to exhibit a plateau region in the stress vs shear rate flow curve [51, 62, 65]. The capillary flow behavior actually closely resembles the oscillatory shear behavior in the sense that the flow curve essentially overlaps on the absolute value of complex modulus G^* vs the oscillation frequency ω [62]. Thus, it appears that the transition-like capillary flow behavior of highly entangled *monodisperse* melts reflects constitutive bulk properties of the melts and is not interfacial in origin. It remains to be explored whether this plateau indeed manifests a real constitutive *instability*, i.e., whether it is double-valued.

A popular line pursued by rheologists involves incorporating some kind of slip boundary condition into a constitutive model. Even a Newtonian model could produce unstable shear flow when a certain slip boundary condition is incorporated [84]. This kind of theoretical efforts generally takes the viewpoint that any model capable of producing unstable flow may offer an explanation for flow instabilities. For example, two recent theoretical studies [44b, 45] investigated sharkskin-like extrusion instability. Instead of considering the existence of such a special region as the die exit, a flow instability is simulated either in a simple shear (parallel plate) geometry [44b] or in an infinitely long slit die without an exit [45]. Unfortunately, none of the predicted instabilities originate from the physical mechanisms revealed by the experimental evidence reported in [69, 81]. On the other hand, a new generation of theoretical modeling will benefit from the recent experimental findings and will have an opportunity to capture the essence of physics involved in these flow problems.

11.2 Gross Extrudate Distortion and Melt Fracture

Gross extrudate distortions have constantly been linked to melt fracture. This alleged connection between a gross extrudate distortion and melt fracture requires clarification. While melt fracture literally refers to a cohesive bulk failure in melt flow, extrudate distortions may not have anything to do with melt fracture. Sharkskin is a prominent example contradicting this correlation. A spiral-like extrudate distortion should be considered to reflect a flow instability rather than melt fracture since the distortion arises from a die entry flow instability which can be eliminated by using a gradually convergent die entry [57], and not from a cohesive breakdown. Such a distortion has been seen in polymers other than polyethylenes and its origin has been studied previously [75, 76, 85]. Direct in situ optical observations show that the spiral-like extrudate distortion originates from development of secondary (vortex) flow near the die entry inside the barrel [86]. These recent observations [86] agree with the previous report on the origin of spiral extrudate distortions [85].

The term "melt fracture" has been applied from the outset [9, 13] to refer to various types of visible extrudate distortion. The origin of sharkskin (often called "surface melt fracture") has been shown in Sect. 10 to be related to a local interfacial instability in the die exit region. The alternating quasi-periodic, sometimes bamboo-like, extrudate distortion associated with the flow oscillation is a result of oscillation in extrudate swell under controlled piston speed due to unstable boundary condition, as discussed in Sect. 8. A third type, spiral like, distortion is associated with an entry flow instability. The latter two kinds have often been referred to as gross "melt fracture." It is clearly misleading and inaccurate to call these three major types of extrudate distortion "melt fracture" since they do not arise from a true melt fracture or bulk failure. Unfortunately, for historical reasons, this terminology will stay with us and be used interchangeably with the phase "extrudate distortion."

A true melt fracture by definition must involve some kind of bulk failure in melt flow. Either massive chain disentanglement or chain scission or both must occur in the *bulk*. Such a real cohesive breakdown *away from* the surface may also produce extrudate distortion. In general, other forms of irregular extrudate distortion do occur at high stresses. Only direct flow visualization may reveal the origin of such extrudate distortions.

Extrudate distortion has been a deceiving character of capillary melt extrusion. It is fair to say that we have not seen any confirmed sign of *true melt fracture* in polyethylene extrusion. For linear polyethylenes, it is clear that a constitutive instability and real melt fracture do not take place in the stress range where the first three leading extrudate distortions occur. Until we have achieved a complete understanding of these prevailing phenomena and found ways to remove them, the subject of constitutive instabilities and true melt fracture may only remain as an academic affair and will not be very useful for industrial polymer processing. Other industrial thermoplastics usually are not sufficiently entangled for an interfacial slip to affect their flow behavior. For these polymers it is important to determine whether entry instability and secondary flow produce any extrudate distortion and how a constitutive instability or melt (cohesive) fracture would arise in capillary flow. Clearly, in situ observation and characterization must be developed and applied to investigate and quantify salient features of a melt fracture or any form of constitutive instabilities as well as secondary flow. In summary, it is important to emphasize that extrudate distortion generally offers useful information about the state of extrusion flow but cannot be used as direct evidence for true melt fracture or constitutive instabilities.

12 Summary

We have surveyed the most recent progress and presented a new molecular level understanding of melt flow instabilities and wall slip. This article can at best be regarded as a partial review because it advocates the molecular pictures emerging from our own work over the past few years [27–29, 57, 62, 69]. Several results from many previous and current workers have been discussed to help illustrate, formulate and verify our own viewpoints. In our opinion, the emerging explicit molecular mechanisms have for the first time provided a unified and satisfactory understanding of the two major classes of interfacial melt flow instability phenomena: (a) sharkskin-like extrudate distortion and (b) stick-slip (flow discontinuity) transition and oscillating flow.

Specifically, we have attempted to extract the generic features of anomalous melt flow behavior from an extensive set of experimental data in [27–29]. The flow anomalies of linear polyethylenes (LPE) summarized in Fig. 1 have now received a coherent explanation in terms of their molecular origins. The *stick-slip* (flow discontinuity) transition arises from a stress induced *coil-stretch transition* of the interfacial chains at the polymer/wall boundary that are different from the unbound chains. The symptoms of *oscillating flow* such as the pressure oscillation and macroscopic quasi-periodic distortion of extrudate are produced by the conformational oscillation of the interfacial chains between their coil and stretch states. This *interfacial instability* exhibits itself under a constant piston speed on a macroscopic time scale rather than a molecular time scale due to long cycles of finite melt compression and decompression in the barrel [64]. The corresponding oscillation of the overall stress level leads to synchronized oscillation of *extrudate swell* in the form of bamboo-like or alternating extrudate distortion. Besides this coil-stretch transition mechanism for wall slip, studies of the influence of surface conditions reveal a second molecular mechanism for wall slip: *stress-induced chain desorption* on weakly adsorbing walls that eliminates chain entanglement at the melt/wall interface and produces wall slip. *Sharkskin*-like extrudate distortion apparently originates from an interfacial molecular instability that modifies the boundary condition of the die exit wall. In particular, the sharkskin dynamics are controlled by melt relaxation dynamics. The fluctuation of the local boundary condition due to the reversible conformational transition of the interfacial chains results in a stress perturbation in the die exit region. Sharkskin shows up explicitly as a small oscillation of the overall extrudate swell due to the stress fluctuation.

Extrudate distortion cannot generally be taken as evidence for *melt fracture*. For example, a commonly observed spiral-like distortion originates from vortex

Table 2. Molecular and macroscopic explanations of three dominant types of extrudate distortion in capillary flow of linear polyethylenes

	Sharkskin	Oscillation	Spirals
Molecular Origin	Coil-stretch transition in the interfacial region near the melt/exit wall boundary	Global coil-stretch transition among the entire die melt/wall interfacial layer	Molecular architecture and MWD influencing viscoelasticity may be important factors
Continuum Depiction	Local oscillation of the melt-wall boundary condition in the exit region causes peturbations on the exit stress and die swell	Oscillation of the overall stress due to unstable boundary condition produces cycles of melt compression and decompression in the barrel and fluctuations in the extrude swell	The extrudate distortion arises from formation of secondary flow (vortices) in the barrel due to the strong converging flow near the die entry

formation near the capillary die entry. As another example, the bamboo-like or alternating extrudate distortion reflects fluctuation of the extrudate swell due to oscillation of the boundary condition and is not at all a real melt fracture phenomenon. The *sharkskin*-like extrudate distortion in linear polyethylenes also has an interfacial origin. The criterion for and characteristics of a true *melt fracture* remain to be established. For linear polyethylenes (LPE), this might happen only at higher stresses, i.e., well beyond the breakdown of interfacial chain entanglement near the PE/wall boundary.

The general framework described in this review should enable us to understand strong melt flow behavior of all polymers. The melt/wall interface is important, and wall slip may influence the flow behavior only when the melt under study is sufficiently entangled in flow. Interfacial flow instabilities such as flow oscillation and sharkskin will not be observable when interfacial slip is immeasurably small. This explains why oscillating flow and sharkskin do not occur in most other commercial polymers including polystyrene, polypropylene, and low density polyethylene. Apparently these polymers do not possess sufficient *effective chain entanglement* at stresses where a coil-stretch transition may occur in the melt/wall interfacial region [62]. Shear thinning originating from chain deformation in flow adversely affect the magnitude of wall slip, as the effective degree of chain entanglement decreases with increasing stress/rate.

In conclusion, it may be useful to highlight our current explanations for the first three types of extrudate distortion (*sharkskin*, *oscillation* and *spirals*) in polymer extrusion in a tabulated manner. In Table 2, using the example of linear polyethylenes, we not only list their molecular origins but their description from a continuum mechanical viewpoint. Of the three, the molecular characteristics of spiral like distortion are least understood. In contrast to bamboo-like and sharkskin-like distortions, the spiral like distortion is a bulk flow phenomenon related to viscoelasticity and is shared by many other polymers that do not suffer from the first two *interfacial* flow instabilities.

At higher stresses yet, extrudate distortion becomes even more distorted. However, it may be misleading and inappropriate to call any of these distortions (from sharkskin to irregular extrudate undulation) "melt fracture" although this term has been arbitrarily employed throughout the history of this field. The real challenges for the future are to verify specifically each of the proposed molecular mechanisms by developing new surface-sensitive experimental methods, and to remove these "extrusion instabilities" by inventing new technological solutions based on the molecular level and fluid dynamic level understanding of these phenomena.

Acknowledgements. The author would like to acknowledge the M.S. research work of P.A. Drda and sharkskin measurements by Y.W. Inn that have been discussed here. Several more discussions were made possible due to the unpublished data on monodisperse PB melts by X. Yang. The author is grateful to H. Ishida for use of the instrumental Monsanto Capillary Rheometer to collect much of the experimental data discussed here. He thanks L. Leibler for pointing out the usefulness to evaluate the value of the critical stress for the stick-slip transition in terms of melt chain statistics. The financial support of the National Science Foundation (Grant No. CTS-9632466) and BP Chemicals is greatly appreciated.

13
References

1. De Gennes (1987) Adv Colloid & Interface Sci 27:1892
2. Lee J-J, Fuller GG (1985) J Colloid & Interface Sci 103:569
3. Frantz P, Granick S (1991) Phys Rev Lett 66: 899; Granick S, Johnson HE (1992) Science 255:966
4. Balazs A, Gempe MC, Zhou Z (1991) Macromolecules 24:4918
5. Dijt JC, Cohen Stuart MA, Fleer GJ (1994) Macromolecules 27:3229
6. Halperin A, Tirrell MV, Lodge TP (1992) Adv Polym Sci 100:31
7. Zheng X, Sauer BB, Vanalsten JG, Scharwz SA, Rafailovich MH, Sokolov J, Rubinstein M (1995) Phys Rev Lett 74:407; Forrest JA, Dalnoki-Veress K, Stevens JR, Dutcher JR (1996) Phys Rev Lett 77:2002
8. Cohen-Addad J-P, Huchot P, Jost P, Pouchelon A (1989) Polymer 30:143
9. (a) Tordella JP (1956) J Appl Phys 27:454; (b) Tordella JP (1963) J Appl Polym Sci 7:215; (c) Tordella JP (1969) In: Rheology, Vol. 5, Eirich FR, Ed, Academic Press:New York, p57
10. (a) Petrie CJS, Denn MM (1976) AIChE J 22:209; (b) Denn MM (1990) Ann Rev Fluid Mech 22:13; (c) Denn MM (1992) In: Moldenaers P, Keunings R (eds) Theoretical and Applied Rheology (Proc XIth Intl Congress on Rheology), Vol. 1, Elsevier, London, pp 45–49. See also Boudreaux E, Cuculo JA (1978) J Macromol Sci C16:39
11. Kalika DS, Denn MM (1987) J Rheol 31:815
12. In the existing literature, extrudate distortion has also been given the term "melt fracture" where the phrase takes its literal meaning of cohesive bulk failure. It will become clear that this practice of associating extrudate distortion with melt fracture is somewhat misleading
13. Bagley EB, Cabott IM, West DC (1958) J Appl Phys 29:109

14. Larson RG (1992) Rheol Acta 31:213
15. Navier M (1823) Mémo Acad Royal Sci Inst France 6:4 14. This reference originates from citation by S. Goldstein in [17] below
16. Stokes GG (1845) Trans Cam Phil Soc 8:299
17. Goldstein S (1938) Modern developments in fluid dynamics, vol 2. Oxford Univ Press, London, pp 676–680
18. Schowalter WR (1988) J Non-Newtonian Fluid Mech 29:25
19. Burton RH, Folkes MJ, Narh KA, Keller A (1983) J Mater Sci 18:315
20. Kissi NEL, Piau JM (1990) J Non-Newtonian Fluid Mech 37:55
21. Larson RG, Patel S Chen Y (1994) Rheol Acta 33:243
22. Kissi NEL, Léger L, Piau JM, Mezghani, A (1994) J Non-Newtonian Fluid Mech 52:249
23. Tritton DJ (1977) Physical Fluid Dynamics. Van Nostrand Reinhold, UK, p 54
24. (a) Lamb H (1932) Hydrodynamics, pp 576 and 586, 6th edn. Dover, New York; (b) Basset AB (1888) A treatise on hydrodynamics. Deighton, Bell and Co., Cambridge
25. De Gennes PG (1979) C R Acad Sci 288B:219
26. De Gennes originally assumed that polymer adsorption would be absent at any stress. It is more realistic to assert that polymer desorption would occur only beyond a certain level of wall stress
27. Wang SQ, Drda PA (1996) Macromolecules 29:4115
28. Wang SQ, Drda PA (1997) Rheol Acta 36:128
29. Wang SQ, Drda PA (1996) Macromolecules 29:2627
30. Brochard F, De Gennes PG (1992) Langmuir 8:3033
31. Bergem N (1976) Proc 7th Intl Congr Rheol, Swedish Soc Rheol, p 50
32. Blyler LL, Hart AC (1970) Polym Sci Eng 10:193
33. Henson DJ, Mackay ME (1995) J Rheol 39:359
34. Cogswell FN (1977) J Non-Newtonian Fluid Mech 2:37
35. Lipscomb GG, Keunings R, Denn MM J Non-Netwonian Fluid Mech (1987) 24:85
36. (a) Huh C, Scriven LE (1971) J Colloid & Interface Sci 35:85; (b) Silliman WJ, Scriven LE (1978) Phys Fluids 21:2115; (c) Silliman WJ, Scriven LE (1980) J Comput Phys 34:287; (d) Kistler SF, Scriven LE (1983) Peason JRA, Richardson SM (eds) Computational analysis of polymer processing. Appl Sci Publisher, London, p243
37. Hatzikiriakos SG, Dealy JM (1991) J Rheol 35:497
38. Migler KB, Hervet H, Leger L (1993) Phys Rev Lett 70:287; MiglerKB, Massey G, Hervet H, Leger L (1994) J Phys – Conden Matt 6:A301
39. Laun HM (1982) Rheol Acta 21:464
40. Macosko CW, Morse DJ (1976) Kalson C, Kubat J (eds) Proceedings of Seventh International Congress on Rheology, Gothenburg, p376
41. Adrian DW, Giacomin AJ (1992) J Rheol 36:1227; Adrian DW, Giacomin AJ (1994) J. Eng Mater Tech 116:446
42. Inn YW, Wang SQ (1994) Rheol Acta 33:108 The observed stress decay at high stresses and slow recovery at lower stresses in PDMS suspensions of glass spheres was thought to arise from interfacial slip and to be controlled by "glassy" chain dynamics. This interpretation turned out to be completely wrong. See ref. [47].
43. Hatzikiriakos SG, Kalogerakis N (1994) Rheol Acta 33:38
44. (a) Graham MD (1995) J Rheol 39:697; (b) Black WB, Graham MD (1996) Phys Rev Lett 77:956
45. Shore JD, Ronis D, Piche L, Grant M (1996) Phys Rev Lett 77:655
46. Petit L, Noetinger B (1988) Rheol Acta 27:437
47. Wang SQ, Inn YW (1995) Polym Intl 37:153
48. Inn YW, Wang SQ (1996) Phys Rev Lett 76:467
49. Kraynik AM, Schowalter WR (1981) J Rheol 25:95
50. Atwood BT, Schowalter WR (1989) Rheol Acta 28:134
51. Lim FJ, Schowalter WR (1989) J Rheol 33:1359
52. Piau JM, Kissi NEl, Tremblay B (1990) J Non-Newtonian Fluid Mech 34:145

53. Hill DA, Hasegawa T, Denn M (1990) J Rheol 34:891
54. Hatzikiriakos SG, Dealy JM (1992) J Rheol 36:703
55. Hatzikiriakos SG, Dealy JM (1992) J Rheol 36:845
56. Piau JM, Kissi NEl, Tremblay B (1994) J Non-Newtonian Fluid Mech 54:121
57. Wang SQ, Drda P (1997) Macromol Chem & Phys 198:673
58. Ui J, Ishimari Y, Morakami H, Fukushma N, Mori Y (1964) SPE Soc Plastics Engr Trans 4:295
59. Uhland E (1979) Rheol Acta 18:1
60. Sabia R (1962) J Appl Polym Sci 6:S42
61. (a) Vinogradov GV (1972) J Polym Sci Part A-2 10:1061; (b) Vinogradov GV, Malkin AY (1980) Rheology of polymers – viscoelasticity and flow of polymers. Springer, Berlin Heidelberg New York; (c) Vinogradov GV, Protasov VP, Dreval VE (1984) Rheol Acta 23:46
62. Yang XP, Wang SQ (1998) Rheol Acta, in press.
63. Lupton JM, Regester JW (1965) Polym Sci Eng 5:235
64. Pearson JRA (1985) Mechanics of polymer processing. Elsevier, London
65. Lin YH (1985) J Rheol 29:506
66. Huseby TW (1966) Trans Soc Rheol 10:181 Similar theoretical studies have been made more recently by Malkus and coworkers
67. (a) McLeish TCB, Ball RC (1986) J Polym Sci B24:1735; (b) McLeish,TCB (1986) J Polym Sci B25:2253
68. Metzger AP, Hamilton CW (1964) SPE Trans 4:107. The authors did not observe a flow instability in a Teflon capillary. However, they were not so sure whether the Teflon capillary retained its original shape and dimensions at 190 °C and high stresses. Consequently a glass-reinforced Teflon capillary die was employed and found to restore the flow instability (i.e., oscillation), thus casting "doubt on the assumed slip mechanism." It is clear to us today that the glass fillers exposed on the die wall may provide sufficient adsorption sites for polyethylene chains, thus restoring the source of the flow oscillation. To our knowledge, this is one of the few early in the scientific literature that studied the surface effect on melt flow instabilities although its conclusion might be misleading
69. Wang SQ, Drda PA, Inn YW (1996) J Rheol 40:875; Barone JR, Plucktaveesak N, Wang SQ (1998) J Rheol 42:813.
70. Kolnaar, JWH, Keller A (1994) Polymer 35:3863; Kolnaar, JWH, Keller A (1995) Polymer 36:821; Kolnaar, JWH, Keller A (1997) Polymer 38:1817
71. Mooney M (1931) Trans Soc Rheol 2:210. It was evident even at the time of Navier that the meaningful way to quantify any level of wall slip is to express it in terms of the extrapolation length *b* [24]. Equation (4a) clearly shows that any correction would enter as the ratio of *b* to a characteristic dimension of the flow apparatus, e.g., the diameter D of capillary dies. It is unfortunate that Mooney abandoned the notion of the extra polation length *b* in favor of the slip velocity v_s
72. Fetters LJ, Lohse DJ, Richter D, Witten TA, Zirkel A (1994) Macromolecules 27:4639
73. De Gennes PG (1996) Private Communication; Brochard-Wyart F, de Gennes PG (1993) C.R. Acad. Sci. Paris t. 317 Série II:13; Ajdari A (1993) C.R. Acad. Sci. Paris t. 317 Série II:1159
74. Drda P and Wang SQ (1995) Phys Rev Lett 75:2698
75. White J (1973) United State – Japan Seminar on Polymer Processing and Rheology. Bogue DC, Yamamoto M, White JL (eds) Wiley, New York, p155
76. Tordella J (1957) Trans Soc Rheol 1:203
77. Bagley EB, Birks AM (1960) J Appl Polym Sci 31:556
78. Ramamurthy AV (1986) J Rheol 30:337
79. Beaufils P, Vergnes B, Agassant JF (1989) Interntl Polym Processing IV:78
80. (a) Kurtz SJ (1984) Advances in Rheology (Proc IX Intl Congr Rheol) Mena B, Garcia-Rejon A, Rangel Nafaile C (eds) vol 3, p 339; (b) Kurtz SJ (1991) 7th Annual Meeting of Polym Proc Soc, Hamilton, Canada, p 54; (c) Kurtz SJ (1992) Theoretical and Applied Rheology (Proc XIth Intl Congr Rheol), vol 1, Moldenaers P, Keunings R (eds), Elsevier, London, p377

81. Moynihan RH, Baird DG, Ramanathan R (1990) J Non-Newtonian Fluid Mech 36:255
82. Sornberger G, Quantin JC, Fajolle R, Vergens B, Agassant JF (1987) J Non-Newtonian Fluid Mech 23:123
83. Tanner RI (1985) Engineering Rheology, Clarendon, Oxford
84. Pearson JRA, Petrie CJS (1968) Whorlow RW (ed) Polymer systems: deformation and flow. Macmillan, New York, p163
85. Bagley EB, Schreiber HP (1961) Trans Soc Rheol 5:341
86. Meijer HEH (1997) Private communication

Received: February 1998

Author Index Volumes 101–138

Author Index Volumes 1–100 see Volume 100

Adolf, D. B. see Ediger, M. D.: Vol. 116, pp. 73-110.

Aharoni, S. M. and *Edwards, S. F.*: Rigid Polymer Networks. Vol. 118, pp. 1-231.

Améduri, B., Boutevin, B. and *Gramain, P.*: Synthesis of Block Copolymers by Radical Polymerization and Telomerization. Vol. 127, pp. 87-142.

Améduri, B. and *Boutevin, B.*: Synthesis and Properties of Fluorinated Telechelic Monodispersed Compounds. Vol. 102, pp. 133-170.

Amselem, S. see Domb, A. J.: Vol. 107, pp. 93-142.

Andrady, A. L.: Wavelenght Sensitivity in Polymer Photodegradation. Vol. 128, pp. 47-94.

Andreis, M. and *Koenig, J. L.*: Application of Nitrogen-15 NMR to Polymers. Vol. 124, pp. 191-238.

Angiolini, L. see Carlini, C.: Vol. 123, pp. 127-214.

Anseth, K. S., Newman, S. M. and *Bowman, C. N.*: Polymeric Dental Composites: Properties and Reaction Behavior of Multimethacrylate Dental Restorations. Vol. 122, pp. 177-218.

Armitage, B. A. see O'Brien, D. F.: Vol. 126, pp. 53-58.

Arndt, M. see Kaminski, W.: Vol. 127, pp. 143-187.

Arnold Jr., F. E. and *Arnold, F. E.*: Rigid-Rod Polymers and Molecular Composites. Vol. 117, pp. 257-296.

Arshady, R.: Polymer Synthesis via Activated Esters: A New Dimension of Creativity in Macromolecular Chemistry. Vol. 111, pp. 1-42.

Bahar, I., Erman, B. and *Monnerie, L.*: Effect of Molecular Structure on Local Chain Dynamics: Analytical Approaches and Computational Methods. Vol. 116, pp. 145-206.

Baltá-Calleja, F. J., González Arche, A., Ezquerra, T. A., Santa Cruz, C., Batallón, F., Frick, B. and *López Cabarcos, E.*: Structure and Properties of Ferroelectric Copolymers of Poly(vinylidene) Fluoride. Vol. 108, pp. 1-48.

Barshtein, G. R. and *Sabsai, O. Y.*: Compositions with Mineralorganic Fillers. Vol. 101, pp.1-28.

Batallán, F. see Baltá-Calleja, F. J.: Vol. 108, pp. 1-48.

Barton, J. see Hunkeler, D.: Vol. 112, pp. 115-134.

Bell, C. L. and *Peppas, N. A.*: Biomedical Membranes from Hydrogels and Interpolymer Complexes. Vol. 122, pp. 125-176.

Bellon-Maurel, A. see Calmon-Decriaud, A.: Vol. 135, pp. 207-226.

Bennett, D. E. see O'Brien, D. F.: Vol. 126, pp. 53-84.

Berry, G.C.: Static and Dynamic Light Scattering on Moderately Concentraded Solutions: Isotropic Solutions of Flexible and Rodlike Chains and Nematic Solutions of Rodlike Chains. Vol. 114, pp. 233-290.

Bershtein, V. A. and *Ryzhov, V. A.*: Far Infrared Spectroscopy of Polymers. Vol. 114, pp. 43-122.

Bigg, D. M.: Thermal Conductivity of Heterophase Polymer Compositions. Vol. 119, pp. 1-30.

Binder, K.: Phase Transitions in Polymer Blends and Block Copolymer Melts: Some Recent Developments. Vol. 112, pp. 115-134.

Binder, K.: Phase Transitions of Polymer Blends and Block Copolymer Melts in Thin Films. Vol. 138, pp. 1-90.

Bird, R. B. see Curtiss, C. F.: Vol. 125, pp. 1-102.
Biswas, M. and *Mukherjee, A.*: Synthesis and Evaluation of Metal-Containing Polymers. Vol. 115, pp. 89-124.
Boutevin, B. and *Robin, J. J.*: Synthesis and Properties of Fluorinated Diols. Vol. 102. pp. 105-132.
Boutevin, B. see Amédouri, B.: Vol. 102, pp. 133-170.
Boutevin, B. see Améduri, B.: Vol. 127, pp. 87-142.
Bowman, C. N. see Anseth, K. S.: Vol. 122, pp. 177-218.
Boyd, R. H.: Prediction of Polymer Crystal Structures and Properties. Vol. 116, pp. 1-26.
Bronnikov, S. V., Vettegren, V. I. and *Frenkel, S. Y.*: Kinetics of Deformation and Relaxation in Highly Oriented Polymers. Vol. 125, pp. 103-146.
Bruza, K. J. see Kirchhoff, R. A.: Vol. 117, pp. 1-66.
Burban, J. H. see Cussler, E. L.: Vol. 110, pp. 67-80.

Calmon-Decriaud, A. Bellon-Maurel, V., Silvestre, F.: Standard Methods for Testing the Aerobic Biodegradation of Polymeric Materials. Vol 135, pp. 207-226.
Cameron, N. R. and *Sherrington, D. C.*: High Internal Phase Emulsions (HIPEs)-Structure, Properties and Use in Polymer Preparation. Vol. 126, pp. 163-214.
Candau, F. see Hunkeler, D.: Vol. 112, pp. 115-134.
Canelas, D. A. and *DeSimone, J. M.*: Polymerizations in Liquid and Supercritical Carbon Dioxide. Vol. 133, pp. 103-140.
Capek, I.: Kinetics of the Free-Radical Emulsion Polymerization of Vinyl Chloride. Vol. 120, pp. 135-206.
Carlini, C. and *Angiolini, L.*: Polymers as Free Radical Photoinitiators. Vol. 123, pp. 127-214.
Casas-Vazquez, J. see Jou, D.: Vol. 120, pp. 207-266.
Chandrasekhar, V.: Polymer Solid Electrolytes: Synthesis and Structure. Vol 135, pp. 139-206
Chen, P. see Jaffe, M.: Vol. 117, pp. 297-328.
Choe, E.-W. see Jaffe, M.: Vol. 117, pp. 297-328.
Chow, T. S.: Glassy State Relaxation and Deformation in Polymers. Vol. 103, pp. 149-190.
Chung, T.-S. see Jaffe, M.: Vol. 117, pp. 297-328.
Connell, J. W. see Hergenrother, P. M.: Vol. 117, pp. 67-110.
Criado-Sancho, M. see Jou, D.: Vol. 120, pp. 207-266.
Curro, J.G. see Schweizer, K.S.: Vol. 116, pp. 319-378.
Curtiss, C. F. and *Bird, R. B.*: Statistical Mechanics of Transport Phenomena: Polymeric Liquid Mixtures. Vol. 125, pp. 1-102.
Cussler, E. L., Wang, K. L. and *Burban, J. H.*: Hydrogels as Separation Agents. Vol. 110, pp. 67-80.

DeSimone, J. M. see Canelas D. A.: Vol. 133, pp. 103-140.
DiMari, S. see Prokop, A.: Vol. 136, pp. 1-52.
Dimonie, M. V. see Hunkeler, D.: Vol. 112, pp. 115-134.
Dodd, L. R. and *Theodorou, D. N.*: Atomistic Monte Carlo Simulation and Continuum Mean Field Theory of the Structure and Equation of State Properties of Alkane and Polymer Melts. Vol. 116, pp. 249-282.
Doelker, E.: Cellulose Derivatives. Vol. 107, pp. 199-266.
Domb, A. J., Amselem, S., Shah, J. and Maniar, M.: Polyanhydrides: Synthesis and Characterization. Vol.107, pp. 93-142.
Dubrovskii, S. A. see Kazanskii, K. S.: Vol. 104, pp. 97-134.
Dunkin, I. R. see Steinke, J.: Vol. 123, pp. 81-126.

Economy, J. and *Goranov, K.*: Thermotropic Liquid Crystalline Polymers for High Performance Applications. Vol. 117, pp. 221-256.
Ediger, M. D. and *Adolf, D. B.*: Brownian Dynamics Simulations of Local Polymer Dynamics. Vol. 116, pp. 73-110.
Edwards, S. F. see Aharoni, S. M.: Vol. 118, pp. 1-231.

Endo, T. see Yagci, Y.: Vol. 127, pp. 59-86.
Erman, B. see Bahar, I.: Vol. 116, pp. 145-206.
Ewen, B. Richter, D.: Neutron Spin Echo Investigations on the Segmental Dynamics of Polymers in Melts, Networks and Solutions. Vol. 134, pp. 1-130.
Ezquerra, T. A. see Baltá-Calleja, F. J.: Vol. 108, pp. 1-48.

Fendler, J.H.: Membrane-Mimetic Approach to Advanced Materials. Vol. 113, pp. 1-209.
Fetters, L. J. see Xu, Z.: Vol. 120, pp. 1-50.
Förster, S. and *Schmidt, M.*: Polyelectrolytes in Solution. Vol. 120, pp. 51-134.
Frenkel, S. Y. see Bronnikov, S. V.: Vol. 125, pp. 103-146.
Frick, B. see Baltá-Calleja, F. J.: Vol. 108, pp. 1-48.
Fridman, M. L.: see Terent´eva, J. P.: Vol. 101, pp. 29-64.
Funke, W.: Microgels-Intramolecularly Crosslinked Macromolecules with a Globular Structure. Vol. 136, pp. 137-232.

Galina, H.: Mean-Field Kinetic Modeling of Polymerization: The Smoluchowski Coagulation Equation. Vol. 137, pp. 135-172.
Ganesh, K. see Kishore, K.: Vol. 121, pp. 81-122.
Geckeler, K. E. see Rivas, B.: Vol. 102, pp. 171-188.
Geckeler, K. E.: Soluble Polymer Supports for Liquid-Phase Synthesis. Vol. 121, pp. 31-80.
Gehrke, S. H.: Synthesis, Equilibrium Swelling, Kinetics Permeability and Applications of Environmentally Responsive Gels. Vol. 110, pp. 81-144.
de Gennes, P.-G.: Flexible Polymers in Nanopores. Vol. 138, pp. 91-106.
Giannelis, E.P., Krishnamoorti, R., Manias, E.: Polymer-Silicate Nanocomposites: Model Systems for Confined Polymers and Polymer Brushes. Vol. 138, pp. 107-148.
Godovsky, D. Y.: Electron Behavior and Magnetic Properties Polymer-Nanocomposites. Vol. 119, pp. 79-122.
González Arche, A. see Baltá-Calleja, F. J.: Vol. 108, pp. 1-48.
Goranov, K. see Economy, J.: Vol. 117, pp. 221-256.
Gramain, P. see Améduri, B.: Vol. 127, pp. 87-142.
Grest, G.S.: Normal and Shear Forces Between Polymer Brushes. Vol. 138, pp. 149-184
Grosberg, A. and Nechaev, S.: Polymer Topology. Vol. 106, pp. 1-30.
Grubbs, R., Risse, W. and *Novac, B.*: The Development of Well-defined Catalysts for Ring-Opening Olefin Metathesis. Vol. 102, pp. 47-72.
van Gunsteren, W. F. see Gusev, A. A.: Vol. 116, pp. 207-248.
Gusev, A. A., Müller-Plathe, F., van Gunsteren, W. F. and *Suter, U. W.*: Dynamics of Small Molecules in Bulk Polymers. Vol. 116, pp. 207-248.
Guillot, J. see Hunkeler, D.: Vol. 112, pp. 115-134.
Guyot, A. and *Tauer, K.*: Reactive Surfactants in Emulsion Polymerization. Vol. 111, pp. 43-66.

Hadjichristidis, N. see Xu, Z.: Vol. 120, pp. 1-50.
Hadjichristidis, N. see Pitsikalis, M.: Vol. 135, pp. 1-138.
Hall, H. K. see *Penelle, J.*: Vol. 102, pp. 73-104.
Hammouda, B.: SANS from Homogeneous Polymer Mixtures: A Unified Overview. Vol. 106, pp. 87-134.
Harada, A.: Design and Construction of Supramolecular Architectures Consisting of Cyclodextrins and Polymers. Vol. 133, pp. 141-192.
Haralson, M. A. see Prokop, A.: Vol. 136, pp. 1-52.
Hedrick, J. L. see Hergenrother, P. M.: Vol. 117, pp. 67-110.
Heller, J.: Poly (Ortho Esters). Vol. 107, pp. 41-92.
Hemielec, A. A. see Hunkeler, D.: Vol. 112, pp. 115-134.
Hergenrother, P. M., Connell, J. W., Labadie, J. W. and *Hedrick, J. L.*: Poly(arylene ether)s Containing Heterocyclic Units. Vol. 117, pp. 67-110.
Hervet, H. see Léger, L.: Vol. 138, pp. 185-226.
Hiramatsu, N. see Matsushige, M.: Vol. 125, pp. 147-186.

Hirasa, O. see Suzuki, M.: Vol. 110, pp. 241-262.
Hirotsu, S.: Coexistence of Phases and the Nature of First-Order Transition in Poly-N-isopropylacrylamide Gels. Vol. 110, pp. 1-26.
Hunkeler, D., Candau, F., Pichot, C., Hemielec, A. E., Xie, T. Y., Barton, J., Vaskova, V., Guillot, J., Dimonie, M. V., Reichert, K. H.: Heterophase Polymerization: A Physical and Kinetic Comparision and Categorization. Vol. 112, pp. 115-134.
Hunkeler, D. see Prokop, A.: Vol. 136, pp. 1-52; 53-74.

Ichikawa, T. see Yoshida, H.: Vol. 105, pp. 3-36.
Ihara, E. see Yasuda, H.: Vol. 133, pp. 53-102.
Ikada, Y. see Uyama, Y.: Vol. 137, pp. 1-40.
Ilavsky, M.: Effect on Phase Transition on Swelling and Mechanical Behavior of Synthetic Hydrogels. Vol. 109, pp. 173-206.
Inomata, H. see Saito, S.: Vol. 106, pp. 207-232.
Irie, M.: Stimuli-Responsive Poly(N-isopropylacrylamide), Photo- and Chemical-Induced Phase Transitions. Vol. 110, pp. 49-66.
Ise, N. see Matsuoka, H.: Vol. 114, pp. 187-232.
Ivanov, A. E. see Zubov, V. P.: Vol. 104, pp. 135-176.

Jaffe, M., Chen, P., Choe, E.-W., Chung, T.-S. and *Makhija, S.*: High Performance Polymer Blends. Vol. 117, pp. 297-328.
Joos-Müller, B. see Funke, W.: Vol. 136, pp. 137-232.
Jou, D., Casas-Vazquez, J. and *Criado-Sancho, M.*: Thermodynamics of Polymer Solutions under Flow: Phase Separation and Polymer Degradation. Vol. 120, pp. 207-266.

Kaetsu, I.: Radiation Synthesis of Polymeric Materials for Biomedical and Biochemical Applications. Vol. 105, pp. 81-98.
Kaminski, W. and *Arndt, M.*: Metallocenes for Polymer Catalysis. Vol. 127, pp. 143-187.
Kammer, H. W., Kressler, H. and *Kummerloewe, C.*: Phase Behavior of Polymer Blends - Effects of Thermodynamics and Rheology. Vol. 106, pp. 31-86.
Kandyrin, L. B. and *Kuleznev, V. N.*: The Dependence of Viscosity on the Composition of Concentrated Dispersions and the Free Volume Concept of Disperse Systems. Vol. 103, pp. 103-148.
Kaneko, M. see Ramaraj, R.: Vol. 123, pp. 215-242.
Kang, E. T., Neoh, K. G. and *Tan, K. L.*: X-Ray Photoelectron Spectroscopic Studies of Electroactive Polymers. Vol. 106, pp. 135-190.
Kato, K. see Uyama, Y.: Vol. 137, pp. 1-40.
Kazanskii, K. S. and *Dubrovskii, S. A.*: Chemistry and Physics of „Agricultural" Hydrogels. Vol. 104, pp. 97-134.
Kennedy, J. P. see Majoros, I.: Vol. 112, pp. 1-113.
Khokhlov, A., Starodybtzev, S. and *Vasilevskaya, V.*: Conformational Transitions of Polymer Gels: Theory and Experiment. Vol. 109, pp. 121-172.
Kilian, H. G. and *Pieper, T.*: Packing of Chain Segments. A Method for Describing X-Ray Patterns of Crystalline, Liquid Crystalline and Non-Crystalline Polymers. Vol. 108, pp. 49-90.
Kishore, K. and *Ganesh, K.*: Polymers Containing Disulfide, Tetrasulfide, Diselenide and Ditelluride Linkages in the Main Chain. Vol. 121, pp. 81-122.
Kitamaru, R.: Phase Structure of Polyethylene and Other Crystalline Polymers by Solid-State ^{13}C/MNR. Vol. 137, pp 41-102.
Klier, J. see Scranton, A. B.: Vol. 122, pp. 1-54.
Kobayashi, S., Shoda, S. and *Uyama, H.*: Enzymatic Polymerization and Oligomerization. Vol. 121, pp. 1-30.
Koenig, J. L. see Andreis, M.: Vol. 124, pp. 191-238.
Kokufuta, E.: Novel Applications for Stimulus-Sensitive Polymer Gels in the Preparation of Functional Immobilized Biocatalysts. Vol. 110, pp. 157-178.
Konno, M. see Saito, S.: Vol. 109, pp. 207-232.

Kopecek, J. see Putnam, D.: Vol. 122, pp. 55-124.
Koßmehl, G. see Schopf, G.: Vol. 129, pp. 1-145.
Kressler, J. see Kammer, H. W.: Vol. 106, pp. 31-86.
Krishnamoorti, R. see Giannelis, E.P.: Vol. 138, pp. 107-148.
Kirchhoff, R. A. and *Bruza, K. J.*: Polymers from Benzocyclobutenes. Vol. 117, pp. 1-66.
Kuchanov, S. I.: Modern Aspects of Quantitative Theory of Free-Radical Copolymerization. Vol. 103, pp. 1-102.
Kuleznev, V. N. see Kandyrin, L. B.: Vol. 103, pp. 103-148.
Kulichkhin, S. G. see Malkin, A. Y.: Vol. 101, pp. 217-258.
Kummerloewe, C. see Kammer, H. W.: Vol. 106, pp. 31-86.
Kuznetsova, N. P. see Samsonov, G. V.: Vol. 104, pp. 1-50.Labadie, J. W. see Hergenrother, P. M.: Vol. 117, pp. 67-110.

Lamparski, H. G. see O´Brien, D. F.: Vol. 126, pp. 53-84.
Laschewsky, A.: Molecular Concepts, Self-Organisation and Properties of Polysoaps. Vol. 124, pp. 1-86.
Laso, M. see Leontidis, E.: Vol. 116, pp. 283-318.
Lazár, M. and RychlΩ, R.: Oxidation of Hydrocarbon Polymers. Vol. 102, pp. 189-222.
Lechowicz, J. see Galina, H.: Vol. 137, pp. 135-172.
Léger, L., Raphaël, E., Hervet, H.: Surface-Anchored Polymer Chains: Their Role in Adhesion and Friction. Vol. 138, pp. 185-226.
Lenz, R. W.: Biodegradable Polymers. Vol. 107, pp. 1-40.
Leontidis, E., de Pablo, J. J., Laso, M. and *Suter, U. W.*: A Critical Evaluation of Novel Algorithms for the Off-Lattice Monte Carlo Simulation of Condensed Polymer Phases. Vol. 116, pp. 283-318.
Lesec, J. see Viovy, J.-L.: Vol. 114, pp. 1-42.
Liang, G. L. see Sumpter, B. G.: Vol. 116, pp. 27-72.
Lin, J. and *Sherrington, D. C.*: Recent Developments in the Synthesis, Thermostability and Liquid Crystal Properties of Aromatic Polyamides. Vol. 111, pp. 177-220.
López Cabarcos, E. see Baltá-Calleja, F. J.: Vol. 108, pp. 1-48.

Majoros, I., Nagy, A. and *Kennedy, J. P.*: Conventional and Living Carbocationic Polymerizations United. I. A Comprehensive Model and New Diagnostic Method to Probe the Mechanism of Homopolymerizations. Vol. 112, pp. 1-113.
Makhija, S. see Jaffe, M.: Vol. 117, pp. 297-328.
Malkin, A. Y. and *Kulichkhin, S. G.*: Rheokinetics of Curing. Vol. 101, pp. 217-258.
Maniar, M. see Domb, A. J.: Vol. 107, pp. 93-142.
Manias, E., see Giannelis, E.P.: Vol. 138, pp. 107-148.
Mashima, K., Nakayama, Y. and *Nakamura, A.*: Recent Trends in Polymerization of a-Olefins Catalyzed by Organometallic Complexes of Early Transition Metals. Vol. 133, pp. 1-52.
Matsumoto, A.: Free-Radical Crosslinking Polymerization and Copolymerization of Multivinyl Compounds. Vol. 123, pp. 41-80.
Matsumoto, A. see Otsu, T.: Vol. 136, pp. 75-138.
Matsuoka, H. and *Ise, N.*: Small-Angle and Ultra-Small Angle Scattering Study of the Ordered Structure in Polyelectrolyte Solutions and Colloidal Dispersions. Vol. 114, pp. 187-232.
Matsushige, K., Hiramatsu, N. and *Okabe, H.*: Ultrasonic Spectroscopy for Polymeric Materials. Vol. 125, pp. 147-186.
Mattice, W. L. see Rehahn, M.: Vol. 131/132, pp. 1-475.
Mays, W. see Xu, Z.: Vol. 120, pp. 1-50.
Mays, J. W. see Pitsikalis, M.: Vol.135, pp. 1-138.
Mikos, A. G. see Thomson, R. C.: Vol. 122, pp. 245-274.
Miyasaka, K.: PVA-Iodine Complexes: Formation, Structure and Properties. Vol. 108. pp. 91-130.
Monnerie, L. see Bahar, I.: Vol. 116, pp. 145-206.
Morishima, Y.: Photoinduced Electron Transfer in Amphiphilic Polyelectrolyte Systems. Vol. 104, pp. 51-96.
Mours, M. see Winter, H. H.: Vol. 134, pp. 165-234.

Müllen, K. see Scherf, U.: Vol. 123, pp. 1-40.
Müller-Plathe, F. see Gusev, A. A.: Vol. 116, pp. 207-248.
Mukerherjee, A. see Biswas, M.: Vol. 115, pp. 89-124.
Mylnikov, V.: Photoconducting Polymers. Vol. 115, pp. 1-88.

Nagy, A. see Majoros, I.: Vol. 112, pp. 1-11.
Nakamura, A. see Mashima, K.: Vol. 133, pp. 1-52.
Nakayama, Y. see Mashima, K.: Vol. 133, pp. 1-52.
Narasinham, B., Peppas, N. A.: The Physics of Polymer Dissolution: Modeling Approaches and Experimental Behavior. Vol. 128, pp. 157-208.
Nechaev, S. see Grosberg, A.: Vol. 106, pp. 1-30.
Neoh, K. G. see Kang, E. T.: Vol. 106, pp. 135-190.
Newman, S. M. see Anseth, K. S.: Vol. 122, pp. 177-218.
Nijenhuis, K. te: Thermoreversible Networks. Vol. 130, pp. 1-252.
Noid, D. W. see Sumpter, B. G.: Vol. 116, pp. 27-72.
Novac, B. see Grubbs, R.: Vol. 102, pp. 47-72.
Novikov, V. V. see Privalko, V. P.: Vol. 119, pp. 31-78.

O'Brien, D. F., Armitage, B. A., Bennett, D. E. and *Lamparski, H. G.*: Polymerization and Domain Formation in Lipid Assemblies. Vol. 126, pp. 53-84.
Ogasawara, M.: Application of Pulse Radiolysis to the Study of Polymers and Polymerizations. Vol.105, pp.37-80.
Okabe, H. see Matsushige, K.: Vol. 125, pp. 147-186.
Okada, M.: Ring-Opening Polymerization of Bicyclic and Spiro Compounds. Reactivities and Polymerization Mechanisms. Vol. 102, pp. 1-46.
Okano, T.: Molecular Design of Temperature-Responsive Polymers as Intelligent Materials. Vol. 110, pp. 179-198.
Okay, O. see Funke, W.: Vol. 136, pp. 137-232.
Onuki, A.: Theory of Phase Transition in Polymer Gels. Vol. 109, pp. 63-120.
Osad'ko, I.S.: Selective Spectroscopy of Chromophore Doped Polymers and Glasses. Vol. 114, pp. 123-186.
Otsu, T., Matsumoto, A.: Controlled Synthesis of Polymers Using the Iniferter Technique: Developments in Living Radical Polymerization. Vol. 136, pp. 75-138.

de Pablo, J. J. see Leontidis, E.: Vol. 116, pp. 283-318.
Padias, A. B. see Penelle, J.: Vol. 102, pp. 73-104.
Pascault, J.-P. see Williams, R. J. J.: Vol. 128, pp. 95-156.
Pasch, H.: Analysis of Complex Polymers by Interaction Chromatography. Vol. 128, pp. 1-46.
Penelle, J., Hall, H. K., Padias, A. B. and *Tanaka, H.*: Captodative Olefins in Polymer Chemistry. Vol. 102, pp. 73-104.
Peppas, N. A. see Bell, C. L.: Vol. 122, pp. 125-176.
Peppas, N. A. see Narasimhan, B.: Vol. 128, pp. 157-208.
Pichot, C. see Hunkeler, D.: Vol. 112, pp. 115-134.
Pieper, T. see Kilian, H. G.: Vol. 108, pp. 49-90.
Pispas, S. see Pitsikalis, M.: Vol. 135, pp. 1-138.
Pitsikalis, M., Pispas, S., Mays, J. W., Hadjichristidis, N.: Nonlinear Block Copolymer Architectures. Vol. 135, pp. 1-138.
Pospíšil, J.: Functionalized Oligomers and Polymers as Stabilizers for Conventional Polymers. Vol. 101, pp. 65-168.
Pospíšil, J.: Aromatic and Heterocyclic Amines in Polymer Stabilization. Vol. 124, pp. 87-190.
Powers, A. C. see Prokop, A.: Vol. 136, pp. 53-74.
Priddy, D. B.: Recent Advances in Styrene Polymerization. Vol. 111, pp. 67-114.
Priddy, D. B.: Thermal Discoloration Chemistry of Styrene-co-Acrylonitrile. Vol. 121, pp. 123-154.
Privalko, V. P. and *Novikov, V. V.*: Model Treatments of the Heat Conductivity of Heterogeneous Polymers. Vol. 119, pp 31-78.

Prokop, A., Hunkeler, D., Powers, A. C., Whitesell, R. R., Wang, T. G.: Water Soluble Polymers for Immunoisolation II: Evaluation of Multicomponent Microencapsulation Systems. Vol. 136, pp. 53-74.
Prokop, A., Hunkeler, D., DiMari, S., Haralson, M. A., Wang, T. G.: Water Soluble Polymers for Immunoisolation I: Complex Coacervation and Cytotoxicity. Vol. 136, pp. 1-52.
Putnam, D. and *Kopecek, J.*: Polymer Conjugates with Anticancer Acitivity. Vol. 122, pp. 55-124.

Ramaraj, R. and *Kaneko, M.*: Metal Complex in Polymer Membrane as a Model for Photosynthetic Oxygen Evolving Center. Vol. 123, pp. 215-242.
Rangarajan, B. see Scranton, A. B.: Vol. 122, pp. 1-54.
Raphaël, E. see Léger, L.: Vol. 138, pp. 185-226.
Reichert, K. H. see Hunkeler, D.: Vol. 112, pp. 115-134.
Rehahn, M., Mattice, W. L., Suter, U. W.: Rotational Isomeric State Models in Macromolecular Systems. Vol. 131/132, pp. 1-475.
Richter, D. see Ewen, B.: Vol. 134, pp.1-130.
Risse, W. see Grubbs, R.: Vol. 102, pp. 47-72.
Rivas, B. L. and *Geckeler, K. E.*: Synthesis and Metal Complexation of Poly(ethyleneimine) and Derivatives. Vol. 102, pp. 171-188.
Robin, J. J. see Boutevin, B.: Vol. 102, pp. 105-132.
Roe, R.-J.: MD Simulation Study of Glass Transition and Short Time Dynamics in Polymer Liquids. Vol. 116, pp. 111-114.
Rozenberg, B. A. see Williams, R. J. J.: Vol. 128, pp. 95-156.
Ruckenstein, E.: Concentrated Emulsion Polymerization. Vol. 127, pp. 1-58.
Rusanov, A. L.: Novel Bis (Naphtalic Anhydrides) and Their Polyheteroarylenes with Improved Processability. Vol. 111, pp. 115-176.
Rychlý, J. see Lazár, M.: Vol. 102, pp. 189-222.
Ryzhov, V. A. see Bershtein, V. A.: Vol. 114, pp. 43-122.

Sabsai, O. Y. see Barshtein, G. R.: Vol. 101, pp. 1-28.
Saburov, V. V. see Zubov, V. P.: Vol. 104, pp. 135-176.
Saito, S., Konno, M. and *Inomata, H.*: Volume Phase Transition of N-Alkylacrylamide Gels. Vol. 109, pp. 207-232.
Samsonov, G. V. and *Kuznetsova, N. P.*: Crosslinked Polyelectrolytes in Biology. Vol. 104, pp. 1-50.
Santa Cruz, C. see Baltá-Calleja, F. J.: Vol. 108, pp. 1-48.
Sato, T. and *Teramoto, A.*: Concentrated Solutions of Liquid-Christalline Polymers. Vol. 126, pp. 85-162.
Scherf, U. and *Müllen, K.*: The Synthesis of Ladder Polymers. Vol. 123, pp. 1-40.
Schmidt, M. see Förster, S.: Vol. 120, pp. 51-134.
Schopf, G. and *Koßmehl, G.*: Polythiophenes - Electrically Conductive Polymers. Vol. 129, pp. 1-145.
Schweizer, K. S.: Prism Theory of the Structure, Thermodynamics, and Phase Transitions of Polymer Liquids and Alloys. Vol. 116, pp. 319-378.
Scranton, A. B., Rangarajan, B. and *Klier, J.*: Biomedical Applications of Polyelectrolytes. Vol. 122, pp. 1-54.
Sefton, M. V. and *Stevenson, W. T. K.*: Microencapsulation of Live Animal Cells Using Polycrylates. Vol.107, pp. 143-198.
Shamanin, V. V.: Bases of the Axiomatic Theory of Addition Polymerization. Vol. 112, pp. 135-180.
Sherrington, D. C. see Cameron, N. R. , Vol. 126, pp. 163-214.
Sherrington, D. C. see Lin, J.: Vol. 111, pp. 177-220.
Sherrington, D. C. see Steinke, J.: Vol. 123, pp. 81-126.
Shibayama, M. see Tanaka, T.: Vol. 109, pp. 1-62.
Shiga, T.: Deformation and Viscoelastic Behavior of Polymer Gels in Electric Fields. Vol. 134, pp. 131-164.
Shoda, S. see Kobayashi, S.: Vol. 121, pp. 1-30.

Siegel, R. A.: Hydrophobic Weak Polyelectrolyte Gels: Studies of Swelling Equilibria and Kinetics. Vol. 109, pp. 233-268.
Silvestre, F. see Calmon-Decriaud, A.: Vol. 207, pp. 207-226.
Singh, R. P. see Sivaram, S.: Vol. 101, pp. 169-216.
Sivaram, S. and *Singh, R. P.*: Degradation and Stabilization of Ethylene-Propylene Copolymers and Their Blends: A Critical Review. Vol. 101, pp. 169-216.
Starodybtzev, S. see Khokhlov, A.: Vol. 109, pp. 121-172.
Steinke, J., Sherrington, D. C. and *Dunkin, I. R.*: Imprinting of Synthetic Polymers Using Molecular Templates. Vol. 123, pp. 81-126.
Stenzenberger, H. D.: Addition Polyimides. Vol. 117, pp. 165-220.
Stevenson, W. T. K. see Sefton, M. V.: Vol. 107, pp. 143-198.
Sumpter, B. G., Noid, D. W., Liang, G. L. and *Wunderlich, B.*: Atomistic Dynamics of Macromolecular Crystals. Vol. 116, pp. 27-72.
Suter, U. W. see Gusev, A. A.: Vol. 116, pp. 207-248.
Suter, U. W. see Leontidis, E.: Vol. 116, pp. 283-318.
Suter, U. W. see Rehahn, M.: Vol. 131/132, pp. 1-475.
Suzuki, A.: Phase Transition in Gels of Sub-Millimeter Size Induced by Interaction with Stimuli. Vol. 110, pp. 199-240.
Suzuki, A. and *Hirasa, O.*: An Approach to Artifical Muscle by Polymer Gels due to Micro-Phase Separation. Vol. 110, pp. 241-262.

Tagawa, S.: Radiation Effects on Ion Beams on Polymers. Vol. 105, pp. 99-116.
Tan, K. L. see Kang, E. T.: Vol. 106, pp. 135-190.
Tanaka, T. see Penelle, J.: Vol. 102, pp. 73-104.
Tanaka, H. and *Shibayama, M.*: Phase Transition and Related Phenomena of Polymer Gels. Vol. 109, pp. 1-62.
Tauer, K. see Guyot, A.: Vol. 111, pp. 43-66.
Teramoto, A. see Sato, T.: Vol. 126, pp. 85-162.
Terent'eva, J. P. and *Fridman, M. L.*: Compositions Based on Aminoresins. Vol. 101, pp. 29-64.
Theodorou, D. N. see *Dodd, L. R.*: Vol. 116, pp. 249-282.
Thomson, R. C., Wake, M. C., Yaszemski, M. J. and *Mikos, A. G.*: Biodegradable Polymer Scaffolds to Regenerate Organs. Vol. 122, pp. 245-274.
Tokita, M.: Friction Between Polymer Networks of Gels and Solvent. Vol. 110, pp. 27-48.
Tsuruta, T.: Contemporary Topics in Polymeric Materials for Biomedical Applications. Vol. 126, pp. 1-52.

Uyama, H. see Kobayashi, S.: Vol. 121, pp. 1-30.
Uyama, Y: Surface Modification of Polymers by Grafting. Vol. 137, pp. 1-40.

Vasilevskaya, V. see Khokhlov, A.: Vol. 109, pp. 121-172.
Vaskova, V. see Hunkeler, D.: Vol.:112, pp. 115-134.
Verdugo, P.: Polymer Gel Phase Transition in Condensation-Decondensation of Secretory Products. Vol. 110, pp. 145-156.
Vettegren, V. I.: see Bronnikov, S. V.: Vol. 125, pp. 103-146.
Viovy, J.-L. and *Lesec, J.*: Separation of Macromolecules in Gels: Permeation Chromatography and Electrophoresis. Vol. 114, pp. 1-42.
Volksen, W.: Condensation Polyimides: Synthesis, Solution Behavior, and Imidization Characteristics. Vol. 117, pp. 111-164.

Wake, M. C. see Thomson, R. C.: Vol. 122, pp. 245-274.
Wang, K. L. see Cussler, E. L.: Vol. 110, pp. 67-80.
Wang, S.-Q.: Molecular Transitions and Dynamics at Melt/Wall Interfaces: Origins of Flow Instabilities and Wall Slip. Vol. 138, pp. 227-276.
Wang, T. G. see Prokop, A.: Vol. 136, pp.1-52; 53-74.
Whitesell, R. R. see Prokop, A.: Vol. 136, pp. 53-74.

Williams, R. J. J., Rozenberg, B. A., Pascault, J.-P.: Reaction Induced Phase Separation in Modified Thermosetting Polymers. Vol. 128, pp. 95-156.
Winter, H. H., Mours, M.: Rheology of Polymers Near Liquid-Solid Transitions. Vol. 134, pp. 165-234.
Wu, C.: Laser Light Scattering Characterization of Special Intractable Macromolecules in Solution. Vol 137, pp. 103-134.
Wunderlich, B. see Sumpter, B. G.: Vol. 116, pp. 27-72.

Xie, T. Y. see Hunkeler, D.: Vol. 112, pp. 115-134.
Xu, Z., Hadjichristidis, N., Fetters, L. J. and *Mays, J. W.*: Structure/Chain-Flexibility Relationships of Polymers. Vol. 120, pp. 1-50.

Yagci, Y. and *Endo, T.*: N-Benzyl and N-Alkoxy Pyridium Salts as Thermal and Photochemical Initiators for Cationic Polymerization. Vol. 127, pp. 59-86.
Yannas, I. V.: Tissue Regeneration Templates Based on Collagen-Glycosaminoglycan Copolymers. Vol. 122, pp. 219-244.
Yamaoka, H.: Polymer Materials for Fusion Reactors. Vol. 105, pp. 117-144.
Yasuda, H. and *Ihara, E.*: Rare Earth Metal-Initiated Living Polymerizations of Polar and Nonpolar Monomers. Vol. 133, pp. 53-102.
Yaszemski, M. J. see Thomson, R. C.: Vol. 122, pp. 245-274.
Yoshida, H. and *Ichikawa, T.*: Electron Spin Studies of Free Radicals in Irradiated Polymers. Vol. 105, pp. 3-36.

Zubov, V. P., Ivanov, A. E. and *Saburov, V. V.*: Polymer-Coated Adsorbents for the Separation of Biopolymers and Particles. Vol. 104, pp. 135-176.

Subject Index

Adhesion 185
-, surface-anchored chains 202
Adhesive strength, normalized enhanced 203, 206
Adsorption, reversible 189
Alexander-de Gennes model 161, 171, 186, 191f
Antisymmetric structure 38–41
Antisymmetric surfaces 59, 79, 81
Ariadne length 99, 100
Aspiration 93, 94
Asymmetric block copolymers 34, 41, 46, 80
Asymmetric mixtures 45, 80
Asymmetric walls 41

Bead-rod model 52
Bead-spring model 161
Bicontinuous ordered phases 80
Binary (AB) mixture 7
Blob ansatz 192
Blob model, polymer brushes 163
Block copolymers 1–82
Bond fluctuation model 43–45, 53–56, 59
Boundary conditions 7, 13, 14, 19, 23, 24, 32, 34, 52, 65, 66, 75, 79, 81
Boundary discontinuities/stress singularities 238
Boundary-less constitutive models 268
Branched polymers 2
Break 103
Brinkman equation 155
Brush/elastomer adhesion 203
- - - energy 205, 208
Brush free energy 192
Brush height 163–167
- -, equilibrium 192
Brush thickness 192
Brushes, blob model 163
-, forces between compressed 168
-, free end distribution 160
-, normal/shear forces 149
-, off-lattice 153
-, polydisperse, half-loops 201
-, polymer 2, 132–142, 191, 201
-, shear force 172
-, static properties 152
Bulk failure 237, 269
Bulk materials 267

Cahn-Hilliard equation 65, 66,
Canonical ensemble 12
Capillary condensation 35
Capillary extrusion instability 249
Capillary flow behavior, anomalous, interfacial origin 261
Capillary waves 49, 72
CEC (cation exchange capacity) 111
Cell dynamics method 66
Chain conformation/shear rates 162
Chain deformation 39
Chain desorption, stress-induced 232
Chain dynamics 227
Chain ends, dangling 258
Chain length 18, 20, 22, 27, 30, 51, 58
Chain stiffness 41, 80
Chain stretching 4, 37
Channel flow 239
Clay minerals 115
Clogging 103
Coarsening 67, 76, 78, 82
Coating flow 239
Coexistence curve 7, 9, 11, 12, 35, 54, 56, 72, 80
Cohesive zone 203
Coil orientation 53, 82
Coil stretching 4, 37, 62, 82
Coil-stretch transition 265
Collective scattering 56
Colloidal suspension, steric stabilization 186
Complete wetting 11, 35
Compressible melts 41
Compression 168
Computer simulation 1, 4, 51, 60, 75, 76

Concentration profiles 2, 16, 72, 80, 190
Configurational statistics 21, 43
Confined interfaces 49
Confinement, strong/weak 100, 101
Conformational asymmetry 46
Connectors 202
Constant-stress viscosity 258
Constructive instabilities 267, 268
Contact angle 2–4
Continuum 179
Continuum solvent case 173
Controlled piston speed rheometers 250
Controlled-pressure capillary rheometer 245
Correlation length 19, 21, 25, 26, 30–33, 48, 50, 55–60, 67, 70–72
Critical amplitude 55, 57
Critical current 98
Critical exponent 55–59
Critical interfacial condition (CIC) 236
Critical point 2, 4, 9–12, 17, 19, 27, 28, 36, 47, 53–55, 71, 81
Critical point shift 28, 54–58, 70, 73
Critical volume fraction 7, 18, 58, 67, 78
Critical wavenumber 65
Critical wetting 50, 51
Criticality, two-dimensional 56–58
Crossover 20, 55–58
Crossover scaling 56, 58
Currents, critical 102

trans-Decalin 157
Density oscillations 9
Derjaguin approximation 170
Dewetting 2–6, 79
Diblock molecules, long 202
Discontinuous flow transition 227
Domains 6, 8, 30, 37, 77
Double-valued plateau 268
Drag flows, wall slip 240
Drumhead model 47
Dumbbells 62
Dynamar 246, 252, 261, 266
Dynamics of phase separation 1, 65, 76

Edge failure/"dynamic slip" 243
Elastomer/brush, adhesion 203
Elastomer/pseudo-brush, adhesion 207
Elastomers, adhesion to solids 202
-, crosslinked 203
-, fluorocarbon 246
Electron microsgraph techniques 75
End-grafted polymers 149
End-to-end distance 166
Entangled melt, interfacial slip 233
Entangled melts/solid surfaces 232
Enthalpic interactions 116
Entropy of confinement 116
Euler-Lagrange equation 13, 24, 29, 32
Excess free energy 10
Excluded volume 44
Exit wall 266
Extrapolation length 19, 232, 258
Extrudate, bamboo-like/alternating 263
Extrudate distortion 263
- - /melt fracture 269
Extrusion anomalies 227

Fast flow, chain dynamics 227
Finite size effects 26, 28, 54
Finite size scaling 25, 54, 55
First-order transition 23, 27, 31–36, 39, 80, 81
Flory theory 43
Flory-Huggins -parameter 7, 9, 15, 22, 44, 47
Flory-Huggins theory 4, 10, 18, 23, 56
Flow discontinuity, interfacial chain disentanglement 251
Flow discontinuity transition 247
Flow instability 229
Flows, pressure-driven 245
Fluctuations 8, 23, 30, 34, 47, 57, 65, 80
Fluid mixtures 35, 77
Fluorocarbon elastomer 246
Fluorohectorite 112, 114, 119–123
Forced permeation 93, 95
Fracture at interface 202
Fracture energy, zero-rate 204
Free energy functional 22, 24, 29, 31, 39, 42, 43, 47
Free interfaces 49
Free surfaces 41, 44
Friction 185
-, interfacial 218
- regimes 213
-, surface-anchored chains 212
Frictional force 149
Frustration 4, 5, 75, 81

Gaussian statistics 22, 38, 41, 42, 53
Gibbs adsorption equation 70
Ginzburg-Landau theory 48
Gradient energy 10, 13, 66
Gradient-square approximation 10
Guiselin's pseudo-brush 193

Gyration radius 4, 21, 22, 25, 44, 53, 55, 58–61, 69–73
Gyration tensor 54

Hard walls 9, 51, 75
Hartree approximation 30
HDPE 242, 245
-, Dynamar 262
-, stick-slip transition 253
Hectorites 111, 115
Hexagonal ordering 41, 80
Horizontal morphology 39, 41
Hydrodynamic boundary condition 227
Hydrodynamic mechanism 67, 69, 77, 78, 82
Hydrodynamic penetration depth 155, 157
Hysteresis effects 150

Incomplete wetting 11
Incompressible melts 41, 44
Instabilities, constructive 267, 268
Interaction range 54
Interdigitation 185, 196
Interface potential 48
Interfaces 3–9, 34, 37, 38, 41, 46, 47, 59, 70, 71, 78–81, 185
Interfacial adsorption, critical stress 256
Interfacial energy 5, 37, 47, 58, 81
Interfacial fluctuations 48, 60
Interfacial friction, modulation 218
Interfacial profile 32, 33, 47, 49, 58, 72, 82
Interfacial slip, ordering in suspensions 244
Interfacial width 1, 47, 50, 59, 60
Interpenetration 168
Intrinsic interfacial profile 47, 49, 60, 72
Ion-beam analysis 78
Ising model 12, 47, 50, 52, 55–58
- - critical exponent 18, 55–58
- -, three-dimensional 55–58

JKR technique 209
Johnson-Kendall-Roberts technique 209

Lamellae, perpendicular orientation 38, 46, 61, 62, 75, 80
Lamellar structure 2, 5, 6, 22, 23, 26–28, 34, 37, 38, 41, 61, 74, 79–81
Landau theory 23
Lateral phase separation 30, 78, 82
Leibler s theory 8, 30, 80
Lennard Jones potential 52
Lennard-Jones interaction 178
Lever rule 9, 12, 47
Lifshitz-Slyozov-law 67
Linear polyethylenes 227, 229, 247
LLDPE 263
Long range surface forces 8, 12, 35, 51, 68–72
Long wavelength concentration fluctuations 65
Long wavelength approximation 4, 22, 69
LPE 227, 229, 247
-, flow curve 230
-, pressure-extrusion 234
Lubrication force 157

Marginal regime 215–217
Mean-field free energy 192
Mean-field theory 1, 4, 8, 14, 15, 21–37, 41–44, 50–57, 70
Melt blobs 197
Melt, entangled, interfacial slip 233
Melt flow anomalies 227, 247
Melt flow instabilities 227
Melt fracture 212, 269
- -, true 270
- -, bamboo-like distortion 264
Melt/solid interfaces 229
Melt/wall interfaces, discontinuity/molecular transition 251
- -, molecular transitions/dynamics 227
Melts, entangled/solid surfaces 232
Membranes 95
Mesh size 100
Metallic surfaces 238
Metastable states 14
Microemulsions 22
Microphase separation 22
Missing neighbors 9, 12, 19, 23, 53
Monomer density 153
Monte Carlo simulations 29, 51, 57, 61, 179
Montmorillonite 111, 115, 118, 130–132
Mushroom regime 191, 197

Nanocomposite hybrids, exfoliated 108, 114, 132, 134
- -, intercalated 108, 114, 119, 124, 130, 131
- -, phase separated 108
Nanocomposites, intercalated 109, 113–122, 131
- properties 109
- rheology 131
Nanopores, flexible polymers 91ff
-, star polymers 96
Navier's slip boundary condition 231

Navier-de Gennes length 236
Navier-de Gennes slip boundary 230
Neutral walls 21, 26, 27, 51, 58, 70
Neutron reflectivity, PDMS layers 200, 201
Neutron reflectometry 70, 74, 76
No-slip boundary condition 233
No-slip condition 231
Nonsymmetric block copolymers 34, 41
Nuclear reaction depth profiling 73, 76
Nylon-6 storage/loss moduli 136

Octadecyltrichlorosilane 213
Octane 194
Order parameter 17–19, 25–32, 37, 53, 54, 59–62, 66–71, 75
Order parameter exponent 17
Order parameter profile 33, 46, 53, 54, 59–62, 66, 68, 76
Order-disorder transition (ODT) 6, 22, 32, 61, 76
Ordering kinetics 76
Oscillating capillary melt flow, molecular mechanism 259
Oscillating flow 227
Overcrowding 150

Packing fraction 43
Parallel orientation of lamellae 38, 46, 61, 75, 80
Parallel plate geometry 239
Partition function 42, 43
Path integral 42
PDMS 186, 242, 249
-, concentration profile 195
-, dry thickness 188, 190
- melts, interfacial velocity 212
Pearls 100
Penetration depth, hydrodynamic 155, 157
Periodic boundary condition 52
Periodic deformation 212
Permeation, critical currents 102
-, forced 93, 95
Phase coexistence 2, 7, 12, 14–17, 32, 34, 47, 65, 71, 72
Phase diagram 2, 7, 11, 12, 17, 18, 26, 27, 36, 46, 52, 56, 65, 71, 78
Phase inversion 2, 79
Phase separation 1, 6, 8, 14, 20, 30, 52, 65–67, 78, 80
Phase transition shift 21
Phyllosilicates 110
-, agglomerates 121
-, hydration 117
PI 209
Pinning 78
Poisson equation 29
Poly(ethylenepropylene) (PEP) 70, 75, 77
Poly(ethylethylene) (PEE) 75
Poly(2-methylvinylcyclohexane) 75
Poly(vinylmethylether) (PVME) 77
Poly(2-vinyl pyridine) (PVP) 75
Polybutadiene (PB) 78, 236, 249
Polycarbonates 93
Polydimethylsiloxane (PDMS) 186, 242, 249
-, , -hydroxyl terminated 186
Polyethylenes, linear 227, 229
-, linear, pressure extrusion 234
Polyisoprene (PI) 76, 209, 249
Polymer blends 1–82
Polymer brushes 2, 132–142
- -, monodisperse 193
- -, polydisperse, half-loops 201
- - /polymeric matrices 196
- -, schematic 191
- -, viscosity 140
Polymer chain rupture 104
Polymer interfaces 227, 229
Polymer melt, irreversible adsorption onto plane surface 193
Polymer mixtures 1–82
Polymer/silicate nanocomposites, PCL end-tethered hybrids 132–136, 138–142
- -, PDMS exfoliated 132
- -, PDMS intercalated 131
- -, PEO intercalated 123–125, 130–131
- -, PS nanocomposites 114, 118–122, 130
- -, PVA intercalated hybrids 115
Polymer/silicate nanocomposites 109–136
Polymer surfaces, interfaces 185
Polymer volume fraction 188
Polymeric matrices, irreversibly adsorbed layers 201
Polymers/organosilicates, miscibility 116
Polymethylmetacrylate (PMMA) 72, 74, 249
Polyolefins 71, 73, 92
Polypropylene 249
Polystyrene (PS) 69, 72–79, 157, 209, 249
-, linear 93
Polystyrene-*co*-4-bromostyrene (PBrS) 69, 72, 78, 79
Pressure driven flows, wall slip 245
Prewetting 10, 11, 35–37, 72, 81
PS 209
PS-PI diblock 209
Pseudo-brushes 187, 193
- /elastomer, adhesion 207
-, Guiselin's 193
-, schematic 194

Pull-out process, schematic 204
Radius of gyration 92
Radius, unperturbed 99
Random phase approximation 22, 23
Regimes, friction 213
-, marginal 215–217
Relaxation times 150
Reptation theory 252
Response function 19, 37
Rheology 92
Rheometers, controlled piston speed 250
Roughening 2, 78
Roughness 2, 70, 72
Rouse friction 217
Rouse time 68
Rupture, polymer chain 104

SANS 114, 140
Scaling 56–58
Scaling law 94
Scanning near-field optimal microscopy (SNOM) 78, 79
SCF 150, 152, 186
- parabolic profile 196
Second-order transition 26, 27, 30, 35, 48, 80
Self-avoiding walk 52, 60
Self-consistent field theory (SCF) 1, 8, 30, 41, 46, 53, 59, 69, 150, 152, 186
Semi-compact structures 103
Semi-grandcanonical ensemble 12
Semi-infinite system 13–17, 20, 25, 34, 36, 66–72, 81
SFA 150, 152
Sharkskin 212, 227, 229, 263, 264
Shear, oscillatory 167
Shear flow, steady-state 149
Shear force 172
Shear stress 174
Shear thinning regime 258
Short range surface forces 8, 51, 67–72
Silica/PDMS systems 187
Silica surface 186
Simulation models 178
Slip, interfacial 240
Slip at the wall 185
Slip velocity, low 220
Small angle scattering 56
Solid binary mixtures 76
Solid surface, shear stress 212
Solvent velocity profiles 158, 159
Spinodal curve 65
Spinodal decomposition 2, 65–68, 76, 78, 82
Squeezing 94
Star polymers in pores 96
Step pyramide structure 6
Steric interactions 8
Stick-slip 249
Stick-slip transition 236
- -, critical condition/molecular characteristics 255
Stiff chains 2, 80
Stress singularities/boundary discontinuities 238
Strong segregation limit 1, 8, 29, 37, 41, 46, 50, 59, 76, 80
Surface density 150
Surface directed spinodal decomposition 2, 66, 68, 76–82
Surface enrichment 2, 8, 9, 16, 17, 21, 44, 46, 69–72, 78–81
Surface excess 7–12, 16, 17, 35, 36, 45, 69–72
Surface excess free energy 12, 13, 20, 23, 82
Surface field 19, 29, 30, 33, 36, 37, 44, 75
Surface forces 8, 67–72, 80
- -, measurements 186
Surface Forces Apparatus 107, 122, 141
Surface layers, formation/characterization 187
- -, internal structure 190
Surface melting 35
Surface transition 20, 26, 28, 36, 37, 72
Surface-anchored chains/polymer melt, interdigitation 196
Surface-anchored polymer chains 185
Surfaces, impenetrable 229
-, metallic 238
-, polymer-bearing 149
Susceptibility 21
Symmetric mixtures 52–57, 79
Symmetry breaking 19

Teflon, glass-reinforced 252
Thin films 1–82
Time-of flight forward recoil spectroscopy (TOF-FRES) 77–79
Tower of Babel 6
Triple point 36
Two-dimensional long range order 26

Unmixing transition 2, 35, 47, 81

Van der Waals forces 12, 35, 73
Velocity field 156
Vertical morphology 39–41
Volume fraction profile 45

Wall slip 214, 238
- -, drag flows 240
- -, interfacial 227
- -, Mooney vs Navier treatment 254
- -, pressure driven flows 245
- - regimes 185

Weak segregation limit 1, 8, 22, 41, 46, 65, 68, 76, 81
Wetting 2–11, 16, 17, 27, 33, 47–50, 67–70, 73, 77–82

Zero-rate fracture energy 204